The Release of Genetically-engineered Micro-organisms

Proceedings of the First International
Conference on the Release of
Genetically-engineered Micro-organisms

The Conference was sponsored by

Society for Applied Bacteriology
Society for General Microbiology
American Society for Microbiology
The International Union of Microbiological Societies

The Release of Genetically-engineered Micro-organisms

Edited by

M. SUSSMAN[1], C. H. COLLINS[2], F. A. SKINNER[3] and D. E. STEWART-TULL[4]

[1]*Department of Microbiology, Medical School, Newcastle upon Tyne NE2 4HH, UK*

[2]*The Ashes, Hadlow, Kent TN11 0AS, UK*

[3]*5 Carisbrooke Road, Harpenden, Hertfordshire AL5 5QS, UK*

[4]*Department of Microbiology, University of Glasgow, Glasgow G12 8QQ, UK*

1988

ACADEMIC PRESS
Harcourt Brace Jovanovich, Publishers
London San Diego New York Berkeley
Boston Sydney Tokyo Toronto

ACADEMIC PRESS LIMITED
24–28 Oval Road,
London NW1 7DX

United States Edition published by
ACADEMIC PRESS INC.
San Diego, CA 92101

British Library Cataloguing in Publication Data

The release of genetically-engineered
micro-organisms.
1. Industrial microbiology. Applications
of genetic engineering of micro-organisms
I. Sussman, M.
660'.62

ISBN 0–12–677521–4 (c)
0–12–677522–2 (p/b)

Typeset by Latimer Trend & Company Ltd
Printed in Great Britain at the Alden Press,
Oxford, London & Northampton

Contents

Participants

E. A. Adelberg	Department of Human Genetics, Yale University, 338 Cedar Street, New Haven, CT 06510, USA
C. J. Allen	NERC Institute of Virology, Mansfield Road, Oxford OX1 3SR, UK
R. M. Atlas	Department of Biology, University of Louisville, Louisville, KY 40292, USA
M. J. Bale	Department of Microbiology, University of Bristol, University Walk, Bristol BS8 1TD, UK
G. F. Barry	Monsanto Company, 700 Chesterfield Village Parkway, St Louis, MO 03198, USA
P. M. Barth	ICI Corporation Bioscience Group, PO Box 11, The Heath, Runcorn, Cheshire WA7 4QD, UK
S. Baumberg	Department of Genetics, University of Leeds, Leeds LS2 9JT, UK
P. M. Bennett	Department of Microbiology, University of Bristol, Medical School, University Walk, Bristol BS8 1TD, UK
J. E. Beringer	Department of Microbiology, University of Bristol, University Walk, Bristol BS8 1TD, UK
D. H. L. Bishop	NERC Institute of Virology, Mansfield Road, Oxford OX1 3SR, UK
E. J. Brandt	Monsanto Company, 700 Chesterfield Village Parkway, St Louis, MO 03198, USA
F. Brown	Wellcome Biotechnology Ltd, Langley Court, Beckenham, Kent BR3 3BS, UK
I. R. Cameron	NERC Institute of Virology, Mansfield Road, Oxford OX1 3SR, UK
C. H. Collins	The Ashes, Hadlow, Kent TN11 0AS, UK
R. R. Colwell	Maryland Biotechnology Institute, 3300 Metzerott Road, Adelphi, MD 20783, USA
D. E. Corpet	Xenobiotiques, INRA, BP 3, F-31931 Toulouse, France
R. Curtiss III	Department of Biology, Washington University, St Louis, MO 63130, USA

S. Cuskey	Gulf Breeze Environmental Research Laboratory, Building 7, Sabine Island, Gulf Breeze, FL 32561, USA
J. E. Davies	Department of Biotechnology, Institut Pasteur, 25 Rue du Docteur Roux, Paris 75724, France
A. De Roeck	Plant Genetic Systems NV, Plateaustraat 22, 3-9000 Ghent, Belgium
P. Desmettre	Rhone Merieux, 254 Rue Marcel Merieux, Lyon Cedex 07, France
D. Drahos	Monsanto Company, 700 Chesterfield Village Parkway, St Louis, MO 03198, USA
D. F. Dwyer	Federal Institute for Biotechnology, Bissendorf, Mascheroder Weg 1, D-3300 Braunschweig, FRG
P. F. Entwistle	NERC Institute of Virology, Mansfield Road, Oxford OX1 3SR, UK
M. J. Gasson	Food Research Institute, Colney Lane, Norwich NR4 7UA, UK
S. W. Glover	Department of Genetics, Medical School, University of Newcastle upon Tyne, Newcastle upon Tyne NE2 4HH, UK
D. T. Gooden	Clemson University, Clemson, SC 29634, USA
E. P. Greenberg	Department of Microbiology, Cornell University, Stocking Hall, Ithaca, NY 14833, USA
R. Grigorova	Institute of Microbiology, Academy of Sciences, Sofia 1113, Bulgaria
N. Gutterson	Advanced Genetic Sciences Inc., 6701 San Pablo Avenue, Oakland, CA 94608, USA
B. C. Hemming	Monsanto Company, 700 Chesterfield Village Parkway, St Louis, MO 03198, USA
M. Hinton	Department of Veterinary Science, Langford House, University of Bristol, Langford, Bristol BS18 7DU, UK
M. G. Höfle	Max Planck Institute of Limnology, Postfach 165, August Thiendmann Strasse 2, D-2320 Plon, FRG
T. A. Hughes	Clemson University, Clemson, SC 29634, USA
J. Hurtado	Leoncio Prado 198, Lima 18, Peru 00184, Peru
J. G. Jones	Freshwater Biological Association, The Ferry House, Ambleside, Cumbria LA22 0LP, UK
H. Joos	Plant Genetic Systems NV, Plateaustraat 22, 3-9000 Ghent, Belgium
S. Kjellberg	Department of Marine Microbiology, University of Goteborg, Carl Skottsbergs Gatan 22, S-413 14, Goteborg, Sweden
E. L. Kline	Clemson University, Clemson, SC 29634, USA

D. A. Kluepfel	Clemson University, Clemson, SC 29634, USA
I. Knight	Department of Microbiology, University of Maryland, College Park, MD 20742, USA
H. Kornberg	Department of Biochemistry, University of Cambridge, Cambridge, CB2 1QW, UK
B. Lambert	Plant Genetic Systems NV, Plateaustraat 22, 3-9000 Ghent, Belgium
B. Languet	Rhone Merieux, 254 Rue Marcel Merieux, Lyon Cedex 07, France
J. L. La Torre	Serrana 665, 1414 Cap Federal, Argentina
S. B. Levy	Department of Molecular Biology, Tufts University, 136 Harrison Avenue, Boston, MA 02111, USA
F. Leyns	Plant Genetic Systems NV, Plateaustraat 22, 3-9000 Ghent, Belgium
A. A. Lindberg	Department of Clinical Microbiology, Karolinska Institute, Huddinge University Hospital, Huddinge S-141 86, Sweden
S. Lindow	Department of Plant Pathology, University of California, Berkeley, CA 94720, USA
B. J. J. Lugtenberg	Department of Plant Molecular Biology, State University of Leiden, Nonnensteeg 3, 2311 B. J. Leiden, Netherlands
B. M. Marshall	Department of Molecular Biology, Tufts University, School of Medicine, Boston, MA 02111, USA
R. V. Miller	2273 Pontiac Circle, Naperville, IL 60565, USA
G. Muyzer	Department of Biochemistry, University of Leiden, Wassenaarsweg 64, Leiden, Netherlands
E. Paoletti	New York State Department of Health, Wadsworth Center Laboratories and Research, Albany, NY 12201, USA
N. J. Panopoulos	Department of Plant Pathology, University of California, Berkeley, CA 94720, USA
G. Pelsey	Directoire Generale D, Alimentation, Département Agricole, 35 Rue St Dominique, 75007 Paris, France
N. J. Poole	Plant Protection, ICI Research Station, Jealott's Hill, Bracknell, Berkshire RG12 3SD, UK
R. D. Possee	NERC Institute of Virology, Mansfield Road, Oxford OX1 3SR, UK
R. A. Prins	Laboratory for Microbiology, Keklaan 30, Haren 9751 NN, Netherlands
H. A. P. Pritchard	EPA Environmental Research Laboratory, Gulf Breeze, FL 32561, USA

M. Riley	Department of Biological Science, State University of New York, Stony Brook, NY 11794, USA
F. Rojo	Department of Medical Biochemistry, Geneva University, Medical Centre, CH-1211, Geneva 4, Switzerland
C. Sanchez-Rivas	Ingebi, Obligado 2490-200 Piso, 1428 Buenos Aires, Argentina
G. Sayler	Graduate Program in Ecology, University of Tennessee, Knoxville, TN 37996, USA
R. Seidler	Terrestrial Biotechnology Program, US EPA, Corvallis, OR 97333, USA
S. Selenska	Institute of Molecular Biology, Academy of Sciences, Sofia 1113, Bulgaria
F. Sinerez	Priomi Mircen, Aveniday Belgrano y, Caseros, 4000 SM De Tucuman, Argentina
F. A. Skinner	5 Carisbrooke Road, Harpenden, Hertfordshire AL5 5QS, UK
H. D. Skipper	Clemson University, Clemson, SC 29634, USA
C. Somerville	Department of Microbiology, University of Maryland, College Park, MD 20742, USA
D. Stahl	Department of Veterinary Pathobiology, University of Illinois, Urbana, IL 61801, USA
R. J. Steffan	Department of Biology, University of Louisville, Louisville, KY 40292, USA
W. Straube	Department of Microbiology, University of Maryland, College Park, MD 20742, USA
M. Sussman	Department of Microbiology, Medical School, Newcastle upon Tyne NE2 4HH, UK
D. E. Stewart-Tull	Department of Microbiology, University of Glasgow, Glasgow G12 8QQ, UK
J. Swings	Plant Genetic Systems NV, Plateaustraat 22, 3-9000 Ghent, Belgium
J. Taylor	New York State Department of Health, Wadsworth Center for Laboratories and Research, Albany, NY 12201, USA
D. Tepfer	Laboratoire de Biologie de la Rhizosphere, Institut National de la Recherche Agronomique, 7800 Versailles, France
J. Tiedje	Department of Crop and Soil Science, Michigan State University, East Lansing, MI 48824, USA
K. N. Timmis	Department of Medical Biochemistry, Geneva University, Medical Centre, CH-1211, Geneva 4, Switzerland
H. Uchida	Teikyo University, 2-11-1 Kaga, Itabashi-ku, Tokyo 171, Japan

G. J. Warren — Advanced Genetic Sciences Inc., 6701 San Pablo Avenue, Oakland, CA 94608, USA

L. S. Watrud — Research Division, Monsanto Agricultural Products Company, 700 Chesterfield Village Parkway, St Louis, MO 63198, USA

E. E. Wedman — Department of Veterinary Medicine, Oregon State University, Corvallis, OR 97331, USA

R. Weinberg — New York State Department of Health, Wadsworth Center for Laboratories and Research, Albany, NY 12201, USA

H. R. Whiteley — Department of Microbiology, University of Washington, Seattle, WA 98195, USA

BUCKINGHAM PALACE.

The development of an entirely new technology inevitably causes anxieties. There is a very natural fear of the unknown, whether it is perceived as potentially good or evil.

The importance of this, the first International Conference on the Release of Genetically Engineered Micro-organisms, is that it will provide an opportunity for open discussion among workers in this new field of activity from all over the world. Such a full and frank exchange of views and opinions between specialists should help to raise the level of understanding and confidence among the general public in the determination of those involved to proceed with responsibility and appropriate caution. I think it is very much to be welcomed that the proceedings of this conference will be made available in a layman's version for the information of the general public.

I welcome all the delegates to the conference and I hope that their deliberations will lead to a better comprehension of the implications of this new and powerful science for the future of all life on this planet.

Philip

Preface

The biotechnical revolution and its potential benefits will become a reality only when the safety of genetically-engineered micro-organisms (GEMs) has been demonstrated and accepted. The First International Conference on the Release of Genetically Engineered Microorganisms—REGEM 1—which took place in Cardiff, South Wales, from 5 to 8 April 1988, was part of the progress towards this ultimate aim. The meeting was more than a scientific conference. Public relations, in terms of the need to communicate with the lay public and governmental bodies about the release of GEMs and the ethical implications of their release were also discussed in some detail.

The proceedings of the conference contained in these pages will, we hope, reflect for the general public as much as for the scientific community, the balance of current opinion about the release of GEMs.

Scientific meetings are more than just occasions to talk about research and its results. New friends are made, old friendships are renewed, new and old places are visited and much is learned. The weather in Cardiff was remarkably good for April—the wet Atlantic fronts stayed away just long enough! It was, therefore, a pleasant microcosm into which the conference participants were released. Even the expected problems of adverse publicity did not materialize. This was probably because all the talk about GEMs made it seem that the venue, St David's Hall, had merely been taken over for the meeting of a Jewellers' Trade Association—a potential hazard only for the purse.

As the Conference Editorial Team we worked, literally, backstage most of the time. Nevertheless, we were able to hear the proceedings of the conference over the public address system as we were labouring to transfer them to paper for posterity. Though exhaustion pursued us most of the time, we enjoyed ourselves. This was in no small way due to the delightful and charming group of secretaries who helped us. First and foremost were Miss Janet Davies, Miss Julie Harvey and Mrs Helen Miller, their smiles from behind mountains of manuscript were a pleasure. The efforts of the Chairmen and Rapporteurs in helping to produce the reports of the Round Tables are also very much appreciated. We are also most grateful to Mr Colin Griffiths, the Conference Organizer, and his secretary Mrs Brenda Maslyn. Their perpetual helpfulness to us in the face of their many other responsibilities places us greatly in their debt. Finally, we wish to express our gratitude to our friends at Academic Press: Mr Rich Cutler, our copy-editor, who has smoothed the path of the manuscript of these proceedings through the press, and Miss Gina Fullerlove who made the arrangements for their rapid publication.

May 1988

M. Sussman
C. H. Collins
F. A. Skinner
D. E. Stewart-Tull

1 Opening Remarks

SIR HANS KORNBERG

Department of Biochemistry, University of Cambridge, Cambridge CB2 1QW, UK

It is a great privilege to be asked to open this Conference and I thank the Organizers for the honour that they have done me in inviting me. Of course, I am aware that my primary reason for being in this honorific but exposed position is that I am currently the Chairman of the Advisory Committee on Genetic Manipulation (ACGM) which, as you know, advises the Health and Safety Executive and the Health and Safety Commission on how these new technologies should be regulated in this country. The position of Chairman is an odd one in that one is expected to know something about the subject under discussion but preferably not too much, since this might be thought to affect one's ability to act in a totally unbiased manner.

There is an old fable that if a newly born child is destined to be a great singer, the Muses kiss it on the lips; if it is destined to be a great thinker, the kiss is implanted on the forehead. I cannot even guess where the Muses kiss those destined to be Chairman.

My pleasure at being here this evening is intensified by my long association with the area of genetic manipulation and with the public perception of these powerful new ways of altering biological matter. This association dates back 14 years. To be precise, it was on 18 June 1974 that Professor Paul Berg of Stanford University wrote to tell me of the meeting that arose from a Gordon Conference and that led a group of concerned colleagues in the USA "to devise a letter to be submitted to *Science* and to *Nature* calling on scientists to defer certain kinds of experiments until some potential hazards could be better evaluated and certainly until there was an opportunity to discuss the issues at a meeting to be called in February 1975". I will not weary you by describing the history of that further meeting that was held in Asilomar, or what went on from there. However, I might just mention that it was as a result of that letter, and in response to Paul Berg's request that I should draw the attention of relevant authorities in the UK

RELEASE OF GENETICALLY-ENGINEERED MICRO-ORGANISMS ISBN 0–12–677521–4

to it, that I asked to give evidence to Sir George Godber's Committee on Dangerous Pathogens, which in turn led to the setting up of Lord Ashby's Working Party which initiated the discussion in the UK of the possible risks that might attend genetic manipulations and how those risks might be minimized.

This Conference is, I believe, not only a further example of this public debate, but is a most important extension of it into a dimension that has not been properly considered previously. If I may echo the epitaph of an author, I hope that you will all be conscious that, "though your sins be scarlet, the proceedings of this Conference will be read".

When one looks back over the past 14 years, one cannot help but be astonished at the rapidity of the progress that has been made. Fourteen years ago we were musing over the first reports that the expression of ribosomal genes of *Xenopus* cloned into *Escherichia coli* could apparently be detected. Today we are long used to seeing not only *E. coli*, but many pro- and eukaryotic micro-organisms successfully harnessed as vectors for the production of many important products on an industrial scale. All this has been done under guidelines and regulations that have gradually been lessened in rigour as experience accrued. I know of no instance in which this technology has harmed the health of any of the workers engaged in it: if true, this must surely be unique amongst all known applications of science to useful ends.

Although we applaud the caution of those who first erected those guidelines in the USA, in Europe and in the UK, we can now be thankful that many of the initial fears that led to their formulation have been shown to be groundless and that we can exploit the wonderful potential of these techniques, at least in the laboratory and in industry, without causing cosmic disaster.

However, we are once again facing an unknown, and as yet unknowable, situation. All our experience has been concerned with containment. It is only now that we are proposing deliberately to introduce into the environment organisms that have been altered in a manner that would not be likely to occur naturally. The expected benefits of such international introductions are well known and no doubt many papers at this Conference will discuss such benefits that range from the hope of improving crop yields by the introduction into plants of genes that confer resistance to disease or particular herbicides, to the use of manipulated micro-organisms as crop protection agents against pests and other environmental stresses.

These developments are becoming a reality because of the precise genetic alterations made possible by recombinant DNA techniques. But let us not forget that inducing genetic change in microbes, plants and animals is by no means a new activity. The ability to cut and splice a gene is but an acceleration of our longstanding capability to tailor organisms to our needs. Cross-breeding and selection are the ancient cornerstones of modern agriculture. Building on these traditional practices, recombinant DNA technology gives us the additional capacity to make precise genetic changes, rather than the more random unpredictable changes of older technologies.

Nevertheless, the potential benefits are accompanied by controversy. It is a controversy being debated on several fronts:

The first is basically a *scientific debate*—will the introduction of recombinant DNA organisms into the environment cause harm? There is no consensus as yet between the views of molecular biologists, who are well used to inducing genetic change; the views of ecologists, who have in mind recorded problems arising from biological invasions such as kudzu vine, gypsy moth and the grey squirrel; and the views of pathologists, who are aware of the many complex criteria that have to be met before a micro-organism can manifest a pathogenic potential. Where there does seem to be consensus, however, is that it is the end *product* destined for release that matters and not the method by which it was produced. We must not lose sight of this sensible and fundamental point in our debate. We will not serve the environment well by singling out for special treatment *one* method of inducing genetic change—we must maintain our perspective.

The second is a *regulatory debate*—the scientific debate is inevitably associated with considerable discussion amongst Government Agencies and bodies such as the OECD and European Commission who are striving to reach the right regulatory framework. The goal here is to achieve balanced regulations so that we do not stifle progress yet are not too lax, with the possibility of allowing unwanted environmental effects and of not carrying public confidence.

And thirdly, but perhaps most importantly, there is the *public perception debate*—here we have an arcane 'new' technology which is capable of generating emotive and polarized views. There is still an air of mystery and dread associated in some minds with recombinant DNA. We must do our utmost to dispel this as recombinant organisms begin to be applied outside contained facilities. We, the scientists, have the responsibility to explain our work and its benefits. And we must be prepared to discuss, rather than merely dismiss, any fears.

So the task before us is to develop a scientifically sound base of knowledge that will also permit the development of a globally acceptable regulatory framework. An essential feature of our task is that it must be capable of addressing concerns, whether real or potential and, most importantly, that it must command public confidence. This requires that in our Conference this week we must engage in a broad scientific debate that accommodates our knowledge of DNA and its properties and, as far as we are able, accommodates our understanding of the manner in which micro-organisms interact with their environment. But we must also not shirk to identify the gaps in our knowledge and suggest how those gaps should be filled.

Here, for example, I am conscious of the fact that for the past seven years, I have spent my summers in teaching a course on microbial ecology at the Marine Biological Laboratory, Woods Hole, but that some people regard this as somewhat less exciting and less 'with it' than the more orthodox courses in embryology, neurobiology or other frontier areas of biology that are also being offered.

This view is something that even the most molecular amongst us must strive to change if we are to move forward in achieving the insight that will be necessary effectively to meet public concerns.

In the absence of a complete understanding of the fate of micro-organisms released into the environment, the Subcommittee of ACGM that particularly deals with applications for such deliberate introductions and that is chaired by John Beringer has been proceeding pragmatically, by a case-by-case approach. So far, there have been five controlled releases in the UK sanctioned by the Committee; there are a number of others in the pipeline. Our firm hope is that, as we gain knowledge and experience and as we further develop our capabilities for risk assessment, this pragmatic approach can be streamlined and we can speed up the procedures. It will also be reasonable to expect that we can begin to identify types of release that pose fewer concerns than others. For example, it may be that we will need to continue to be more watchful over the release of genetically-manipulated microbes into the environment than about plants or animals, simply because microbes are less easy to recall. On the other hand, there is increasing evidence that genetically-manipulated micro-organisms are less able to compete with their more robust wild-type brothers and less able to survive than would, for example, a rabbit or a kudzu vine introduced into an environment that did not already contain these organisms. I must also confess that, whenever I consider the ability of Man to alter to any significant and longstanding extent the interplay of the myriad genes already present in the environment, I remember Aesop's fable of the fly that sat on the axle of the wheel of the chariot and said: "Look at the dust I raise". Only time, and the results of the type of experiments that we are already carrying out and that will be the substance of our discussions here, will tell.

I hope that we will soon be in a position to give advice to Government, to industry, to employees and to the public at large that is even more securely based on hard scientific data than is the advice we give at present. But I believe that in giving such advice, we should also aim to satisfy three main conditions.

First, the advice we give should *not* be solely derived from peer review: we should aim also for watchdog review. We cannot expect public confidence unless we bring together all elements of reasonable debate on the subject.

Secondly, our advice must be derived from a transparent system. This means that we must ensure widespread public dissemination of our regulatory activities and full local consultations on the part of those proposing to do the work.

Thirdly, the regulatory body set up to oversee genetic manipulations must have enforceable powers to require the notification of planned release proposals, and must also be able to require notification of the outcome and data generated by each release. The experience gained from each release is a vital part of our learning process, of development of risk assessment criteria, and of movement away from a rigid case-by-case approach.

Let me repeat something that I have already said. This Conference gives us a unique opportunity to begin to make the public aware of the science that underlies our work, to make the public aware of the care and responsible concern that those engaged in the work devote to it, and to make the public aware that scientists are as concerned in safeguarding our environment as are other members of the community. After all, we must all be concerned about the environment since all of us will spend the rest of our lives in it.

I therefore wish this Conference scientific success. I also hope that its proceedings will attract wide public notice and that they will rightly be regarded as an important step forward in applying new knowledge to the benefit of mankind.

2 Engineering Organisms for Safety: What is Necessary?

ROY CURTISS III

Department of Biology, Washington University, St Louis, MO 63130, USA

HISTORICAL BACKGROUND

In considering the risks and benefits of using the new recombinant DNA technology, the Asilomar Conference proposed the use of two means of reducing the consequences of any perceived hazards (Berg *et al.*, 1975). *Physical containment* standards were proposed to reduce the likelihood that the genetically-modified organisms might escape from the research laboratory while *biological containment* was designed to reduce the probability of the survival of recombinant DNA in the environment if the organism or vector containing it were to escape from the laboratory. These recommendations were eventually adopted in a modified form and promulgated as the National Institutes of Health Guidelines for Recombinant DNA Molecule Research (National Institutes of Health, 1976). We (Curtiss *et al.*, 1976a,b, 1977b) and others (Armstrong *et al.*, 1977; Blattner *et al.*, 1977; Bolivar *et al.*, 1977; Leder *et al.*, 1977; Dougan *et al.*, 1987) worked diligently during the year after Asilomar to develop *Escherichia coli* hosts and phage and plasmid vectors to provide a level of biological containment called for by the newly adopted National Institutes of Health Guidelines. During the ensuing years, risk assessment experiments were designed and conducted by a large number of individuals expert in a diversity of subject areas of little familiarity to the early recombinant DNA researchers. Meetings were also held to consider various scenarios according to which recombinant DNA research might cause harm (National Academy of Sciences, 1977; Proceedings, 1978). With each passing year, it became clearer that the use of recombinant DNA techniques to deduce organization and regulation of genetic information from a diversity of organisms was exceedingly safe and this

RELEASE OF GENETICALLY-ENGINEERED MICRO-ORGANISMS ISBN 0–12–677521–4

led to successive revisions in the National Institutes of Health Guidelines to decrease containment conditions for the conduct of most cloning experiments (Guidelines, 1982). With the exception of extending the host-range of genes conferring antibiotic resistance to pathogenic micro-organisms that did not already express them and the cloning of genes for potent polypeptide toxins—two concerns identified before the drafting of any guidelines (Berg *et al.*, 1974), other experiments were judged to be of relatively low risk. During the last few years, concerns have shifted away from the safety of conducting recombinant DNA experiments in laboratories, to the risk and consequence associated with the intentional release of genetically-modified organisms. It also seems that the less visible the organism, the greater the perceived risk and, hence, the emphasis of the present Symposium on the release of genetically-engineered micro-organisms.

GOALS IN RELEASING MICRO-ORGANISMS

Numerous potential objectives may be achieved by releasing micro-organisms into new ecological niches. This is so whether one is dealing with pure cultures of unmodified wild-type micro-organisms or introducing microbes that have been altered by standard genetic means, such as by mutation, or engineered by recombinant techniques and gene cloning. Micro-organisms are released to serve as commensals to benefit health and or productivity of animals and plants. We certainly employ and can develop microbes to use as fungicides, insecticides and even to deal with mollusc, bird and small rodent pests. Since our environment seems to be accumulating pollutants deleterious to the well-being of organisms on which we depend, as well as to ourselves, the release of micro-organisms to destroy or otherwise remove a diversity of pollutants is becoming increasingly likely. Micro-organisms are also used to leach out noxious contaminants from ores or to retrieve and concentrate heavy metals that in and of themselves have value. A principal objective in my own laboratory group is the design and construction of genetically-modified micro-organisms that can be used as live vaccines to induce protective immunity against a diversity of microbial pathogens.

In all of these endeavours, the concept of physical containment as proposed for recombinant DNA activities in laboratory settings no longer holds. One must, therefore, ask whether micro-organisms to be released should be engineered to possess some biological containment attributes to reduce their longevity in the ecological niche to which they are released or to alter their ability to transmit genetic material to other micro-organisms encountered in the new habitat. Such modifications cannot be excessive since the genetically-engineered micro-organism must persist long enough in its new niche to be effective in achieving the task to be performed. At the same time, one must consider the economics of the use of genetically-modified micro-organisms. Obviously, if the micro-organism survives so poorly that it must be repeatedly released, the costs for use are likely to price it out of the market. On the other hand, very long-term persistence would diminish the likelihood of repeat sales and might dissuade commercial development and sales of the genetically-modified micro-organism. Some middle ground is necessary if society is to achieve maximal benefit. However, this must clearly be chosen with due consideration for the potential to cause harm.

ASSESSMENT OF POTENTIAL CONCERNS

Fourteen years ago the gene-cloning scientific community and their adversaries made endless lists of possible genetic constructs and considered a diversity of points of view on the potential risk of conducting each experiment and the consequences and severity of the harm if the experiment created 'new life forms' and if these were to escape the laboratory and 'run wild' (Sinsheimer, 1975; Chargaff, 1976; Curtiss, 1976; Wald, 1976; Cohen, 1977; Curtiss *et al.*, 1977a,b; Davis, 1977; Grobstein, 1977; Krimsky, 1977; National Academy of Sciences, 1977; US House of Representatives, 1977; Curtiss, 1978a,b; Gorbach, 1978a,b; US Senate, 1978). This exercise led to inflammatory rhetoric but, in the end, was beneficial because after all the argument, the scientific community gradually reached a consensus that most of the speculated harmful consequences were pretty much figments of our collective imaginations. In the present era, it would therefore be useful to consider scenarios for causing harm when specific genetically-modified micro-organisms are intentionally released and then to evaluate the probability of harm and to examine the possible consequences of that harm. Such an assessment needs to be subdivided so that one first considers the probabilities and consequences of harm associated with the specific micro-organism *per se* and second, considers the consequences of transfer of genetic information to indigenous micro-organisms. In the latter case, the likelihood for transfer and the consequences of that transfer both require attention. It must be admitted that there are a great number of generalizations but few, if any, specific examples of potential harm associated with the release of genetically-modified micro-organisms designed to fulfil a specific need.

MEANS BY WHICH THE PROBABILITY FOR SURVIVAL AND GENE TRANSMISSION IS DECREASED

On the assumption that it might be useful, in some situations, to introduce into a given environment micro-organisms that would survive less well than the wild-type of the species, one can consider means by which this may be accomplished. It has often been suggested that the engineering of a micro-organism to produce something not originally produced by the wild-type, would serve to debilitate the microbe and thus lessen the likelihood for its survival. Although this may be true in some situations, and this would be testable, there are situations in which the expressed cloned gene trait might actually enhance the survival and persistence of the released micro-organism.

One means of reducing the likelihood for long-term survival could, therefore, be to introduce mutations such as those that abolish the ability to synthesize cyclic AMP (cAMP) and the cyclic AMP receptor protein (CRP), since strains of enteric bacteria with *cya* and *crp* mutations have a generation time 50% longer than their wild-type parents (Curtiss and Kelly, 1987). Micro-organisms that have the cAMP-CRP means for positive gene regulation use this regulatory system to regulate the expression of a diversity of genes and operons for amino-acid and carbohydrate transport and utilization (Perlman and Pastan, 1969; Alper and Ames, 1970; Pastan and Perlman, 1970; Yokota and Gots, 1970; Rickenberg, 1974; Gemmil *et al.*, 1984) and for the synthesis of glycogen (Dietzler *et al.*, 1977, 1979) and cell-surface structures such as fimbriae (Saier *et al.*, 1978), flagella (Yokota and Gots, 1970; Komeda *et al.*, 1975) and outer-membrane proteins (Movva *et al.*, 1981; Bremer *et al.*, 1988).

A second means of reducing the probability for microbial persistence would be to introduce a gene such as *hok* which encodes a 52-amino acid polypeptide that mediates a lethal collapse of the trans-membrane potential (Gerdes *et al.*, 1986a,b) in such a manner that it occasionally switches on and so kills some fraction of cells in the population. Molin *et al.* (1987) have developed such a system with the flip-flop invertible system for pilus expression in *E. coli* (Abraham *et al.*, 1985; Klemm, 1986) in which the *hok* gene is fused to the *fimA* promoter. Since a certain fraction of bacterial cells die each generation, the effect is effectively to prolong the mean generation time and, thus, diminish the likelihood for establishing a population larger than the wild-type species could achieve in the same ecological niche.

As a third means, it should be possible to engineer a genetically-modified micro-organism so that it would be dependent for growth or survival on a substance present only in the ecological niche into which the micro-organism is introduced. This could be achieved in a variety of ways, perhaps by having a plant which is much more sessile than a microbe, produce the required substance, especially by its roots, so that the microbe would colonize the rhizosphere and there express a trait of benefit to the plant.

If one has concluded that transfer of a recombinant vector is likely to occur and that such transfer would have deleterious consequences, then it is possible to develop means to restrict the vector to the genetically-engineered micro-organism or, alternatively, to endow the vector with a 'kill' function such that it would, upon entrance into another micro-organism, cause its demise. Two means to accomplish this were considered more than 12 years ago. The first approach was to use a plasmid vector with a nonsense mutation in a gene indispensable for plasmid replication and/or maintenance and a suppressor mutation in the chromosome that would allow translational read-through of the message of the gene (Curtiss *et al.*, 1976b). This scenario was predicated in the belief, substantiated by a survey (Robeson *et al.*, 1980), that suppressor-containing *E. coli* were relatively rare (i.e. less than 0.5%) in nature. Unfortunately, such a system has never successfully been developed. Today this approach should be much more feasible, given the substantial information now available about vector genes essential for replication and maintenance. An alternate strategy would be to endow a plasmid vector with a gene for a restriction enzyme where the ability to synthesize the modification methylase would be specified by a chromosomal gene. In this case, transfer of the plasmid to a recipient cell would lead to the synthesis of the restriction enzyme that would digest the DNA of the recipient and cause death. Although the workers at Herbert Boyer's laboratory came close to achieving a functional system of this type, it was never put into practice, in part because of difficulties at the time in down-regulating the expression of the restriction enzyme gene.

If either of the above means to restrict a vector to its host were not easily achievable in the bacterial species to be modified, it would be possible to associate the cloned gene trait with the bacterial chromosome. In this case, the ability of the microbe to persist in the environment would be the only measure for evaluating its potential risks. Several means exist for inserting genes into the chromosome of the host (Grinter, 1983; Barry, 1986; Obukowicz *et al.*, 1986, 1987). For example, a modified Tn*5* vector (Grinter, 1983; Barry, 1986) has been used to introduce the gene for *Bacillus thuringiensis* (BT) toxin into the chromosome of *Pseudomonas fluorescens* (Barry, 1986). It should be noted, however, that this approach eliminates the benefit of high cloned gene expression that is achievable when the cloned gene is on a high-copy number plasmid vector.

One can also consider design of host–vector systems in which the survival of the host is dependent on the vector and where either the maintenance of the vector is dependent on the host or the transfer of the vector would kill the recipient. For example, in terms of host-dependence on the vector, we have developed a balanced lethal construction with a deletion mutation in the chromosome that blocks synthesis of an essential metabolite which is not readily available in nature. We then endowed the cloning vector with a non-homologous gene sequence to complement the chromosomal gene defect (Nakayama *et al.*, 1988). We used a gene involved in diaminopimelic acid (DAP) biosynthesis. DAP is a unique constituent of the rigid layer of the cell walls of Gram-negative bacteria (Schleifer and Kandler, 1972). Inability to synthesize DAP during growth results in DAP-less death with cell lysis and release of cell contents (Bauman and Davis, 1956; Meadow and Work, 1956; Rhuland, 1957; Meadow *et al.*, 1957). Since DAP is only synthesized by bacteria and a few plant species, a bacterial strain unable to synthesize DAP because of a deletion mutation and containing a plasmid vector with a complementing *dap*$^+$ gene constitutes a balanced lethal system in which the bacterium is unable to survive in the absence of the vector. To achieve this we used deletions of the gene for aspartate β-semialdehyde dehydrogenase (*asd*) absence of which results in obligate requirements for threonine and methionine, which are prevalent in nature and DAP, which is rare in nature. We then made plasmid cloning vectors with the *asd*$^+$ gene cloned from *Streptococcus mutans*, a principal causative agent of dental caries. The *S. mutans asd*$^+$ gene is expressed at exceedingly high levels in enteric bacteria (Jagusztyn-Krynicka *et al.*, 1982; Nakayama and Curtiss, unpublished) because of the high affinity of the *asd* promoter for RNA polymerase (Cardineau and Curtiss, 1987). To limit the vector to the host one could endow the *asd*$^+$ vector with a kill-function; perhaps a *hok* gene could be fused to a promoter such as P*lac*, that would be repressed in a bacterial host with a chromosomal *lac*Iq gene. Transfer of such a high-copy number plasmid to wild-type strains with *lac*I$^+$ genes would lead to derepression of the P*lac-hok* fusion and death of the recipient. Alternatively, the vector might have a gene for a restriction enzyme when the gene for the modification methylase is on the chromosome. If one endowed either of these organisms with a diminished ability to grow or persist in a given habitat, the result would be a safe, contained microbe to accomplish a specific objective.

GENETICALLY-MODIFIED MICRO-ORGANISMS AS SAFE, LIVE VACCINES

Since use of live bacterial strains for oral immunization of animals and humans constitutes intentional release of micro-organisms and since these live vaccine strains are invariably avirulent mutants of pathogens that are, by using recombinant DNA techniques, endowed with abilities to express colonization and/or virulence antigens from other pathogens, it is appropriate to discuss in a specific way how we have limited the ability of these strains to persist and/or cause harm. *Salmonella typhimurium*, upon ingestion, attaches to, invades and persists in the gut-associated lymphoid tissue (GALT or Peyer's patches) before reaching deep tissues such as the liver and spleen to cause a fatal bacteraemia in mice (Carter and Collins, 1974). Since delivery of antigens to the GALT stimulates a generalized secretory immune response (Cebra *et al.*, 1976; Bienenstock *et al.*, 1978; Weisz-Carrington *et al.*, 1979; McCaughan and Basten, 1983; Lefever and Joel, 1984) and leads to humoral and cellular immunity (Elson and Ealding, 1984; Tramont *et al.*, 1984; Stevenson and Manning, 1985; Clements *et al.*, 1986; Dougan *et al.*, 1986; Brown *et al.*, 1987; Curtiss *et al.*, 1987), and since *Salmonella* can be rendered avirulent by a diversity of mutations (Bacon *et al.*, 1950, 1951; Germanier and Furer, 1971, 1975; Hoiseth and Stocker, 1981; Curtiss and Kelly, 1987; McFarland and Stocker, 1987; Curtiss *et al.*, 1987) without impairing its tissue tropism to the GALT (Clements and El-Morshidy, 1984; Curtiss *et al.*, 1987; Curtiss and Kelly, 1987; Dougan *et al.*, 1987; Maskell *et al.*, 1987), such avirulent *Salmonella* can be genetically-engineered to express colonization or virulence antigens from other pathogens (Clements and El-Morshidy, 1984; Stevenson and Manning, 1985; Clements *et al.*, 1986; Dougan *et al.*, 1986, 1987; Maskell *et al.*, 1986; Curtiss *et al.*, 1987, 1988a,b) and serve as vectors to deliver those antigens to the GALT to induce secretory, humoral and cellular immune responses against both the foreign antigens and the *Salmonella* vector. We and others have demonstrated that absence of a plasmid contributing to the virulence of *Salmonella* (Jones *et al.*, 1982) is without effect on the ability of *Salmonella* to attach to, invade and persist in the GALT but reduces the ability of the *Salmonella* to reach and/or survive in the liver and spleen (Hackett *et al.*, 1986; Pardon *et al.*, 1986; Gulig and Curtiss, 1987). Similarly, we have shown that Δ*cya* Δ*crp* *Salmonella* mutants are likewise able to attach to, invade and persist in the GALT, but also have a greatly diminished ability to reach and/or survive in the liver and spleen (Curtiss and Kelly, 1987). Either of these means of attenuation is effective in reducing the virulence of the *Salmonella* relative to the virulence of the wild-type parent from 500-fold in the case of plasmid loss, to 10 000-fold in the case of the Δ*cya* and Δ*crp* mutations (Curtiss and Kelly, 1987; Curtiss *et al.*, 1988b) mutations. More importantly, mice surviving infection with these mutant derivatives achieved a high level of long-lasting immunity to subsequent infection with wild-type virulent *Salmonella* (Curtiss *et al.*, 1988a,b).

Since the US Food and Drug Administration does not permit the use of micro-organisms that express antibiotic resistance as live vaccines, we have developed Asd^+ vectors as described above for use in Δ*cya* Δ*crp* Δ*asd* *S. typhimurium* strains (Nakayama *et al.*, 1988). In this case, the plasmid was selectively maintained and 100% of the micro-organisms expressed the cloned gene trait whenever the micro-organisms were recovered from an immunized animal, even after several weeks. It should be noted that the Δ*asd* mutation renders *S. typhimurium* completely avirulent but that introduction of the Asd^+ vector into such mutants leads to wild-type virulence. These Δ*cya* Δ*crp* Δ*asd* hosts with Asd^+ vectors which express colonization and virulence antigens from various pathogens should, therefore, be maximally effective in inducing secretory, humoral, and cellular immune responses against a diversity of pathogens, since they are especially effective against pathogens that colonize or invade through a mucosal surface.

The Δ*cya* Δ*crp* Δ*asd* *Salmonella* vaccine strains with the Asd^+ vector have several safety features. First, all of the mutations are deletions and cannot revert. However, suppressor mutations do occur in Δ*cya* Δ*crp* strains but are exceedingly rare. These suppressor mutations result in promoters that are less dependent on activation of transcription by the CRP protein but the suppressor-containing revertants so-far tested remained avirulent and are immunogenic. Strains with either Δ*cya* Δ*crp* mutations are completely avirulent and immunogenic. Transduction of Δ*cya* and Δ*crp* mutations to cya^+ or crp^+, respectively, occurs at reduced frequencies because of the sizes of the deletions. The Δ*cya* and Δ*crp* mutations are 11 min apart on the chromosome (Sanderson and Hurley, 1987) and could only be restored to the wild-type state by a mating with a Hfr-type donor; such donors being as yet undiscovered in either *E. coli* or *Salmonella* in nature. The pStSR100 virulence plasmid is not self-transmissible and could only be re-introduced into the *Salmonella* vaccine strain by conjugation with a *Salmonella* possessing the virulence plasmid and a conjugative plasmid. It is important that the basis for avirulence cannot be phenotypically reversed by anything in the diet or anything supplied by the host. The strains are easy to grow and store and they exhibit very high-level expression of foreign genes with high stability of the Asd^+ vectors, even in the presence of DAP. Significantly, the absence of drug resistance genes eliminates any selective advantage due to the widespread use of antibiotics, especially in agriculture.

Since oral immunization of animals with live vaccine strains leads to the persistence of the organism by colonization of the small and large bowel and invasion and persistence in the GALT, it should be evident that microorganisms will be shed in the faeces. We have shown that such avirulent *Salmonella* strains can be fed to mice and induce immunity. Therefore, it can be surmised that the survival of the vaccine strains and their distribution might lead to immunization of animals that were not initially vaccinated. In many instances, especially in the control of diseases in farm animal populations or in human populations in parts of the world, such secondary spread might be desirable. In fact, effective control of poliomyelitis is partially due to continued excretion of attenuated poliovirus and involuntary vaccination of all those who were not vaccinated (Salk, 1980). When contending with a disease such as rabies, one might even desire to achieve a ubiquitous spread of the live vaccine strain throughout populations of all the animal reservoirs for rabies (Fenner *et al.*, 1974). Nevertheless, it will be important to investigate the ability of the avirulent Δ*cya* Δ*crp* Δ*asd* *S. typhimurium* strain with the Asd^+ vector to survive in appropriate environments and to determine whether animal-to-animal spread is possible.

SUMMARY

After establishing the intended purpose for releasing a genetically-modified micro-organism, one must assess the potential hazards in contrast to the potential or real benefits. The microbe can be designed and constructed to achieve the established goals and then, if perceived risks are contemplated, the microbe can be evaluated for its likelihood of causing harm. If none can be envisioned, then it is appropriate to proceed cautiously.

Although it is prudent to proceed in the manner noted and scientifically inappropriate to do otherwise, it would be remiss if it were not stated that not a single, specific example of a possible harmful consequence in the use of any genetically-modified micro-organism designed to achieve a beneficial purpose has yet come to light. Because of this, there is no need, unless valid perceived risks are apparent, to require individuals to consider introducing into genetically-modified micro-organisms attributes not necessary to achieve the beneficial purpose of the construct but rather solely to provide perceived, enhanced biological containment and/or safety.

ACKNOWLEDGEMENTS

I thank Kathleen Eaton and Sandra Kelly for editorial assistance in the preparation of the manuscript.

REFERENCES

Abraham, J. M., Freitag, C. S., Clements, J. R. and Eisenstein, B. I., 1985. An invertible element of DNA controls phase variation of type 1 fimbriae of *Escherichia coli*. *Proceedings of the National Academy of Sciences, USA* **82,** 5724–5727.

Alper, M. D. and Ames, B. N., 1970. Transport of antibiotics and metabolite analogs by systems under cyclic AMP control: positive selection of *Salmonella typhimurium cya* and *crp* mutants. *Journal of Bacteriology* **133,** 149–157.

Armstrong, K. A., Hershfield, V. and Helinski, D., 1977. Gene cloning and containment properties of plasmid ColEl and its derivatives. *Science* **196,** 172–174.

Bacon, G. A., Burrows, T. W. and Yates, M., 1950. The effects of biochemical mutation on the virulence of *Bacterium typhosum*: the virulence of mutants. *British Journal of Experimental Pathology* **32,** 714–724.

Bacon, G. A., Burrows, T. W. and Yates, M., 1951. The effects of biochemical mutation on the virulence of *Bacterium typhosum*: the loss of virulence of certain mutants. *British Journal of Experimental Pathology* **32,** 85–96.

Barry, G. F., 1986. Permanent insertion of foreign genes into the chromosomes of soil bacteria. *Biotechnology* **4,** 446–449.

Bauman, N., Davis, B. D., 1957. Selection of auxotrophic bacterial mutants through diaminopimelic acid or thymine deprival. *Science* **126,** 170.

Berg, P., Baltimore, D., Boyer, H. W., Cohen, S. N., Davis, R. W., Hogness, D. S., Nathans, D., Roblin, R., Watson, J. D., Weissman, S. and Zinder, N. D., 1974. Potential biohazards of recombinant DNA molecules. *Science* **185,** 303.

Berg, P., Baltimore, D., Brenner, S., Roblin, R. O. and Singer, M., 1975. Summary statement of the Asilomar conference on recombinant DNA molecules. *Proceedings of the National Academy of Sciences, USA* **72,** 1981–1984.

Bienenstock, J., McDermott, M., Befus, D. and O'Neil, L. M., 1978. A common mucosal immunologic system involving the bronchus, breast, and bowel. *Advances in Experimental Medicine and Biology* **107,** 53–59.

Blattner, F. R., Williams, B. C., Blechl, A. E., Denniston-Thompson, K., Faber, H. E., Furlong, L. A., Granwald, D. J., Kiefer, D. O., Moore, D. D., Shumm, J. W., Sheldon, E. L. and Smithies, O., 1977. Charon phages: safer derivatives of bacteriophage lambda for DNA cloning. *Science* **196,** 163–169.

Bolivar, F., Rodriguez, R. L., Greene, P. J., Betlach, M. C., Heynecker, H. L., Boyer, H. W., Crosa, J. H. and Falkow, S., 1977. Construction and characterization of new cloning vehicles, II. A multi-purpose cloning system. *Gene* **2,** 95–113.

Bremer, E., Gerlach, P. and Middendorf, A., 1988. Double negative and positive control of *tsx* expression in *Escherichia coli. Journal of Bacteriology* **170,** 108–116.

Brown, A., Hormaeche, C. E., Demarco de Hormaeche, R., Winther, M., Dougan, G., Maskell, D. J. and Stocker, B. A. D., 1987. An attenuated *aroA Salmonella typhimurium* vaccine elicits humoral and cellular immunity to cloned β-galactosidase in mice. *Journal of Infectious Diseases* **155,** 86–92.

Cardineau, G. A. and Curtiss, R. III., 1987. Nucleotide sequence of the *asd* gene of *Streptococcus mutans*: identification of the promoter region and evidence for attenuator like sequences preceding the stuctural gene. *Journal of Biological Chemistry* **262,** 3344–3353.

Carter, P. B. and Collins, F. M., 1974. The route of enteric infection in normal mice. *Journal of Experimental Medicine* **139,** 1189–1203.

Cebra, J. J., Gearhart, P. J., Kamat, R., Robertson, S. M. and Tseng, J., 1976. Origin and differentiation of lymphocytes involved in the secretory IgA response. *Cold Spring Harbor Symposia on Quantitative Biology* **41,** 201–221.

Chargaff, E., 1976. On the dangers of genetic meddling. *Science* **192,** 938–940.

Clements, J. D. and El-Morshidy, S., 1984. Construction of a potential live oral bivalent vaccine for typhoid fever and cholera-*Escherichia coli*-related diarrheas. *Infection and Immunity* **46,** 564–569.

Clements, J. D., Lyon, F. L., Lowe, K. L., Farrand, A. L. and El-Morshidy, S., 1986. Oral immunization of mice with attenuated *Salmonella enteritidis* containing a recombinant plasmid which encodes for production of the B subunit of heat-labile *Escherichia coli* enterotoxin. *Infection and Immunity* **53,** 685–692.

Cohen, S. N., 1977. Recombinant DNA—fact or fiction? *Science* **195,** 654–657.

Curtiss, R. III, 1976. Genetic manipulation of microorganisms: potential benefits and biohazards. *Annual Review of Microbiology* **30,** 507–533.

Curtiss, R. III, 1978a. A scientist's perspective on regulation of recombinant DNA activities. In *Genetics of Industrial Microorganisms*, Sebek, O. K. and Laskin, A. I. (eds), pp. 242–250. Washington, D.C.: American Society for Microbiology.

Curtiss, R. III, 1978b. Biological containment and cloning vector transmissibility. *Journal of Infectious Diseases* **137,** 668–675.

Curtiss, R. III and Kelly, S. M., 1987. *Salmonella typhimurium* deletion mutants lacking adenylate cyclase and cyclic AMP receptor protein are avirulent and immunogenic. *Infection and Immunity* **55,** 3035–3043.

Curtiss, R. III, Pereira, D. A., Hsu, J. C., Hull, S. C., Clark, J. E., Maturin, L. J., Sr., Goldschmidt, R., Moody, R., Inoue, M. and Alexander, L., 1976a. Biological containment: the subordination of *Escherichia coli* K-12. In *Recombination molecules: Impact on Science and Society*, Beers, R. F. and Bassett, E. G. (eds), pp. 45–56. New York: Raven Press.

Curtiss, R. III, Szybalski, W., Helsinki, D. R. and Falkow, S., 1976b. Workshop on design and testing of safer prokaryotic vehicles and bacterial hosts for research on recombinant DNA molecules. *ASM News* **42,** 134–138.

Curtiss, R. III, Clark, J. E., Goldschmidt, R., Hsu, J. C., Hull, S. C., Inoue, M., Maturin, L. J., Sr., Moody, R. and Pereira, D. A., 1977a. Biohazard assessment of recombinant DNA molecule research. In *Plasmids: Medical and Theoretical Aspects*, Mitsuhashi, S., Rosival, L. and Kremery, V. (eds), pp. 375–387. Prague: Avicenum, Czechoslovak Medical Press.

Curtiss, R. III, Inoue, M., Pereira, D., Hsu, J. C., Alexander, L. and Rock, L., 1977b. Construction and use of safer bacterial host strains for recombinant DNA research. In *Molecular Cloning of Recombinant DNA*, Scott, W. A. and Werner, R. A. (eds), pp. 99–114. New York: Academic Press.

Curtiss, R. III, Goldschmidt, R. M., Kelly, S. M., Lyons, M., Michalek, S., Pastian, R. and Stein, S., 1987. Recombinant avirulent *Salmonella* for oral immunization to induce mucosal immunity to bacterial pathogens. In *Vaccines: New Concepts and Developments*, Kohler, H. and LoVerde, P. T. (eds), pp. 261–271, *Proceedings of the 10th International Convocation on Immunology*. Harlow: Longman Scientific and Technical.

Curtiss, R. III, Goldschmidt, R. M., Fletchall, N. B. and Kelly, S. M., 1988a. Avirulent *Salmonella typhimurium* Δ*cya* Δ*crp* oral vaccine strains expressing a streptococcal colonization and virulence antigen. *Vaccine*, in press.

Curtiss, R. III, Kelly, S. M., Gulig, P. A., Gentry-Weeks, C. R. and Galan, J. E., 1988b. Avirulent *Salmonella* expressing virulence antigens from other pathogens for use as orally-administered vaccines. In *Virulence Mechanisms*, Roth, J. (ed.), pp. 311–328. Washington DC: American Society for Microbiology.

Davis, B. D., 1977. Epidemiological and evolutionary aspects of research on recombinant DNA. In *Research with Recombinant DNA: An Academy Forum*, pp. 124–135. Washington DC: National Academy of Sciences.

Dietzler, D. N., Leckie, M. P., Sternheim, W. L., Taxman, T. L., Unger, J. M. and Porter, S. E., 1977. Evidence for the regulation of bacterial glycogen synthesis by cyclic AMP. *Biochemical and Biophysical Research Communications* **77,** 1468–1477.

Dietzler, D. N., Leckie, M. P., Magnini, J. L., Sughrue, M. J., Bergstein, P. E. and Sternheim, W. L., 1979. Contribution of adenosine cyclic 3′ : 5′-monophosphate to the regulation of bacterial glycogen synthesis *in vivo*. *Journal of Biological Chemistry* **254,** 8308–8317.

Dougan, G., Sellwood, R., Maskell, D. J., Sweeney, K., Liew, F. Y., Beesley, J. and Hormaeche, C., 1986. *In vivo* properties of a cloned K88 adherence antigen determinant. *Infection and Immunity* **52,** 344–347.

Dougan, G., Hormaeche, C. E. and Maskell, D. J., 1987. Live oral *Salmonella* vaccines: potential use of attenuated strains as carriers of heterologous antigens to the immune system. *Parasite Immunology* **9,** 151–160.

Elson, C. O. and Ealding, W., 1984. Generalized systemic and mucosal immunity in mice after mucosal stimulation with cholera toxin. *Journal of Immunology* **132,** 2736–2741.

Fenner, F., McAuslan, B. R., Mims, C. A., Sambrook, J. and White, D. O., 1974. *The Biology of Animal Viruses*, 2nd edn. NewYork: Academic Press.

Gemmill, R. M., Tripp, M., Friedman, S. B. and Calvo, J. M., 1984. Promoter mutation causing catabolite repression of the *Salmonella typhimurium* leucine operon. *Journal of Bacteriology* **158,** 948–953.

Gerdes, K., Bech, F. W., Jorgensen, S. T., Lobner-Olesen, A., Rasmussen, P. B., Atlung, T., Boe, L., Karlstrom, O., Molin, S. and Von Meyenburg, K., 1986a. Mechanism of postsegregational killing by the *hok* gene product of the *parB* system of plasmid R1 and its homology with the *relF* gene product of the *E. coli relB* operon. *The EMBO Journal* **5,** 2023–2029.

Gerdes, K., Rasmussen, P. B. and Molin, S., 1986b. Unique type of plasmid maintenance function: Postsegregational killing of plasmid-free cells. *Proceedings of the National Academy of Sciences, USA* **83,** 3116–3120.

Germanier, R. and Furer, E., 1971. Immunity in experimental salmonellosis. II. Basis for the avirulence and protective capacity of *galE* mutants of *Salmonella typhimurium. Journal of Infectious Diseases.*

Germanier, R. and Furer, E., 1975. Isolation and characterization of GalE mutant Ty21a of *Salmonella typhi*: a candidate strain for a live, oral typhoid vaccine. *Journal of Infectious Diseases* **131,** 553–558.

Gorbach, S. L., 1978a. Recombinant DNA: an infectious disease prospective. *Journal of Infectious Diseases* **137,** 615–623.

Gorbach, S. L., 1978b. Risk assessment protocols for recombinant DNA experimentation. *Journal of Infectious Diseases* **137,** 704–708.

Grinter, N. J., 1983. A broad-host-range cloning vector transposable to various replicons. *Gene* **21,** 133–143.

Grobstein, C., 1977. The recombinant DNA debate. *Scientific American* **237,** 22–33.

Guidelines for Research Involving Recombinant DNA Molecules and Actions under Guidelines, 1982. *Federal Register* **47,** 17166–17198.

Gulig, P. A. and Curtiss, R. III, 1987. Plasmid-associated virulence of *Salmonella typhimurium. Infection and Immunity* **55,** 2891–2901.

Hackett, J., Kotlarski, I. K., Mathan, V., Francki, K. and Rowley, D., 1986. The colonization of Peyer's patches by a strain of *Salmonella typhimurium* cured of the cryptic plasmid. *Journal of Infectious Diseases* **153,** 1119–1125.

Hoiseth, S. and Stocker, B. A. D., 1981. Aromatic-dependent *Salmonella typhimurium* are nonvirulent and effective as live vaccines. *Nature* **291,** 238–239.

Jagusztyn-Krynicka, E. K., Smorawinska, M. and Curtiss, R. III, 1982. Expression of *Streptococcus mutans* aspartate semialdehyde dehydrogenase gene cloned into plasmid pBR322. *Journal of General Microbiology* **128,** 1135–1145.

Jones, G. W., Rabert, D. K., Svinarich, D. M. and Whitfield, H. J., 1982. Association of adhesive, invasive, and virulent phenotypes of *Salmonella typhimurium* with autonomous 60-megadalton plasmids. *Infection and Immunity* **338,** 476–486.

Klemm, P., 1986. Two regulatory *fim* genes, *fimB* and *fimE*, control the phase variation of type 1 fimbriae in *Escherichia coli. The EMBO Journal* **5,** 1389–1393.

Komeda, Y., Suzuki, H., Ishidsu, J. and Iino, T., 1975. The role of cAMP in flagellation of *Salmonella typhimurium. Molecular and General Genetics* **142,** 289–298.

Krimsky, S., 1977. Recombinant DNA research: public must regulate. *Chemical Engineering News* 30 May 1977, 36–41.

Leder, P., Tiemeier, D., Tilghman, S. and Enquist, L., 1977. Use of an EK-2 vector for the cloning of DNA from higher organisms. In *Molecular Cloning of Recombinant DNA*, Scott, W. H. and Werner, R. A. (eds), pp. 205–217. New York: Academic Press.

Lefever, M. E. and Joel, D. D., 1984. Peyer's patch epithelium: an imperfect barrier. In *Intestinal Toxicology*, Schiller, C. M. (ed.), pp. 45–56. New York: Raven Press.

Maskell, D., Liew, F. Y., Sweeney, K., Dougan, D. and Hormaeche, C., 1986. Attenuated *Salmonella typhimurium* as live oral vaccines and carriers for delivering antigens to the secretory immune system. In *Vaccines 86*, pp. 213–217. Cold Spring Harbor: Cold Spring Harbor Laboratory.

Maskell, D. J., Sweeney, K. J., O'Callaghan, D., Hormaeche, C. E., Liew, F. Y. and Dougan, G., 1987. *Salmonella typhimurium aroA* mutants as carriers of the *Escher-*

chia coli heat-labile enterotoxin B subunit to the murine secretory and systemic immune systems. *Microbial Pathogenesis* **2,** 211–221.

McCaughan, G. and Basten, A., 1983. Immune system of the gastrointestinal tract. *International Review of Physiology* **28,** 131–157.

McFarland, W. C. and Stocker, B. A. D., 1987. Effect of different purine auxotrophic mutations on mouse-virulence of a Vi-positive strain of *Salmonella dublin* and of two strains of *Salmonella typhimurium. Microbial Pathogenesis* **3,** 109–116.

Meadow, P. and Work, E., 1956. Interrelationships between diaminopimelic acid, lysine and their analogues in mutants of *Escherichia coli. Biochemical Journal* **64,** 11.

Meadow, P., Hoare, D. S. and Work, E., 1957. Interrelationships between lysine and αε-diaminopimelic acid and their derivatives and analogues in mutants of *Escherichia coli. Biochemical Journal* **66,** 270–282.

Molin, S., Klemm, S., Poulsen, L. K., Biehl, H., Gerdes, K. and Andersson, P., 1987. Conditional suicide system for containment of bacteria and plasmids. *Biotechnology* **5,** 1315–1318.

Movva, R. N., Green, P., Nakamura, K. and Inouye, M., 1981. Interaction of cAMP receptor protein with the *ompA* gene, a gene for a major outer membrane protein of *Escherichia coli. FEBS Letters* **128,** 186–190.

Nakayama, K., Kelly, S. M. and Curtiss, R. III, 1988. Construction of an Asd$^+$ expression-cloning vector: stable maintenance and high level expression of cloned genes in a *Salmonella* vaccine strain. *Bio/Technology*, in press.

National Academy of Sciences, USA:, 1977. *Research with Recombinant DNA: An Academy Forum.* Washington DC: National Academy of Sciences.

National Institutes of Health, 1976. Recombinant DNA research: guidelines for recombinant molecule research. *Federal Register* **41,** 37902–37943.

Obukowicz, M. G., Perlak, F. J., Bolten, S. L., Kusano-Kretzmer, K., Mayer, E. J. and Watrud, L. S., 1987. IS*50L* as a non-self transposable vector used to integrate the *Bacillus thuringiensis* delta-endotoxin gene into the chromosome of root-colonizing pseudomonads. *Gene* **51,** 91–96.

Obukowicz, M. G., Perlak, F. J., Kusano-Kretzmer, K., Mayer, E. J., Bolten, S. L. and Watrud, L. S., 1986. Tn*5*-mediated integration of the delta-endotoxin gene from *Bacillus thuringiensis* into the chromosome of root-colonizing Pseudomonads. *Journal of Bacteriology* **168,** 982–989.

Pardon, P., Popoff, M. Y., Coynault, C., Marly, J. and Miras, I., 1986. Virulence-associated plasmids of *Salmonella* serotype typhimurium in experimental murine infection. *Annales de Microbiologie (Institut Pasteur)* **137B,** 47–60.

Pastan, I. and Perlman, R. L., 1970. Cyclic adenosine monophosphate in bacteria. *Science* **169,** 339–344.

Perlman, R. L. and Pastan, I., 1969. Pleiotropic deficiency of carbohydrate utilization in an adenyl cyclase deficient mutant of *Escherichia coli. Biochemical and Biophysical Research Communications* **37,** 151–157.

Proceedings from a Workshop on Risk Assessment of Recombinant DNA Experimentation with *Escherichia coli* K-12, 1978. *Journal of Infectious Diseases* **137,** 613–714.

Rhuland, L. E., 1957. Role of αε-diaminopimelic acid in the cellular integrity of *Escherichia coli. Journal of Bacteriology* **73,** 778–783.

Rickenberg, H. V., 1974. Cyclic AMP in prokaryotes. *Annual Reviews of Microbiology* **28,** 353–369.

Robeson, J. P., Goldschmidt, R. and Curtiss, R. III, 1980. Potential of *Escherichia coli* isolated from nature to propagate cloning vectors. *Nature* **283,** 104–106.

Saier, M. H., Schmidt, M. R. and Leibowitz, M., 1978. Cyclic AMP-dependent synthesis of fimbriae in *Salmonella typhimurium*: effects of *cya* and *pts* mutations. *Journal of Bacteriology* **134,** 356–358.

Salk, D., 1980. Eradication of poliomyelitis in the United States. III. Polio vaccines—practical considerations. *Reviews of Infectious Diseases* **2,** 258.

Sanderson, K. E. and Hurley, J. A., 1987. Linkage map of *Salmonella typhimurium*. In *Escherichia coli and Salmonella typhimurium: Cellular and Molecular Biology,* **Vol. 2,** Neidhardt, F. C. (ed.), pp. 877–918. Washington DC: American Society for Microbiology.

Schleifer, K. H. and Kandler, O., 1972. Peptidoglycan types of bacterial cell walls and their taxonomic implications. *Bacteriological Reviews* **36,** 407–477.

Sinsheimer, R. L., 1975. Troubled dawn for genetic engineering. *New Scientist* 148–151.

Stevenson, G. and Manning, P. A., 1985. Galactose epimeraseless (*galE*) mutant G30 of *Salmonella typhimurium* is a good potential live oral vaccine carrier for fimbrial antigens. *FEMS Microbiology Letters* **28,** 317–321.

Tramont, E. C., Chung, R., Berman, S., Keren, D., Kapfer, C. and Formal, S. B., 1984. Safety and antigenicity of typhoid-*Shigella sonnei* vaccine (Strain 5076-1C). *Journal of Infectious Diseases* **149,** 133–136.

US House of Representatives, 1977. Science policy implications of DNA recombinant molecule research. Hearings before the Subcommittee on Science, Research and Technology of the Committee on Science and Technology. Washington DC: US Government Printing Office.

US Senate, 1978. Regulation of recombinant DNA research. Hearings before the Subcommittee on Science, Technology, and Space of the Committee on Commerce, Science, and Transportation. Washington DC: US Government Printing Office.

Wald, G., 1976. The case against genetic engineering. *Transactions of the New York Academy of Sciences* **16,** 6–11.

Weisz-Carrington, P., Roux, M., McWilliams, M., Phillips-Quagliata, J. M. and Lamm, M. E., 1979. Organ and isotype distribution of plasma cells producing specific antibody after oral immunization: evidence for a generalized secretory immune system. *Journal of Immunology* **123,** 1705–1708.

Yokota, R. and Gots, J. S., 1970. Requirement of adenosine 3′, 5′-cyclic phosphate for flagella formation in *Escherichia coli* and *Salmonella typhimurium. Journal of Bacteriology* **103,** 513–516.

3 Engineering Organisms for Use

JULIAN DAVIES

Département des Biotechnologies, Institut Pasteur, 75724 Paris, Cedex 15, France

One of the principal commercial goals of genetic engineering has been to develop microbes, bacteria and yeasts as 'factories' to produce large amounts of otherwise rare proteins. From a scientific and commercial point of view, this enterprise has been extremely successful and a number of mammalian proteins has been obtained in quantities sufficient for pharmaceutical development (Table 1). However, this is not the only goal of genetic engineering for profit and considerable effort has gone into attempts to improve the capacity of microbes to over-produce amino acids, vitamins or antibiotics, or to ameliorate their natural ability to degrade and metabolize diverse and exotic substrates. In

Table 1. Some mammalian proteins produced by genetic engineering in microbes.*

Protein	Application
Insulin†	Diabetes
Interferon α† Interferon β† Interferon γ	Treatment of various forms of cancer
Human growth hormone†	Dwarfism
Bovine growth hormone	Growth, milk production
Interleukin 1 (a + b) Interleukin 2	Cancer
Colony stimulating factors	Adjuvants, cancer, etc. Graft therapy
Hepatitis B surface protein† (and other viral antigens)	Vaccine

*Not complete.
†Current commercial products.

RELEASE OF GENETICALLY-ENGINEERED MICRO-ORGANISMS ISBN 0–12–677521–4

addition, the microbial production of enzymes for food processing and other applications has been the target of extensive recombinant DNA studies. The net result is that numerous engineered, improved micro-organisms with a variety of different phenotypes have been produced for potential industrial use. In some of these cases the change in phenotype confers important new properties in relationship to the environment, including the overproduction of proteases, amylases, cellulases, etc., or the capability to metabolize or produce large amounts of complex organic molecules. In such industrial applications the whole organism is employed in a given conversion or in manufacturing processes with specific bioreactor conditions.

However, in other cases a dramatic change in phenotype does not occur. For example, a microbe which overproduces interferon does not differ phenotypically from its non-interferon producing parent, apart from growth rate (see below). The application here is primarily to produce new and needed therapeutic agents on a scale that permits commercial expoitation in the pharmaceutical market. Lack of a clearly distinguishable phenotype after *in vitro* genetic manipulation can be somewhat of a disadvantage, as we shall discuss later.

What is required for the hyper-production of a given protein in a microbial host? If we put aside for a moment the role of the host, improved gene expression is attained routinely by cloning the open-reading frame for a protein into an 'expression'. The latter, a plasmid, possesses optimal nucleotide sequences for transcription and translation of an appropriately inserted gene (Rodriguez and Denhardt, 1987). Depending on the promoter system used, transcription may be constitutive or regulated. Table 2 lists promoters that have been used in the production of heterologous proteins in microbes (Reznikoff and McClure, 1986). Some of these promoters such as *tac*, are synthetic and have

Table 2. Microbial gene expression systems for foreign protein production.

	Promoter	Characteristics
Bacteria	lac	Regulated or non-regulated
	trp	Regulated or non-regulated
	tac, trc	Hybrids
	lpL	Regulated (ts) or non-regulated
	lpp	Strong, associated with secreted protein
	rrn	Strong non-regulated; can be used in hybrid form
Yeasts	glycolytic enzymes (PK, ADH, GAPDH)	Non-regulated } and hybrids
	gal	Regulated } and hybrids
	matα	Associated with secreted protein

been designed to combine the most desirable features of two different promoters in attempts to optimize transcription activity. 'Desirable features' are usually deduced by the comparison of oligonucleotide sequences from different transcription systems and subsequent definition of so-called consensus sequences, which represent common or near-universal features involved in control functions. With respect to translation most attention has been given to ribosome binding sites (Shine-Dalgarno sequences); considerable trial-and-error effort has led to identification of consensus oligonucleotide sequences in 5′ untranslated regions of messenger RNA that have varying degrees of complementarity with the 3′ end of 16S ribosomal RNA (Stormo, 1986). In addition, for efficient translation initiation, the distance between the ribosome binding site and the initiation codon is an important variable which often depends on the gene to be expressed (Looman *et al.*, 1986).

When a protein is synthesized in large amounts in a microbe, it frequently accumulates in the form of granules or particles (inclusion bodies) in the cell cytoplasm (Taylor *et al.*, 1986). These inclusion bodies are essentially pure protein in a fully reduced, denatured and insoluble form. Their formation, which may well have deleterious effects on the host, can be advantageous, since the protein so produced may lend itself to convenient physical and chemical methods of purification. However, since the protein in inclusion bodies is denatured and biologically inactive, serious problems are created. Considerable effort has to be expended to solubilize the protein with mild or strong chaotropic agents and then to remove the chaotrope under redox conditions appropriate for subsequent refolding and formation of correct disulphide bridges (Marston, 1987).

The solubilization and refolding process can be complex and is often arrived at by trial and error; our knowledge of the processes of protein folding is still somewhat primitive. The larger and more complex the protein, the more difficult it is to obtain high-efficiency renaturation. In addition, the use of substantial quantities of solvents and chaotropic agents creates environmental disposal problems. Large amounts of urea or guanidinium hydrochloride may have to be employed and their recycling may be only a partial solution. An early initiative in protein engineering involved the replacement of cysteine residues in certain proteins by site-directed mutagenesis *in vitro* to obtain molecules in which the formation of undesirable inter- and intramolecular disulphide bridges would be minimized during the refolding process (Mark *et al.*, 1984).

Because of the difficulties inherent in the solubilization of protein inclusion bodies and their subsequent renaturation and reactivation, and also the problem of protein toxicity, attempts have been made to use the capability of certain micro-organisms to secrete proteins. The coincident signal peptide processing also eliminates the problem of heterogeneity, which is due to the presence of extra *N*-terminal methionine residues, often found on heterologous proteins expressed cytoplasmically in micro-organisms (Nicaud *et al.*, 1986). The *N*-terminal methionine is present as a consequence of the requirement in the translation stage for an initiation codon (ATG). A pharmacologically-active mammalian protein with an extra *N*-terminal amino acid may have altered biological properties and also modified antigenic characteristics. For human therapeutic use it is desirable to use a product as close to the native material as possible. Although methionylaminopeptidases (Ben-Bassat *et al.*, 1987) are present in bacteria such as *Escherichia coli*, the enzymes show certain sequence specificities and can remove only the *N*-terminal methionine when the amino acid residues adjacent to the methionine are compatible. Several methionylaminopeptidases have been developed for post-translational processing, inside or outside the bacterial cell.

Secretion vectors have been developed that promote the export of heterologous proteins into the periplasm of *E. coli*, *Bacillus subtilis* and *Saccharomyces cerevisiae* and these have been used for the secretion of genetically-engineered proteins into the culture medium. It has even been possible to devise systems for *E. coli* that release proteins into the culture medium rather than the periplasm (Abrahmsen *et al.*, 1986). The yields of the heterologous proteins exported or secreted from microbes are variable and, in some cases, processing of the signal peptide during the secretion process may be heterogenous leading to variable *N*-terminal sequences. The latter is obviously a bad feature, since such heterogeneity is to be avoided if the product is to be used pharmaceutically. However, secretion is an extremely valuable approach since the proteins so produced are, by definition, soluble and in their natural active conformation. Purification from culture media presents decided advantages over purification from cell extracts and, in addition, microbial protein secretion systems lend themselves to the development of immobilized cell 'factories'. In some cases full biological activity of a protein requires post-translational modifications, such as glycosylation, which occur during the normal process of export through the endoplasmic reticulum in eukaryotic cells. Since bacteria are not capable of such modifications protein secretion does not solve this particular problem. Should more attention, therefore, be focused on eukaryotic microbes?

High-level expression of genes in microbes can be deleterious to the host and often an effect on growth parameters is evident and, occasionally, marked toxic effects result. The latter are usually due to the production of extremely hydrophobic proteins but the mechanisms for these toxic effects have not been biochemically defined. It is generally assumed that toxicity of hydrophobic proteins is due to non-specific attachment and interference with the function of the host-cell membranes, although other explanations cannot be excluded.

The net result is that, when a micro-organism produces a growth-limiting compound, the spontaneous selection of variants free of this impediment ensues. This may occur because of loss of the plasmid encoding the offending product, or by deletion or some other mechanism which reduces or eliminates its production. Relatively few studies have analysed this problem in detail, even though it must be a fairly common occurrence. A great deal of effort has gone into the construction of host–vector combinations to ensure that the expression plasmid–host combination is *not* disturbed. Several approaches have been used (Loison *et al.*, 1986), usually involving some form of positive selection to maintain the plasmid in the host, or by 'suicide' systems which would eliminate the non-plasmid containing cells that arise spontaneously during growth (Table 3). Unfortunately, no procedure is available to ensure that the heterologous gene itself is not deleted or mutated in an otherwise stably maintained vector.

In any event, to minimize toxic effects of over-expressed proteins in microbes while, at the same time, allowing good production yields, it is advisable and sometimes necessary to use promoters whose function can be regulated. Under these circumstances the expression of the gene product is normally repressed and can be turned-on at an appropriate time, for example when cell mass (fermentation yield) is appropriately high. Since the induction and synthesis of the required protein usually leads to loss of cell viability, the organisms essentially 'die with their boots on'. Various inducible promoters have been developed for this purpose (Table 2); the use of induction methods on a large scale can involve additional and expensive mechanical engineering design.

Of the heterologous genes tested for protein production in micro-organisms, some have provided only low and commercially non-acceptable quantities of the desired product. The reasons for these 'failures' are, in most instances, not well understood. Sometimes extreme toxicity of the product to the host bacterium, even when expressed under carefully regulated conditions, may be the reason. In other cases proteolytic degradation of the heterologous protein in the host cytoplasm contributes to poor recovery, which can occasionally be prevented by using protease-deficient hosts (Goldberg and Goff, 1986). Various microbial host strains have different characteristics with respect to their capacity to maintain plasmids, to transcribe and translate genes and to produce heterologous proteins in the form of inclusion bodies. Very little solid experi-

Table 3. Systems for maintaining plasmids in cells.

Antibiotic resistance
Metal resistance
Antibiotic sensitivity/dependence
Bacteriophage repressor/lysogen
Immunity/toxin
Prototrophy/auxotrophy
Repressor/lysis
Suppressor/nonsense

mental evidence bears on the properties of different hosts, their interactions with different vector systems, and/or the expression of different heterologous proteins.

Finally, what is the potential for untoward characteristics in micro-organisms engineered for the production of heterologous proteins for application in the pharmaceutical, food and other industries? How might such genetically engineered organisms interact with the environment in the event of dispersion? To begin with, recombinant organisms engineered for protein production often incorporate antibiotic resistance genes used for selection and maintenance of the chimeric plasmid in its microbial host. The wisdom of using antibiotic selection for this purpose is questionable and it is obvious that regulatory agencies look askance at production processes requiring the use of antibiotics such as tetracycline and ampicillin at levels of 100 μg ml^{-1} in 10 000 l, or larger, fermentors. The antibiotics can be destroyed by chemical or enzymatic treatment of the spent media but it is preferable that the antibiotics themselves be used only during construction of the producing clone and perhaps for its maintenance in culture collections. More research on methods to prevent spontaneous segregation of expression plasmids during production should be encouraged. What about the accidental dissemination of bacteria carrying plasmids encoding antibiotic resistance? Does this provide another potential source of antibiotic resistance genes in nature? Present guidelines require that industrial processes involving recombinant organisms incorporate procedures for killing of the micro-organism. The likelihood of adding to the existing antibiotic resistance gene pool is very small by comparison with the effects of antibiotic use in hospitals and in agriculture, particularly because transfer must also take place.

Recombinant micro-organisms are engineered, and their products made, under conditions for strict containment. The organisms themselves are ill-fitted for survival outside the laboratory since they do not obviously offer any positive competitive traits. They are extremely unlikely to be capable of establishing themselves in a natural, highly-competitive environment. Most of the recombinant organisms engineered for large scale protein production possess only few additional genes and it is difficult to imagine conditions in which the ability to overproduce a given mammalian protein would be an advantage in competition with healthy wild-type organisms in nature. One can consider 'protein-overproducers' in the same light as any other organism improved for production of a given product; they are usually fickle and fragile beasts. Far too much emphasis has been placed on the methods used to obtain such 'mutant' micro-organisms rather than consideration of the important applications to which they will be put. Admittedly, most of this discussion is only supposition but we have very little knowledge of all the characteristics necessary to allow a 'new' microbe to establish house in a new ecological niche. It is certain that multiple characteristics are necessary and cytokine overproducing *E. coli* HB 101 is not likely to possess these!

Is any hazard associated with the dispersion of bacteria overproducing biologically active mammalian proteins into the environment? There is no easy answer to this question; as always, a hypothetical scenario can be created in which a quantity of, say, interleukin 2 ends up in animal, resulting in an immunological disturbance. Nonetheless, it should be remembered that cytokines and similar proteins are injectable agents. Indeed, the development of effective means of drug delivery remains one of the problems in the use of pharmaceutically-active proteins and, therefore, the chances of potent, biologically active molecules escaping from their hosts in active form and arriving at appropriate receptor sites is *vanishingly small.* The protein, even if made by the producing organism outside the laboratory, would probably be degraded, and highly expensive cytokines, etc., would end up as cheap sources of amino acids for other micro-organisms in nature!

It should be remembered that in these cases we are considering the question of *accidental* dissemination of protein-producing organisms; they are designed to be used in large scale fermentors under carefully controlled conditions. They are not end products designed for environmental use, but biological 'factories' which will be used in large scale fermentations as part of an industrial production process. The environmental issue is that of any waste disposal problem. However, in many other potential beneficial applications of genetic engineering, the recombinant organism is the product: for example, engineered organisms in the conversion of biomass to fuel and food, to degrade pesticides, or to remove environmental pollutants, etc. In these cases, their use outside of a contained laboratory environment is required and this raises different kinds of questions, adequately discussed in other parts of this Symposium.

REFERENCES

Abrahamsen, L., Moks, T., Nilsson, B. and Uhlen, M., 1986. Secretion of heterologous gene products to the culture medium of *Escherichia coli. Nucleic Acids Research* **14,** 7487–7500.

Ben-Bassat, A., Bauer, K., Chang, S. Y., Myambo, K., Boosman, A. and Chang, S., 1987. Processing of the initiation methionine from proteins properties of the *Escherichia coli* methionine peptidase and its gene structure. *Journal of Applied Bacteriology* **169,** 751–757.

Goldberg, A. L. and Goff, S. A., 1986. The selective degradation of abnormal proteins in bacteria. In *Maximising Gene Expression*, Reznikoff, W. and Gold, L. (eds), pp. 287–314. Boston: Butterworth.

Loison, G., Nguyen-Juilleret, M., Alouani, S. and Marquet, M., 1986. Plasmid-transformed *ura3 fur1* double-mutants of *S. cerevisiae*: an autoselection system applicable to the production of foreign proteins. *Bio/Technology* **4,** 433–437.

Looman, A. C., Bodlaender, J., de Gruyter, M., Vogellar, A. and van Knippenberg, P. H., 1986. Secondary structure as primary determinant of the efficiency of ribosome binding sites in *Escherichia coli*. *Nucleic Acids Research* **14,** 5481–5497.

Mark, D. F., Lu, S. D., Greasey, A. A., Yamamoto, R. and Lin, L. S., 1984. Site-specific mutagenesis of the human fibroblast interferon gene. *Proceedings of National Academy of Sciences, USA* **81,** 5662–5666.

Marston, F. A. O., 1987. The purification of eukaryotic proteins expressed in *Escherichia coli*. In *DNA Cloning*, Vol. III, Glover, D. M. (ed.), pp. 59–88. Oxford: IRL Press.

Nicaud, J.-M., Mackman, N. and Holland, I. B., 1986. Current status of secretion of foreign proteins by microorganisms. *Journal of Biotechnology* **3,** 255–270.

Reznikoff, W. S. and McClure, W. R., 1986. *E. coli* promoters. In *Maximising Gene Expression*, Reznikoff, W. and Gold, L. (eds), pp. 1–33. Boston: Butterworth.

Rodriguez, R. L. and Denhardt, D. T., 1987. *Vectors: A Survey of Molecular Cloning Vectors and Their Uses*, pp. 179–225. Boston: Butterworth.

Stormo, G. D., 1986. Translation Initiation. In *Maximising Gene Expression*, Reznikoff, W. and Gold, L. (eds), pp. 195–224. Boston: Butterworth.

Taylor, G., Hoare, M., Gray, D. R. and Marston, F. A. O., 1986. Size and density of protein inclusion bodies. *Bio/Technology* **4,** 553–557.

4 The Survival and Persistence of Genetically-engineered Micro-organisms

JOHN E. BERINGER and MARK J. BALE

Microbiology Department, University of Bristol, Bristol, UK

INTRODUCTION

The only examples of the intentional release of genetically-engineered micro-organisms (GEMs) into the environment have been recent and are few in number. As a result we have little experience of their ability to survive and persist. The basis for the construction of a GEM is that only one or very few defined genes are added or deleted: consequently the GEM differs little from the parent strain (*ca* 1 gene in 2000) and should behave in the same way once released into the environment, unless the added gene, or genes, modify the ecological properties of the strain. It is, therefore, relevant to examine existing information about the release of micro-organisms into the environment.

Two concerns often raised about GEMs are that they might be pathogenic and that they may colonize the environment and displace existing species. Because pathogenic bacteria are released into the environment continuously in sewage they provide a good example of our ability to handle such releases without harm. Millions of hectares of land are inoculated with the bacterium *Rhizobium* each year to improve the growth of leguminous crops. Because of its commercial importance there have been many studies of the survival and persistence of *Rhizobium* stretching back to the beginning of this century. We will consider how existing knowledge about these releases is relevant to the release of GEMs.

THE RELEASE OF POTENTIALLY PATHOGENIC MICRO-ORGANISMS

Released GEMs are highly likely to enter freshwater or marine environments via agricultural run-off (van Donsel *et al.*, 1967), drainage of industrial effluent, or

RELEASE OF GENETICALLY-ENGINEERED MICRO-ORGANISMS ISBN 0-12-677521-4

faecal contamination by birds, animals or man (Edmonds, 1976; White and Godfree, 1985). Pollution by faecal bacteria entering by such routes is routinely monitored in order to maintain potable water quality standards (Olson and Nagy, 1984). Much information concerning the survival and persistence of various enteric bacteria is therefore available. Evidence of faecal contamination is obtained by the detection of various indicator species (Elliot and Colwell, 1985), including *Escherichia coli*, total coliforms (*Escherichia*, *Klebsiella* and *Enterobacter* spp.), anaerobic bacteria (*Bacteroides* and *Bifidobacterium* spp., Allsop and Stickler, 1985; Carrillo *et al.*, 1985) faecal streptococci (enterococci) and spore-forming bacteria, such as *Clostridium perfringens*. The pathogenic fungus *Candida albicans* has also been studied (Buck, 1978). Bacteriophage specific for faecal bacteria has been used to indicate the survival and persistence of enteric viruses (Stetler, 1984; Tartera and Jofre, 1987). Some of these groups of bacteria also have commercial uses and are potential sources of GEMs.

Survival studies in the laboratory and *in situ* have been done using batch chambers, diffusion chambers and dialysis bags. More realistic studies have traced the movement of bacteria from sewage outfall pipes and other discharges (Niemela and Vaatanen, 1982; Evison, 1985; White and Godfree, 1985). The survival times quoted in these studies vary widely and depend on the precise conditions employed. Early studies by McFeters and Stuart (1972) showed survival of *E. coli* in fresh water for more than 5 days. McFeters *et al.* (1974) compared pathogen survival in well water and showed decreases in population of 50% in 2–27 h. Granai and Sjogren (1981) used a membrane-filter sandwich containing *E. coli* and *Streptococcus faecalis*, and demonstrated that populations declined by 95% in 48 and 96 h, respectively, in lake water. Survival was better at 12 than at 16°C. Davenport *et al.* (1976) showed that ice covering a river enhanced survival of faecal and total coliforms and enterococci. Between 8 and 33% survived for 7 days, following a rapid initial reduction in the first two days.

Enhanced survival of micro-organisms at low temperatures has been reported by many authors (Faust *et al.*, 1975; Niemela and Vaatanen, 1982; Lessard and Sieburth, 1983; Flint, 1987). In contrast, Anderson *et al.* (1983) found that in filtered estuarine water higher temperatures gave longer micro-organism survival times (over 15 days), but in the presence of the eucaryotic microbiota, survival times were reduced to 6–9 days and were affected by temperature. There is ample evidence that *E. coli*, *Klebsiella* spp., *Vibrio* spp. and *Candida albicans* will survive indefinitely tropical waters and may grow (Carrillo *et al.*, 1985; Lopez-Torres *et al.*, 1987; Valdes-Collazo *et al.*, 1987; Perez-Rosas and Hazen, 1988).

Sunlight is also known to inactivate enteric bacteria in freshwater and marine environments. Perhaps the first convincing demonstration of this was by Gameson and Saxon (1967) who exposed bottles containing suspensions of sewage isolates and found a strong correlation between die-off and intensity of incident light. McCambridge and McMeekin (1980) found that light and predators affected survival of several coliforms. In autoclaved seawater 50% survival after 10 days was obtained, whereas predators and light led to a 10^9-fold reduction in less than 8 days. They suggested that light induced superoxide formation which damaged cells and enhanced the effects of predation. Sunlight induced non-lethal injury (Kapuscinksi and Mitchell, 1981) and strongly influenced survival of *E. coli* and enterococci in seawater, reducing 90% survival rates from several days to 30–90 and 60–80 min, respectively (Fujioka *et al.*, 1981). Similar effects were seen on the survival of a *Serratia* bacteriophage, where the time to reach 90% survival was reduced from over 300 to 30 min in bright sunlight (Drury and Wheeler, 1982): sunlight appears not to be as effective in fresh water (Fujuoka *et al.*, 1981; Niemela and Vaatanen, 1982). The mortality of *E. coli* was inversely proportional to salinity in illuminated coastal, estuarine and river water samples (Pike, 1975). Similarly, Lessard and Sieburth (1983) showed that light had little effect on *E. coli* and enterococci in an estuary but showed complex effects in a salt marsh. Sunlight is not confined to the immediate surface layer. Pike (1975) states that the lethal effects can be demonstrated at depths of 5 m, but is influenced by turbidity; 90% survival times of *E. coli* in North Sea water increased from 6.6 to 20 h when the water was turbid. It has been suggested that multiple-resistant, plasmid-containing coliforms may be selected for by high levels of sunlight (Meckes, 1982; Bell *et al.*, 1983), and that UV-resistance plasmids may enhance the survival of laboratory bacteria in shallow rivers (Bale *et al.*, 1988).

Little is known about how pH affects survival of bacteria in aquatic habitats. McFeters and Stuart (1972) showed that the optimal pH for survival lay between 5.5 and 7.5 and that survival declined rapidly outside these limits. Sjogren and Gibson (1981) found that some bacteria, such as *Klebsiella* species, survived better at pH 5.5 than, for instance, *Acinetobacter*, *Escherichia coli*, *Erwinia* and *Pseudomonas* spp. They suggest that low pH could be used by certain bacteria as an electrochemical gradient to produce energy. Reduction of counts of coliforms in retention lagoons may be enhanced by high pH (pH 9) produced by algal activity (Pike, 1975). Studies of streams receiving acid mine drainage with low pH values showed rapid inactivation of *E. coli* and other indicators whilst aciduduric *Candida albicans* was relatively unaffected (DePasquale *et al.*, 1987).

Estuarine and coastal aquatic habitats contain large amounts of silt and suspended solids. These are extremely important in affecting the survival of viruses (Gerba, 1984) and bacteria (Roper and Marshall, 1974; Faust *et al.*, 1975). The binding of *Bdellevibrio* and bacteriophages to *E. coli* was found to be reduced by the absorption of colloidal clay particles to the bacteria (Roper and Marshall, 1974). These effects have been extensively reviewed elsewhere (Stotzky and Krasovsky, 1981; Stotzky and Babich, 1986). The absorption to suspended solids often leads to the accumulation of introduced (allochthonous) bacteria in sediments. Sediments enhance the survival of bacteria in laboratory systems containing sediment and seawater, for example, levels of *E. coli* declined to undetectable levels in 4 days in the water and 8 days in the sediment (Gerba and McLeod, 1976). The same strains grew faster in autoclaved sediment–water systems than in water. Whilst studying plasmid transfer in sediments Stewart and Koditscheck (1980) showed enhanced survival in sediments compared to the overlying seawater.

Similar results were found by Laliberte and Grimes (1982) who proposed that the enhanced survival observed in non-sterile sediment was due to higher nutrient levels. The ability to grow in autoclaved sediment was thought to be due to reduced competition. Sediments also enhanced the survival of *Salmonella* spp. relative to *E. coli* (van Donsel and Geldrich, 1971). Counts of *E. coli* and enterococci in intertidal sediments were 3–4 log values higher than in the overlying water despite the variations in moisture, temperature and light caused by tides (Shiaris *et al.*, 1987). Goyal and Adams (1984) isolated drug-resistant *E. coli* from a sludge dumping site in 45–60 m of water 30 months after sludge was last dumped. A possible reason for these observations is that deep-sea conditions enhance the survival of some enteric bacteria. Baross *et al.* (1975) found that pressures of up to 500 atm enhanced survival of *E. coli* and enterococci, and enterococci were not affected by 1000 atm. Conversely, the survival of *Clostridium perfringens* and *Vibrio* spp. was reduced by presures over 1 atm. Enhanced survival of *E. coli* in sediments was shown to lead to large numbers in the overlying water following dredging operations (Grimes, 1980).

The survival and persistence of introduced GEMs will largely depend on interaction with the indigenous microflora. Many studies have shown that competition, parasitism and predation strongly influence survival (Mitchell, 1968). The biological nature of competition is suggested by the reduced mortality of *E. coli* in autoclaved water. Klein and Alexander (1986) found that a toxin in sterile lake water inhibited *Klebsiella pneumoniae* but not *Micrococcus flavus* and could be removed by heating, dialysis and cation exchange resins. Lessard and Sieburth (1983) suggested that the lethal effect of light was related to inhibitory products from algae. One such substance, acrylic acid is known to affect *E. coli* in sewage and lake water (Mitchell, 1968; Brown *et al.*, 1977; Rowbury and Hicks, 1987). McCambridge and McMeekin (1980, 1981) also linked light and predation and suggested that light injured the prey and enhanced predation. The survival of *E. coli* in this instance depended on the population of predators and exhibited a lag whilst the predator population increased. Using marked *E. coli* minicells as prey, Wikner *et al* (1986) studied the effect of protozoa on cell numbers and found that they consumed 60–115% of the bacterial biomass produced. Roper and Marshall (1977) showed that myxobacteria were responsible for the lysis of *E. coli* in seawater, whereas Enzinger and Cooper (1976) found that protozoa, not lytic bacteria, were responsible for the mortality of *E. coli* in an estuary. Filtration of river water through Whatman filter paper did not significantly influence the survival of *E. coli* (Flint, 1987) and it was concluded that protozoa were not important as predators but increased survival after the water had been Millipore-filtered (0.45 μm pore size) suggested that the removal of competing bacteria was more important (Flint, 1987). Once decay began, however, small, motile, non-culturable cells were seen; these may have been *Bdellovibrio* parasites (Starr and Seidler, 1971; Shilo, 1983) but Fry and Staples (1974) showed that bdellovibrios were not responsible for the decline of *E. coli* in dialysis sacs in fresh water, and they concluded that it was due to bacterial competition.

Bacteriophages are thought to be unimportant in reducing bacterial numbers (Varon and Zeigler, 1978; Wiggins and Alexander, 1985) except at very high population densities, and often bacteriophages themselves do not survive well in lake water and sewage. Fujuoka *et al.* (1980) suggest that micro-organisms are responsible for virus destruction and Ward *et al.* (1986) showed that this was due to proteases and nucleases digesting the capsid and genome. Many types of bacteriophage can, however, be isolated from sediments where they may be protected (Tatera and Jofre, 1987). The complex nature of interactions between predators and prey are reviewed extensively elsewhere (Alexander, 1981).

Indigenous (autochthonous) bacteria usually prevent introduced bacteria from becoming established. This is often via competition for nutrients (Liang *et al.*, 1982). The role of resistance to starvation as a survival strategy has been reviewed (Stevenson, 1978; Morita, 1982; Roszak and Colwell, 1987). Starvation results in changes in microbial cells; changes similar to those induced by other stresses (Groat and Martin, 1986; Sterkenberg *et al.*, 1986; Albertson *et al.*, 1987) which include the presence of specific proteins and antigens. Resistance to starvation does not, itself, appear to guarantee persistence in an environment (Sinclair and Alexander, 1984). There is increasing evidence that bacteria can actively adapt to starvation conditions by a series of biochemical (Palmer *et al.*, 1984; Munro *et al.*, 1987) and morphological changes (Morita, 1982; Kjelleberg *et al.*, 1987).

Using alternative techniques of assessing viability, it has been shown that starved populations of *E. coli*, *Salmonella enteritidis*, *Campylobacter jejuni* and *Vibrio cholerae* enter a non-culturable resting stage whilst remaining able to infect animals (Xu *et al.*, 1982; Baker *et al.*, 1983; Roszak *et al.*, 1984; Colwell *et al.*, 1985; Rollins and Caldwell, 1986). Thus the apparent rapid die-off of indicator strains as measured by selective media may reflect the emergence of non-culturable forms. It is well known that conditions in natural habitats lead to sub-lethal stresses (Rhodes *et al.*, 1983).

THE RELEASE OF *RHIZOBIUM*

Strains of *Rhizobium* have been grown under laboratory conditions and released into the environment since the turn of the century (Date and Roughley, 1977; Thompson, 1980) when their role in inducing the formation of nitrogen-fixing root nodules was first demonstrated. Because leguminous plants do not nodulate if appropriate rhizobia are not present in soils, the use of *Rhizobium* inoculants has been intimately linked with the introduction of legume crops, such as the soybean, into regions in which they are not native. As a result, in the USA alone in 1980 almost 4×10^6 kilograms of inoculant were produced for use on 15.5×10^6 hectares of land (Burton, 1982).

Most *Rhizobium* inoculants contain between 10^8 and 10^{10} rhizobia per gram, and should contain fewer than 0.1% of contaminating micro-organisms (i.e. 10^5–10^7 per gram; Thompson, 1980). Unfortunately these standards are not adhered to in all countries and, as a result, a large proportion of inoculants sold will contain extremely large numbers of unknown micro-organisms. Despite the lack of knowledge of what contaminants are present in most samples of inoculants, we can find no reference to health or environmental problems arising from their use. Recommendations for use stress the need to protect the inoculum and to ensure that the appropriate strains for the crop are used (Date and Roughley, 1977; Thompson, 1980; Burton, 1982).

From the point of view of human health and possible risks to the environment this experience is valuable because it demonstrates clearly that the release of micro-organisms *per se* is not inherently dangerous or environmentally damaging. Work with *Rhizobium* has also provided us with a wealth of experience about the ability of released bacteria to survive in the environment, compete with indigenous strains and retain their original characteristics. Before discussing our knowledge of these properties it is useful to summarize problems associated with the use of *Rhizobium* inoculants:

(1) It is difficult to market inoculants in regions where rhizobia nodulating relevant crops are indigenous because the introduced strains will not necessarily compete successfully for nodule formation.
(2) Optimal growth responses arise from compatible combinations of host and *Rhizobium* strains, but these can seldom be exploited because of problems associated with competition from indigenous strains.
(3) We know so little about the biochemical determinants of competitiveness that we are unable to utilize modern genetic techniques to produce more competitive inoculants.
(4) Quality control standards vary and are often optional (e.g. in the USA), therefore inoculants are often of little value.
(5) When a non-native crop is introduced, for which indigenous rhizobia are not available, rhizobia can be introduced without difficulty. If the strains used in the inoculant are not particularly efficient at nitrogen fixation they can establish a competitive indigenous population of low agronomic value. There are no regulations to prevent the introduction of such strains

which have the potential to reduce the value of symbiotic nitrogen fixation for a very long time.

(6) Problems of competition can be overcome by adding very large numbers of inoculant bacteria to produce a temporary numerical advantage for the inoculum (usually at least 103:1 required). Unfortunately it is common to find soil populations of indigenous rhizobia of at least 10^4–10^7 per gram which makes this a very expensive option because large amounts of inoculum are required.

Most of our information about the introduction of *Rhizobium* strains has come from work in Australia and the USA because native rhizobia for many agriculturally-important legumes are not present. In Europe the formation of polders in the Netherlands provided an excellent opportunity for studies of the colonization of soils by bacteria. In all three examples rhizobia were introduced intentionally and were also distributed by natural means and inadvertently on seeds.

The Dutch experience has been documented (van Schreven and Harmsen, 1968) and, perhaps not surprisingly, colonization reflected the observed ability of different *Rhizobium* species to persist in European soils. For example, the biovars *trifolii* and *viciae* of *R. leguminosarum*, which are found in large numbers (10^1–10^7) in most soils (Nutman and Ross, 1969; de Escuder, 1972) quickly became established in polder soils whether or not inoculants were used. Conversely, strains of *R. meliloti*, which are often absent or present in very low numbers in agricultural soils (Nutman and Ross, 1969; de Escuder, 1972) were very slow to colonize and lucerne usually needed to be inoculated. We do not know why *R. meliloti*, which grows much more vigorously on most laboratory media than biovars of *R. leguminosarum*, is such a poor colonizer of most soil types.

The colonization of North American soils by non-indigenous rhizobia has not been studied in as much detail as might be expected, with the possible exception of studies of soybean rhizobia (*Bradyrhizobium japonicum*) in the north-central region where a single serotype (123) has become so firmly established that inoculation with other strains is usually futile (Reyes and Schmidt, 1979; Stacey and Upchurch, 1984). Soybeans are not indigenous and appropriate rhizobia are required in virgin soils. Where the existing serotype (123) came from is unknown. Also we do not know whether existing strains of this serotype are derivatives of a single ancestral strain or not. If they are not derivatives of a single strain there must be at least one characteristic of serotype 123 that favours its persistence and competitiveness. Unfortunately we have little or no idea what might be involved. Studies of this serotype should provide extremely important information about the ability of strains of bacteria to colonize soils, persist and compete with incoming strains.

Australian studies of introduced strains of *Rhizobium leguminosarum* biovar *trifolii* have indicated that the serotype can change and distinct strains evolve within a few years of introduction (Brockwell *et al.*, 1981; Gibson *et al.*, 1976; Roughley *et al.*, 1976). Because these studies were initiated before good techiques were available for identifiying strains at the level of DNA sequence, the serotype was the basis for recognition and we cannot be sure that the novel strains arose from the original population, although the absence of indigenous rhizobia capable of nodulating clover makes this the most likely interpretation.

There have been many studies of the ability of rhizobia to survive and persist in the environment (Table 1) and several useful reviews have been published

Table 1. Factors affecting the survival and persistence of *Rhizobium* in soils.

Antagonism and stimulation by other organisms
Virtanen and Linkola (1948), Hattingh and Louw (1966a,b), Holland and Parker (1966), Schwinghamer (1971), Patel (1974)

Competition with other rhizobia
Franco and Vincent (1976), Gibson *et al.* (1976), Roughley *et al.* (1976), Brockwell *et al.* (1982)

Moisture
Vincent *et al.* (1962), Hamdi (1971), Chatel and Parker (1973), Danso and Alexander (1974), Bushby and Marshall (1977), Read (1953)

Organic matter
van Schreven and Harmsen (1969), Danso and Alexander (1974)

pH
Wilson (1930), Robson and Loneragan (1970a,b), Ham *et al.* (1971), Obaton (1971), McLoughlin and Dunican (1981), Brockwell *et al.* (1982), Wood *et al.* (1985)

Presence of host and other plants
Rovira (1961), Robinson (1967), van Schreven and Harmsen (1969), Brockwell and Robinson (1970), Reyes and Schmidt (1979), Brockwell *et al.* (1982), Jensen and Sørensen (1987)

Predation
Parker and Grove (1970), Habte and Alexander (1977), Evans *et al.* (1979), Ramirez and Alexander (1980)

Soil Properties
Vojinovic (1961), Marshall and Roberts (1963), Robson (1969), Brockwell and Robinson (1970), Ham *et al.* (1971), Obaton (1971), Bhardwaj (1974), Marshall (1975), Brockwell *et al.* (1982), Wood *et al.* (1985), Corman *et al.* (1987)

Temperature
Wilson (1930), Read (1953), Bowen and Kennedy (1959), Marshall (1964), Wilkins (1967), Ek-Jander and Fahraeus (1971)

Other observations
Dudman and Brockwell (1968), Chatel and Greenwood (1973a,b), Chatel *et al.* (1973), Diatloff (1977)

(Wilson, 1930; Purchase and Vincent, 1949; Vincent, 1954, 1962; Bushby, 1982; Date, 1982; Stacey, 1985). The most important findings have been the major importance of moisture, temperature and organic matter and the relatively unimportant role for predators and antagonists. Not unexpectedly, salinity and extremes of pH are very detrimental to rhizobia, and they are also important factors in preventing healthy growth of host plants. The rhizobia appear to be perfectly normal soil micro-organisms which will survive and maintain healthy populations for extremely long periods of time if temperatures are moderate and there is sufficient moisture and organic matter present. A serious problem faced by suppliers of inoculants is that of introducing strains into environments in which rhizobia are already present.

An important aspect of persistence is the ability to survive in dry soils under adverse conditions. Sen and Sen (1956) dried samples of Indian soils in air and stored them for 19 years. At the end of this time two of the 19 samples contained rhizobia capable of nodulating test plants. Jensen (1961) reported studies in which soils were autoclaved and amended by the addition of nutrients. After 30–45 years storage, 40 out of 40 soils examined contained rhizobia. The implication from these studies is that even though *Rhizobium* does not form spores or known resting stages, it has the potential to survive extremely long periods under adverse conditions.

CONCLUSIONS

We have presented a brief overview of literature relating to the release of organisms into the environment. It is clear from the information that has been gathered during the last century that many factors influence survival and persistence. It is also clear that with the exception of extreme environments, it is extremely difficult to make predictions about the potential of a given organism to become established and to maintain high populations in a given environment.

We have been unable to comment on published studies of the release of GEMs because such releases have only occurred recently. With *Rhizobium*, however, mutant strains have been isolated and these facilitate identification (Bergersen *et al.*, 1971; Schwinghamer and Dudman, 1973; Brockwell *et al.*, 1977; Kuykendall and Weber, 1978). In all these reports the bacteria were spontaneous mutants selected in the laboratory. Spontaneous resistance to antibiotics has been particularly useful for these studies, but strain identification has often been confused by the presence of intrinsically resistant rhizobia in soil. Over the last few years rhizobia carrying transposable genetic elements conferring drug resistance have been released to field soils to study the fate of genes carried on plasmids (P. R. Hirsch, J. Spokes and J. E. Beringer and others) but publication of detailed results from these experiments is still awaited.

Experience shows us that very large numbers of micro-organisms can be released into the environment without causing harm. Previous studies also indicate that nothing can be predicted with certainty. It therefore appears that the release of GEMs into the environment should not be discouraged, but that care should be taken to ensure that those organisms that are released are not derived from species capable of causing harm. A great deal more research will be needed before it will be possible to make a simple assessment as to whether or not a previously unknown organism which has been manipulated should be released.

REFERENCES

Albertson, N. H., Jones, G. W. and Kjelleberg, S., 1987. The detection of starvation-specific antigens in two marine bacteria. *Journal of General Microbiology* **133,** 2225–2231.

Alexander, M., 1981. Why microbial predators and parasites do not eliminate their prey and hosts. *Annual Review of Microbiology* **35,** 113–133.

Allsop, K. and Stickler, D. J., 1985. An assessment of *Bacteriodes fragilis* group organisms as indicators of human faecal pollution. *Journal of Applied Bacteriology* **58,** 95–99.

Anderson, I. C., Rhodes, M. W. and Kator, H. I., 1983. Seasonal variation in survival of *Escherichia coli* exposed *in situ* in membrane diffusion chambers containing filtered and nonfiltered estuarine waters. *Applied and Environmental Microbiology* **45,** 1877–1883.

Baker, R. M., Singleton, F. L. and Hood, M. A., 1983. Effect of nutrient deprivation on *V. cholerae. Applied and Environmental Microbiology* **46,** 930–940.

Bale, M. J., Fry, J. C. and Day, M. J., 1988. Transfer and occurrence of large mercury resistance plasmids in river epilithon. *Applied and Environmental Microbiology*, in press.

Baross, J. A., Hanus, F. J. and Morita, R. Y., 1975. Survival of human enteric and other sewage microorganisms under simulated deep-sea conditions. *Applied Microbiology* **30,** 309–318.

Bell, J. B., Elliott, G. E. and Smith, D. W., 1983. Influence of sewage treatment and urbanization on selection of multiple resistance in faecal coliform populations. *Applied and Environmental Microbiology* **46,** 227–232.

Bhardwaj, K. K. R., 1974. Growth and symbiotic effectiveness of indigenous *Rhizobium* species in a saline–alkali soil. *Proceedings of the Indian National Science Academy* **40,** 540–543.

Bowen, G. D. and Kennedy, M., 1959. Effect of high soil temperatures on *Rhizobium* spp. *Queensland Journal of Agricultural Science* **16,** 177–197.

Brockwell, J., Gault, R. R., Mytton, L. R., Schwinghamer, E. A. and Gibson, A. H., 1981. Intrastrain instability in symbiotic effectiveness in *Rhizobium* spp. In *Current Perspective in Nitrogen Fixation*, Gibson, A. H. and Newton, W. E. (eds), pp. 433. Canberra: Australian Academy of Science.

Brockwell, J., Gault, R. R., Zorin, M. and Roberts, M. J., 1982. Effects of environmental variables on the competition between inoculum strains and naturalized populations of *Rhizobium trifolii*, for nodulation of *Trifolium subterraneum* L. and on rhizobia persistence in the soil. *Australian Journal of Agricultural Research* **33,** 803–815.

Brockwell, J. and Robinson, A. C., 1970. Observations on the natural distribution of *Rhizobium* spp. relative to physical features of the landscape. In *Proceedings of the XI International Grassland Congress*. Brisbane: University of Queensland Press.

Brockwell, J., Schwinghamer, E. A. and Gault, R. R., 1977. Ecological studies of root-nodule bacteria introduced into field environments—5. A critical examination of the stability of antigenic and streptomycin-resistance markers for identification of strains of *Rhizobium trifolii*. *Soil Biology and Biochemistry* **9,** 19–24.

Brown, R. K., McMeekin, T. A. and Balis, C., 1977. Effect of some unicellular algae on *Escherichia coli* population in seawater and oysters. *Journal of Applied Bacteriology* **43,** 129–136.

Buck, J. D., 1978. Comparison of *in situ* and *in vitro* survival of *Candida albicans* in seawater. *Microbial Ecology* **4,** 291–302.

Burton, J, C., 1982. Modern concepts in legume inoculation. In *Biological Nitrogen Fixation Technology for Tropical Agriculture*, Graham, P. H. and Harris, S. C. (eds), pp. 105–114. Cali: Centro Internacional de Agricultura.

Bushby, H. V. A., 1982. Ecology. In *Nitrogen Fixation*, Vol. 2. *Rhizobium*, Broughton, J. W. (ed.), pp. 35–75. Oxford: Clarendon Press.

Bushby, H. V. A. and Marshall, K. C., 1977. Some factors affecting the survival of root-nodule bacteria on desiccation. *Soil Biology and Biochemistry* **9,** 143–147.

Carrillo, M., Estrada, E. and Hazen, T. C., 1985. Survival and enumeration of the faecal indicators *Bifidobacterium adolescientis* and *Escherichia coli* in a tropical rain forest watershed. *Applied and Environmental Microbiology* **50,** 468–476.

Chatel, D. L. and Greenwood, R. M., 1973a. The colonization of host-root and soil by rhizobia. II. Strain differences in the species *Rhizobium trifolii*. *Soil Biology and Biochemistry* **5,** 433–440.

Chatel, D. L. and Greenwood, R. M., 1973b. Differences between strains of *Rhizobium trifolii* in ability to colonize soil and plant roots in the absence of their specific host plants. *Soil Biology and Biochemistry* **5,** 809–813.

Chatel, D. L. and Parker, C. A., 1973. Survival of field-grown rhizobia over the dry summer period in Western Australia. *Soil Biology and Biochemistry* **5,** 415–423.

Chatel, D. L., Shipton, W. A. and Parker, C. A., 1973. Establishment and persistence of *Rhizobium trifolii* in western Australian soils. *Soil Biology and Biochemistry* **5,** 815–824.

Colwell, R. R., Brayton, P. R., Grunes, D. J., Roszak, D. B., Huq, S. A. and Palmer, L. M., 1985. Viable but non-culturable *Vibrio cholerae* and related pathogens in the environment: implications for release of genetically engineered microorganisms. *Biotechnology* **3,** 817–820.

Corman, A., Crozat, Y. and Cleyet-Marel, J. C., 1985. Modelling of survival kinetics of some *Bradyrhizobium japonicum* strains in soils. *Biology and Fertility of Soils* **4,** 79–84.

Danso, S. K. A. and Alexander, M., 1974. Survival of two strains of *Rhizobium* in soil. *Soil Science Society of America Proceedings* **38,** 86–89.

Date, R. A., 1982. Assessment of rhizobial status of the soil. In *Nitrogen Fixation in Legumes*, Vincent, J. M. (ed.), pp. 85–94. Sydney: Academic Press.

Date, R. A. and Roughley, R. J., 1977. Preparation of legume seed inoculants. In *A Treatise on Dinitrogen Fixation, Section IV: Agronomy and Ecology*, Hardy, R. W. F. and Gibson, A. H. (eds), pp. 243–273. New York: Wiley

Davenport, C. V., Sparrow, E. B. and Gordon, R. C., 1976. Fecal indicator bacteria persistence under natural conditions in an ice-covered river. *Applied and Environmental Microbiology* **32,** 527–536.

Depasquale, D. A., Law, C. B. and Bissonnette, G. K., 1987. Comparative survival and injury of *Candida albicans* and bacterial indicator organisms in streams receiving acid mine drainage. *Water Research* **21,** 1525–1530.

Diatloff, A., 1977. Ecological studies of root-nodule bacteria introduced into field environments—6. Antigenic and symbiotic stability of *Lotononsis* rhizobia over a 12-year period. *Soil Biology and Biochemistry* **9,** 85–88.

van Donsel, D. J., Geldrich, E. E. and Clarke, N. A., 1967. Seasonal variation in survival of indicator bacteria in soil and their contribution to storm-water pollution. *Applied Microbiology* **15,** 1362–1370.

van Donsel, D. J. and Geldrich, E. E., 1971. Relationships of salmonella to fecal coliforms in bottom sediments. *Water Research* **5,** 1079–1087.

Drury, D. F. and Wheeler, D. C., 1982. Applications of a *Serratia marcescens* bacteriophage as a new microbial tracer of aqueous environments. *Journal of Applied Bacteriology* **53,** 137–142.

Dudman, W. F. and Brockwell, J., 1968. Ecological studies of root-nodule bacteria introduced into field environments. 1. A survey of field performance of clover inoculants by gel immune diffusion serology. *Australian Journal of Agricultural Research* **19,** 739–745.

Edmonds, R. L., 1976. Survival of coliform bacteria in sewage sludge applied to a forest clearcut and potential movement into groundwater. *Applied and Environmental Microbiology* **32,** 537–546.

Ek-Jander, J. and Fahraeus, G., 1971. Adaption of *Rhizobium* to subarctic environment in Scandinavia. *Plant and Soil*, Special Volume, pp. 129–137.

Elliot, E. L. and Colwell, R. R., 1985. Indicator organisms for estuarine and marine waters. *FEMS Microbiology Reviews* **32,** 61–79.

Enzinger, R. M. and Cooper, R. C., 1976. Role of bacteria and protozoa in the removal of *Escherichia coli* from estuarine waters. *Applied and Environmental Microbiology* **31,** 758–763.

de Escuder, A. M. Q., 1972. A survey of rhizobia in farm soils at Wye College, Kent. *Journal of Applied Bacteriology* **35,** 109–118.

Evans, J., Barnet, Y. M. and Vincent, J. M., 1979. Effect of a bacteriophage on the colonization and nodulation of clover roots by a strain of *Rhizobium trifolii. Canadian Journal of Microbiology* **25,** 968–973.

Evison, L. M., 1985. Bacterial pollution of coastal waters in the UK and Mediterranean. In *Microbial Aspects of Water Management*, White, W. R. and Passmore, S. M. (eds). Society for Applied Bacteriology Symposium Series, Vol. 14, pp. 815–935. Oxford: Blackwell Scientific.

Faust, M. A., Aotaky, A. E. and Hargadon, M. T., 1975. Effect of physical parameters of the *in situ* survival of *Escherichia coli* MC-6 in an estuarine environment. *Applied Microbiology* **30,** 800–806.

Flint, K. P., 1987. The long term survival of *Escherichia coli* in river water. *Journal of Applied Bacteriology* **63,** 261–270.

Franco, A. A. and Vincent, J. M., 1976. Competition among rhizobial strains for colonization and nodulation of two tropical legumes. *Plant and Soil* **45,** 27–48.

Fry, J. C. and Staples, D. G., 1974. The occurrence and role of *Bdellovibrio bacteriovorus* in a polluted river. *Water Research* **8,** 1029–1035.

Fujuoka, R. S., Loh, P. C. and Lau, S., 1980. Survival of human enterovirus in the Hawaiian ocean environment: evidence for virus-inactivating microorganisms. *Applied and Environmental Microbiology* **39,** 1105–1110.

Fujuoka, R. S., Hashimoto, H. H., Siwak, E. B. and Young, R. H. F., 1981. Effect of sunlight on survival of indicator bacteria in seawater. *Applied and Environmental Microbiology* **41,** 690–696.

Gameson, A. L. H. and Saxon, J. R., 1967. Field studies on the effect of daylight on mortality of coliform bacteria. *Water Research* **1,** 279–295.

Gerba, C. P., 1984. Applied and theoretical aspect of virus absorption to surfaces. *Advances in Applied Microbiology* **30,** 133–168.

Gerba, C. P. and McLeod, J. S., 1976. Effect of sediments on the survival of *Escherichia coli* in marine waters. *Applied and Environmental Microbiology* **32,** 114–120.

Gibson, A. H., Date, R. A., Ireland, J. A. and Brockwell, J., 1976. A comparison of competitiveness and persistence amongst five strains of *Rhizobium trifolii*. *Soil Biology and Biochemistry* **8,** 395–401.

Goyal, S. M. and Adams, W. N., 1984. Drug resistant bacteria in continental shelf sediment. *Applied and Environmental Microbiology* **48,** 861.

Granai, C. and Sjogren, R. E., 1981. *In situ* and laboratory studies of bacterial survival using a microporous membrane sandwich. *Applied and Environmental Microbiology* **41,** 190–195.

Grimes, D. J., 1980. Bacteriological water quality effects of hydraulically dredging contaminated upper Mississippi River bottom sediment. *Applied and Environmental Microbiology* **39,** 782–789.

Groat, R. G. and Martin, A., 1986. Synthesis of unique proteins at the onset of carbon starvation in *Escherichia coli*. *Journal of Industrial Microbiology* **1,** 69–73.

Habte, M. and Alexander, M., 1977. Further evidence for the regulation of bacterial populations in soil by protozoa. *Archives of Microbiology* **113,** 181–183.

Ham, G. E., Frederick, L. R. and Anderson, I. C., 1971. Serogroups of *Rhizobium japonicum* in soybean nodules sampled in Iowa. *Agronomy Journal* **63,** 69–72.

Hamdi, Y. A., 1971. Soil-water tension and the movement of rhizobia. *Soil Biology and Biochemistry* **3,** 121–126.

Hattingh, M. J. and Louw, H. A., 1966a. The antagonistic effects of soil micro-organisms, isolated from the root region of clovers, on *Rhizobium trifolii*. *South African Journal of Agricultural Science* **9,** 239–252.

Hattingh, M. J. and Louw, H. A., 1966b. The incidence of micro-organisms, stimulating *Rhizobium trifolii*, in the soil and root region of clovers. *South African Journal of Agricultural Science* **9,** 453–460.

Holland, A. A. and Parker, C. A., 1966. Studies on microbial antagonism in the establishment of clover pasture. 2. The effect of saprophytic soil fungi upon *Rhizobium trifolii* and the growth of subterranean clover. *Plant and Soil* **25,** 329–340.

Jensen, H. L., 1961. Survival of *Rhizobium meliloti* in soil culture. *Nature* **192,** 682–683.

Jensen, E. S. and Sørensen, L. H., 1987. Survival of *Rhizobium leguminosarum* in soil after addition as inoculant. *FEMS Microbiology Ecology* **45,** 221–226.

Kapuscinski, R. B. and Mitchell, R., 1981. Solar radiation induced sublethal injury in *Escherichia coli* in seawater. *Applied and Environmental Microbiology* **41,** 670–674.

Kjelleberg, S., Hermansson, M. and Marden, P., 1987. The transient phase between growth and nongrowth of heterotrophic bacteria, with emphasis on the marine environment. *Annual Review of Microbiology* **41,** 25–49.

Klein, T. M. and Alexander, M., 1986. Bacterial inhibitors in lake water. *Applied and Environmental Microbiology* **52,** 114–118.

Kuykendall, L. D. and Weber, D. F., 1978. Genetically marked *Rhizobium* identified as inoculum strain in nodules of soybean plants grown in fields populated with *Rhizobium japonicum*. *Applied and Environmental Microbiology* **36,** 915–919.

Laliberte, P. and Grimes, D. J., 1982. Survival of *Escherichia coli* in lake bottom sediment. *Applied and Environmental Microbiology* **43,** 623–628.

Lessard, E. J. and Sieburth, J. M., 1983. Survival of natural sewage populations of enteric bacteria in diffusion and batch chambers in the marine environment. *Applied and Environmental Microbiology* **45,** 950–959.

Liang, L. N., Sinclair, J. L., Mallor, L. M. and Alexander, M., 1982. Fate in model ecosystems of microbial species of potential use in genetic engineering. *Applied and Environmental Microbiology* **44,** 707–714.

Lopez-Torres, A. J., Hazen, T. C. and Toranzo, G. A., 1987. Distribution and *in situ* survival and activity of *K. pneumoniae* and *Escherichia coli* in a tropical rain forest watershed. *Current Microbiology* **15,** 213.

McCambridge, J. and McMeekin, T. A., 1980. Relative effects of bacterial and protozoan predators on survival of *Escherichia coli* in estuarine water samples. *Applied and Environmental Microbiology* **40,** 907–911.

McCambridge, J. and McMeekin, T. A., 1981. Effect of solar radiation and predacious microorganisms on survival of fecal and other bacteria. *Applied and Environmental Microbiology* **41,** 1083–1087.

McFeters, G. A. and Stuart, D. G., 1972. Survival of coliform bacteria in natural waters: field and laboratory studies with membrane filter chambers. *Applied Microbiology* **24,** 805–811.

McFeters, G. A., Bissonnette, G. K., Jezeski, J. J., Thomson, C. A. and Stuart, D. G., 1974. Comparative survival of indicator bacteria and enteric pathogens in well water. *Applied Microbiology* **27,** 823–829.

McLoughlin, T. and Dunican, L. K., 1981. An ecological study of marked *Rhizobium trifolii* strains on the host plant *Trifolium repens* var. Huai in an acidic peat and a neutral mineral soil. *Journal of Applied Bacteriology* **50,** 65–72.

Marshall, K. C., 1964. Survival of root-nodule bacteria in dry soils exposed to high temperatures. *Australian Journal of Agricultural Research* **15,** 273–281.

Marshall, K. C., 1975. Clay mineralogy in relation to survival of soil bacteria. *Annual Review of Phytopathology* **13,** 357–373.

Marshall, K. C. and Roberts, F. J., 1963. Influence of fine particle materials on survival of *Rhizobium trifolii* in sandy soils. *Nature* **198,** 410–411.

Meckes, M. C., 1982. Effect of UV light disinfection on antibiotic-resistant coliforms in wastewater effluents. *Applied and Environmental Microbiology* **43,** 371–377.

Mitchell, R., 1968. Factors affecting the decline of non-marine micro-organisms in seawater. *Water Research* **2,** 535–543.

Morita, R. Y., 1982. Starvation-survival of heterotrophs in the marine environment. *Advances in Microbial Ecology* **6,** 171–198.

Munro, P. M., Gauthuer, M. J. and Laumond, F. M., 1987. Changes in *Escherichia coli* cells starved in seawater or grown in seawater–wastewater mixtures. *Applied and Environmental Microbiology* **53,** 1476–1481.

Niemela, S. I. and Vaatanen, P., 1982. Survival in lake water of *K. pneumoniae* discharged by a paper mill. *Applied and Environmental Microbiology* **44,** 264–269.

Nutman, P. S. and Ross, G. J. S., 1969. Rhizobium in the soils of the Rothamsted and Woburn farms. *Report of Rothamsted Experimental Station for 1969*, pp. 148–169.

Obaton, M., 1971. Influence de la composition chimique du sol sur l'utilité de l'inoculation des graines de luzerne avec *Rhizobium meliloti. Plant and Soil*, Special volume, pp. 273–285.

Olson, B. H. and Nagg, L. A., 1984. Microbiology of potable water. *Advances in Applied Microbiology* **30,** 73–132.

Palmer, L. M., Baya, A. M., Grunes, D. J. and Colwell, R. R., 1984. Molecular genetic and phenotypic alteration of *Escherichia coli* in natural water microcosms containing toxic chemicals. *FEMS Microbiology Letters* **21**, 169–173.

Parker, C. A. and Grove, P. L., 1970. *Bdellovibrio bacteriovorus* parasitizing *Rhizobium* in Western Australia. *Journal of Applied Bacteriology* **33**, 253–255.

Patel, J. J., 1974. Antagonism of actinomycetes against rhizobia. *Plant and Soil* **41**, 395–402.

Perez-Rosas, N. and Hazen, T. C., 1988. *In situ* survival of *Vibrio cholerae* and *Escherichia coli* in tropical coral reefs. *Applied and Environmental Microbiology* **52**, 1–9.

Pike, E. B., 1975. Aerobic bacteria. In *Ecological Aspects of Used Water Treatment, Vol. 1. The organisms and their ecology*, pp. 1–65. London: Academic Press.

Purchase, H. F. and Vincent, J. M., 1949. A detailed study of the field distribution of strains of clover nodule bacteria. *Proceedings of the Linnean Society of New South Wales* **74**, 227–236.

Ramirez, C. and Alexander, M., 1980. Evidence suggesting protozoan predation on *Rhizobium* associated with germinating seeds and in the rhizosphere of beans (*Phaseolus vulgaris* L.) *Applied and Environmental Microbiology* **40**, 492–499.

Read, M. P., 1953. The establishment of serologically identifiable strains of *Rhizobium trifolii* in field soils in competition with the native microflora. *Journal of General Microbiology* **9**, 1–14.

Reyes, V. G. and Schmidt, E. L., 1979. Population densities of *Rhizobium japonicum* strain 123 estimated directly in soil and rhizospheres. *Applied and Environmental Microbiology* **37**, 854–858.

Rhodes, M. W., Anderson, I. C. and Kator, H. I., 1983. *In situ* development of sublethal stress in *Escherichia coli*: effects on enumeration. *Applied and Environmental Microbiology* **45**, 1870–1876.

Robinson, A. C., 1967. The influence of host on soil and rhizosphere populations of clover and lucerne root-nodule bacteria in the field. *Journal of the Australian Institute of Agricultural Science* **33**, 207–209.

Robson, A. D., 1969. Soil factors affecting the distribution of annual *Medicago* species. *Journal of the Australian Institute of Agricultural Science* **35**, 154–167.

Robson, A. D. and Loneragan, J. F., 1970a. Nodulation and growth of *Medicago truncatula* on acid soils. I. Effect of calcium carbonate and inoculation level on the nodulation of *Medicago truncatula* on a moderately acid soil. *Australian Journal of Agricultural Research* **21**, 427–434.

Robson, A. O. and Loneragan, J. F., 1970b. Nodulation and growth of *Medicago truncatula* on acid soils. II. Colonization of acid soils by *Rhizobium meliloti*. *Australian Journal of Agricultural Research* **21**, 435–445.

Rollins, D. M. and Colwell, R. R., 1986. Viable but nonculturable stage of *Campylobacter jejuni* and its role in survival in the natural aquatic environment. *Applied and Environmental Microbiology* **52**, 531–538.

Roper, M. M. and Marshall, K. C., 1974. Modification of the interaction between *Escherichia coli* and bacteriophage in saline sediment. *Microbial Ecology* **1**, 1–13.

Roper, M. M. and Marshall, K. C., 1977. Lysis of *Escherichia coli* by a marine myxobacter. *Microbial Ecology* **3**, 167–171.

Roper, M. M. and Marshall, K. C., 1978. Effects of a clay mineral on microbial predation and parasitism on *Escherichia*. *Microbial Ecology* **4**, 279–290.

Roszak, D. B., Grimes, D. J. and Colwell, R. R., 1984. Viable but nonrecoverable state of *Salmonella enteritidis* in aquatic systems. *Canadian Journal of Microbiology* **30**, 334–338.

Roszak, D. B. and Colwell, R. R., 1987. Survival strategies of bacteria in the natural environment. *Annual Review of Microbiology* **51,** 365–379.

Roughley, R. J., Blowes, W. M. and Herridge, D. F., 1976. Nodulation of *Trifolium subterraneum* by introduced rhizobia in competition with naturalized strains. *Soil Biology and Biochemistry* **8,** 403–407.

Rovira, A. D., 1961. *Rhizobium* numbers in the rhizospheres of red clover and paspalum in relation to soil treatment and numbers of bacteria and fungi. *Australian Journal of Agricultural Research* **12,** 77–82.

Rowbury, D. B. and Hicks, S. J., 1987. Resistance of attached *Escherichia coli* to acrylic acid and its significance for the survival of plasmid bearing organisms in water. *Annals Institut Pasteur/Microbiologie* **138,** 359–363.

van Schreven, D. A. and Harmsen, G. W., 1968. Soil bacteria in relation to the development of polders in the region of the former Zuidersee. In *The Ecology of Soil Bacteria: an International Symposium*, Gray, T. R. G. and Parkinson, D. (eds), pp. 474–499. Liverpool: Liverpool University Press.

Schwinghamer, E. A., 1971. Antagonism between strains of *Rhizobium trifolii* in culture. *Soil Biology and Biochemistry* **3,** 355–363.

Schwinghamer, E. A. and Dudman, W. F., 1973. Evaluation of spectinomycin resistance as a marker for ecological studies with *Rhizobium* species. *Journal of Applied Bacteriology* **36,** 263–272.

Sen, A. and Sen, A. N., 1956. Survival of *Rhizobium japonicum* in stored air dry soils. *Journal of the Indian Society of Soil Science* **4,** 215–220.

Shiaris, M. P., Rex, A. C., Pettibone, G. W., Keay, K., McManus, P., Rex, M. A., Ebersole, J. and Gallagher, E., 1987. Distribution of indicator bacteria and *Vibrio parahaemolyticus* in sewage-polluted intertidal sediments. *Applied and Environmental Microbiology* **53,** 1756–1761.

Shilo, M., 1983. *Bdellovibrio* as a predator. In *Current Perspectives in Microbial Ecology*, Klug, M. J. and Reddy, C. A. (eds), pp. 334–339. Washington DC: American Society for Microbiology.

Sinclair, J. L. and Alexander, M., 1984. Role of resistance to starvation in bacterial survival in sewage and lake water. *Applied and Environmental Microbiology* **48,** 410–415.

Sjogren, R. E. and Gibson, M. J., 1981. Bacterial survival in a dilute environment. *Applied and Environmental Microbiology* **41,** 1331–1336.

Stacey, G., 1985. The *Rhizobium* experience. In *Engineered Organisms in the Environment: Scientific Issues*, Halvorson, H. O., Pramer, D. and Rogul, M. (eds), pp. 109–121. Washington DC: American Society for Microbiology.

Stacey, G. and Upchurch, R. G., 1984. Rhizobium inoculation of legumes. *Trends in Biotechnology* **2,** 65–70.

Starr, M. P. and Seidler, R. J., 1971. The *Bdellovibrios. Annual Review of Microbiology* **25,** 649–675.

Sterkenberg, A., Vlegels, E. and Wouters, J. T., 1986. Global control in *Salmonella typhimurium*: two dimensional electropheretic analysis of starvation-, anaerobiosis-, and heat shock-inducible proteins. *Journal of Bacteriology* **168,** 420–424.

Stetler, R. E., 1984. Coliphages as indicators of enteroviruses. *Applied and Environmental Microbiology* **47,** 319–401.

Stevenson, L. H., 1978. A case for bacterial dormancy in aquatic systems. *Microbial Ecology* **4,** 127–133.

Stewart, K. R. and Koditscheck, L., 1980. Drug resistance transfer in *E. coli* in New York Bight sediment. *Marine Pollution Bulletin* **11,** 130–133.

Stotzky, G. and Krasovsky, V. N., 1981. Ecological factors that affect the survival, establishment, growth and genetic recombination of microbes in natural habitats. In *Molecular Biology, Pathogenicity and Ecology of Bacterial Plasmids*, Levy, R. C., Clowes, S. B. and Koenig, E. L. (eds), pp. 31–42. London: Plenum Press.

Stotzky, G. and Babich, H., 1986. Survival of, and genetic transfer by, genetically engineered bacteria in natural environments. *Advances in Applied Microbiology* **31,** 93–138.

Tartera, C. and Jofre, J., 1987. Bacteriophages active against *Bacteriodes fragilis* in sewage-polluted waters. *Applied and Environmental Microbiology* **53,** 1632–1637.

Temple, K. L., Camper, A. K. and McFeters, G. A., 1980. Survival of two *Enterobacteria* in faeces buried in soil under field conditions. *Applied and Environmental Microbiology* **40,** 794–797.

Thompson, J. A., 1980. Pollution and quality control of legumes inoculants. In *Methods for Evaluating Biological Nitrogen Fixation*, Bergersen, F. J. (ed.), pp. 489–533. Brisbane: Wiley.

Thornton, H. G. and Gangulee, N., 1926. The life cycle of the nodule organisms, *Bacillus radicicola* (Beij.), in soil and its relation to the infection of the host plant. *Proceedings of the Royal Society* **B99,** 427–451.

Valdes-Collazo, L., Schultz, A. J. and Hazen, T. C., 1987. Survival of *Candida albicans* in tropical marine and freshwaters. *Applied and Environmental Microbiology* **53,** 1762–1767.

Varon, M. and Zeigler, B. P., 1978. Bacterial predator–prey interaction at low prey density. *Applied and Environmental Microbiology* **36,** 11–17.

Vincent, J. M., 1954. Root nodule bacteria of pasture legumes. *Proceedings of the Linnean Society of New South Wales* **79,** 4–32.

Vincent, J. M., 1962. Australian studies of the root-nodule bacteria. A review. *Proceedings of the Linnean Society of New South Wales* **87,** 8–38.

Vincent, J. M., Thompson, J. A. and Donovan, K. O., 1962. Death of root-nodule bacteria on drying. *Australian Journal of Agricultural Research* **13,** 258–270.

Virtanen, A. I. and Linkola, H., 1948. On the antibacterial effect of spore-forming soil bacteria on the legume bacteria. *Suomen Kemistilehti B* **21,** 12–13.

Vojinović, Z. D., 1961. Investigation of the distribution of some important nodule bacteria in the soils of Serbia. *Journal for Scientific Agricultural Research* **45,** 1–17.

Ward, R. L., Knowlton, D. R. and Winston, P. E., 1986. Mechanism of inactivation of enteric viruses in fresh water. *Applied and Environmental Microbiology* **52,** 450–459.

White, W. R. and Godfree, A. F., 1985. Pollution of freshwater and estuaries. In *Microbial Aspects of Water Management*, White, W. R. and Passmore, S. M. (eds). Society for Applied Bacteriology Symposium Series, Vol. 14, pp. 675–795. Oxford: Blackwell Scientific.

Wiggins, B. A. and Alexander, M., 1985. Minimum bacterial density for bacteriophage replication, implications for significance of bacteriophages in natural ecosystems. *Applied and Environmental Microbiology* **49,** 19–23.

Wikner, J., Andersson, A., Normark, S. and Hagstrom, A., 1986. Use of genetically marked minicells as a probe in measurement of predation on bacteria in aquatic environments. *Applied and Environmental Microbiology* **52,** 4–8.

Wilkins, J., 1967. The effects of high temperatures on certain root-nodule bacteria. *Australian Journal of Agricultural Research* **18,** 299–304.

Wilson, J. K., 1930. Seasonal variation in the number of two species of *Rhizobium* in soil. *Soil Science* **30,** 289–296.

Wood, M., Cooper, J. E. and Campbell, D. S., 1985. A survey of clover and *Lotus* rhizobia in Northern Ireland pasture soils. *Journal of Soil Science* **36**, 357–365.

Xu, H.-S., Roberts, N., Singleton, F. L., Attwell, R. W., Grimes, D. J. and Colwell, R. R., 1982. Survival and viability of nonculturable *Escherichia coli* and *Vibrio cholerae* in the estuarine and marine environment. *Microbial Ecology* **8**, 313–323.

5 Detection and Monitoring of Genetically-engineered Micro-organisms

RITA R. COLWELL, C. SOMERVILLE, I. KNIGHT and W. STRAUBE

Department of Microbiology, University of Maryland, College Park, MD 20742, USA

INTRODUCTION

Advances in the application of recombinant DNA (rDNA) techniques have extended the possibilities for the exploitation of genetic change, which is in itself a fundamental biological phenomenon and the basic process of evolution. As a result, there are now many opportunities to address a variety of socio-economic problems, such as pollution control, the fight against disease and the improvement of food production. The extensive research efforts in molecular biology during the past 30 years have yielded some astonishing discoveries, which carry the potential of tremendous contributions in advancing human health and welfare. Nevertheless, because of the relative dearth in funding for microbial ecology and systematics during the same period, the full potential of the discoveries and their possible applications have not been realized. As a result of the lack of information from which to make predictions of the fate of genetically-engineered micro-organisms (GEMs) and their effects when introduced into the environment, the exploitation of the recombinant DNA discoveries has been essentially halted.

If the major ecological issues that arise from introduction of GEMs to the environment are reduced to their essential points, it can be concluded that the following represent information gaps, that is, areas of research needed: (1) detection and monitoring, (2) horizontal transfer of the genetic information of the GEM, (3) fate of the GEMs after release into the environment, e.g. survival and dispersion, and (4) effects of GEMs on the environment. Subsidiary to these, but equally important, are containment and recovery, i.e. 'recall', if adverse effects occur after release of GEMs. Such a significant proportion of the potential

RELEASE OF GENETICALLY-ENGINEERED MICRO-ORGANISMS ISBN 0–12–677521–4

applications of GEMs involve introduction into the environment, there is a clear need to assess risks associated with the introduction and monitoring of GEMs in the environment. One of the requirements is to ensure public health and environmental safety. As a result of this need, information about the survival, dispersion, fate and effects of GEMs in the environment must be gathered before any planned introduction. Here, only one of the major issues will be considered, that is monitoring of GEMs, including detection and enumeration, in the environment after planned introductions.

Methods frequently employed to detect GEMs require culture of the micro-organisms. Culture methods can be useful for detection and monitoring of micro-organisms in the environment but a significant disadvantage is that culture methods do not necessarily track the genes of interest. In addition, there are a host of problems associated with the isolation and culture of micro-organisms from samples collected from air, water or soil.

Direct methods for the detection, enumeration and monitoring of GEMs, on the other hand, provide relatively greater efficiency of detection. However, the direct methods generally do not permit differentiation of viable and dead micro-organisms. Modification of direct microscopic methods to include immunological procedures, i.e. highly specific, fluorescent-labelled monoclonal antibodies, it is possible to target specific micro-organisms precisely but the method does not track the genes. Gene probes that allow tracking of the genome have been developed. The probes recognize species-specific sequences and, more importantly, the genetically-modified sequences of GEMs. Chemical methods of detection, such as spent-medium analysis and chemical analysis of characteristic structural and metabolic compounds or elements have also been proposed. In this case, the metabolic or structural and functional properties of the GEM of interest serve as a 'tag'; however, based on available data, it can be concluded that an accurate and reliable approach to the problem of the detection and monitoring of GEMs in the environment requires a combination of fluorescent-labelled monoclonal antibody and epifluorescent microscopy, together with a gene probe. However, the technology of detection and monitoring is moving swiftly and improvements both in sensitivity and selectivity in all methodologies are to be expected. Ultimately, the least expensive and most reliable method will prevail.

In seeking the ideal method to detect and monitor for GEMs, certain conditions must be met. For example, survival rates of GEMs in the range of conditions likely to occur in the release area and in the surrounding environment must be determined. It is also important to know: (1) the reproduction rate of GEMs in the environment into which they have been introduced, (2) whether the GEM of interest survives and reproduces on target organisms or sites and/or on non-target organisms or sites in the environment, (3) whether the GEM is able to spread from the release area and, if so, by what mechanisms, (4) whether the GEM can establish itself in non-target species or substances and (5) the sensitivity of the method employed to monitor the populations of modified, target, and non-target organisms. Such information is essential for the assessment of risk.

CULTURE METHODS

Culture methods that employ selective media, colorimetric reactions and spent medium analysis require growth of the GEM. Unfortunately, these methods do not necessarily track the genome. To accomplish this, gene probes, restriction enzyme analysis, nucleic acid sequencing or insertion of marker elements are the methods of choice. Furthermore, culture methods may fail if the microorganisms, once released into the environment, become dormant, i.e. remain viable but cannot be cultured by any of the currently available methods for enrichment, selection, or non-specific culture (Colwell *et al.*, 1985) (Fig. 1). Such

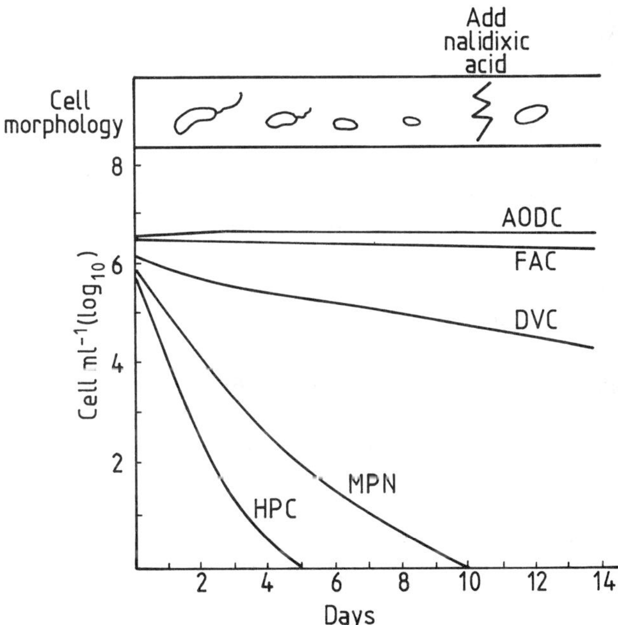

Fig. 1. Enumeration by several methods of enteric pathogens exposed to river, bay or seawater. Results are generalized from experiments providing evidence for viable but non-culturable organisms (Colwell *et al.*, 1985): AODC, acridine orange direct count; FAC, fluorescent antibody count; DVC, direct viable count; MPN, most probable number; HPC, heterotrophic plate count (Brayton and Colwell, 1987).

organisms can best be detected by direct immunofluorescent–epifluorescent microscopy, with confirmation by gene probe or other genetic methods. Thus, direct microscopy, coupled with use of a specific gene probe offers the greatest sensitivity and specificity at the present time (Table 1).

A useful example for the discussion of the merits of culture methods is that of the coliforms and, in particular, *Escherichia coli*. Examination of environmental samples for faecal indicators and, directly, for *E. coli* has been done since the turn of the century. Detection of *E. coli* is based mainly on the ability of this species to grow at 44°C, to produce acid and gas from lactose and indole from tryptophane (Anon., 1983, 1985). Isolation and enumeration of *E. coli* by traditional methods usually requires an enrichment step in liquid media or membrane filtration and incubation on a solid medium, in each case followed by subculture and additional testing of positive cultures or colonies. Improvements in this procedure have been suggested by many investigators. Among the many suggestions in the voluminous literature on this subject, one of the most recent is a direct enumeration method, based exclusively on the ability of the organism to produce indole on a selective medium containing bile salts at 44°C (Delaney *et al.*, 1962). Direct plating has been adapted for analysis of water samples by Havelaar and During (1988) who found that it gave better recoveries of *E. coli* than a membrane filter method with 0.1% sodium lauryl sulphate agar. McClure and Roberts (1987) recently examined the effect of incubation time and temperature on growth of *E. coli* on two-dimensional gradient plates and staining, laser densitometry and computer graphics. The results of these recent studies, combined with the vast historical literature on the subject, show that a well-studied organism such as *E. coli* is recovered with variable success by culture methods. The efficiency and accuracy of recovery depends on the temperature, nutrients, salt concentration and a host of other related environmental factors.

A comparison of media and of spread with pour plate methods for the enumeration of heterotrophic bacteria in drinking water, recently reported by Gibbs and Hayes (1988), showed that the pour plate method with yeast extract agar and a 3-day incubation period, which is the standard method in the UK for enumeration of heterotrophic bacteria in drinking water, yielded significantly lower bacterial numbers than the R2A spread-plate method with a 7-day incubation period. The latter method yielded 520 times the bacterial numbers than the standard method. Thus, standard procedures for non-engineered bacteria are highly inefficient at best. This, and the well-known phenomenon of phenotypic dissonance of bacteria growing in the laboratory compared with strains of the same species in the natural environment, an example of which is the occurrence of atypical forms of *E. coli* in streams (Rychert and Stephenson, 1981), casts serious doubt on the usefulness of culture methods for detection of GEMs.

As compared with culturing methods, greater precision in detection can be achieved by direct microscopy and use of fluorescent monoclonal antibodies to purified cell components. With this approach, specific organisms in environmental samples can be detected directly. This procedure has been used to identify *Rhizobium* spp. in soil (Bohlool and Schmidt, 1980), to assess the serological diversity of marine ammonia- and nitrite-oxidizing bacteria in seawater (Ward and Carlucci, 1985) and to detect *Sphaerotilis natans* sufficiently to stain the cells individually. Furthermore, the fluorescent antiserum stained the cells brightly and they could be observed through activated sludge flocs and other suspended matter.

The direct staining method has been criticized for its inability to differentiate between living and non-living micro-organisms. However, the use of nalidixic acid and yeast extract, a procedure which results in enlargement of viable cells, allows identification of living cells (Kogure *et al.*, 1979; Brayton and Colwell, 1987).

The degree of cross-reactivity between antisera prepared against one strain and other related strains of the same physiological type or species can be a problem in the immunological approach. However, Belser and Schmidt (1978) found very little inter-genus cross-reactivity between different strains of the same genus in their studies. Josserand and Cleyet-Marel (1979) were able to identify at least four unique serotypes among nitrite oxidizers of identical morphology. Ward and Carlucci (1985) included a variety of discernibly different morphological types under similar culture conditions and found a greater cross-reactivity between antisera prepared against one strain and other related strains of the same physiological type. It should be noted that the marine ammonium-oxidizing population is less diverse than the soil population of these organisms. It is possible that the water column environment in the sea may present less habitat diversity than the soil environment, resulting in a relatively less diverse microbial assemblage. In any case, the immunofluorescent–epifluorescent microscopy approach is of some value in the direct detection, enumeration and monitoring of micro-organisms in the environment.

The distribution of *Vibrio cholerae* in the natural environment of a cholera endemic region has been studied with the direct microscopy procedure and a highly specific monoclonal antibody (Brayton and Colwell, 1987). On the basis of O antigens, *V. cholerae* is subdivided into serovars and 83 O serogroups have been recognized, the cholera vibrio being assigned to O1 (Sakazaki and Donovan, 1984). Two antigenic forms of *V. cholerae*, Owaga and Inaba, with the antigenic formulae AB(C) and AC, respectively, are recognized. Shimada and Sakazaki (1988) reported a serogroup of non-O1 *V. cholerae*, Hakata, which possess the C (Inaba) factor but not B (Ogawa) and A factors of *V. cholerae* O1. Shimada *et al.* (1987) reported a group of marine vibrios that possess the B and C factors of *V. cholerae* O1 as well as their own major antigen factor but the A

Table 1. Methods for detection, identification, enumeration and monitoring of GEMs in environmental samples.

Methods of Detection	Advantages	Disadvantages	Cost	Application
Plating/culturing	Fairly sensitive Selective/colorimetric media can be employed Operationally feasible and relatively simple Follows the genome Amenable to statistical analysis Enrichments can be done Spent media can be analysed	Organisms must be culturable Inserted markers may be unstable Limited by media, temperature, etc.	Relatively inexpensive	Gerhardt (1981) Barry (1986)
Direct microscopy/ monoclonal antibody tagging	High specificity Permits direct detection in environmental samples High sensitivity for samples that can be concentrated Acridine orange direct staining, direct viable count (nalidixic acid) procedure, and fluorescent antibody	Low sensitivity for soil or samples that cannot be concentrated No differentiation between dead and living cells if the DVC method does not apply* Time consuming if not automated Organic slimes may prevent binding	Relatively inexpensive If automated, cost can be high Preparation of monoclonals is time-consuming and expensive	Bohlool and Schmidt (1980) Irby *et al.* (1985) Brayton and Colwell (1987) Zelibor *et al.* (1987)

	labelling of cells can be employed Permits detection of organisms that cannot be cultured	Antigen gene may be unstable in environment		
Protein product fingerprinting	Highly specific	Low sensitivity	Relatively expensive, requiring radiolabelled proteins and polyacrilamide gel electrophoresis	Summers and Hoops (1980)
Gene probes	Detection based on genotype not phenotype	Low sensitivity (but improving)	Relatively expensive	Barkay *et al.* (1985) Holben *et al.* (1988)
rRNA sequencing and fingerprinting	High specificity	Low sensitivity	Expensive Requires extensive data base	Loh *et al.* (1982) MacDonell and Colwell (1984) Stahl *et al.* (1985) Giovannoni *et al.* (1988)
Flow cytometry	Highly promising in ability to automate individual cell analysis	Little work done to date to assess sensitivity	Expensive	

*Kogure *et al.*, 1979.

factor has not, so far, been found in any vibrios other than *V. cholerae* O1. Thus, the A-subunit-specific monoclonal antibody used by Brayton and Colwell (1987) allows detection and enumeration of this organism in environmental samples when culture methods fail (Fig. 1). The conclusions are that the problem of species cross-reaction and intra-species antigenic variation will reduce both the specificity and sensitivity of the method but these problems are not entirely insurmountable. Use of a particle counter, such as the Elzone unit, permits automation of the microscopy (Zelibor *et al.*, 1987) and improves precision, not to mention relief from the tedium and the time saved when counting large numbers of environmental samples.

GENETIC METHODS

Hybridization with gene probes relies on base pairing between homologous nucleic acid sequences, which allows identification of specific sequences homologous with the labelled DNA probe. This method has been used to detect particular organisms in environmental samples (Loh *et al.*, 1982; Barkay *et al.*, 1985; Holben *et al.*, 1988).

Specific DNA probes for identification of *Campylobacter jejuni* (Korolik *et al.*, 1988), enterotoxigenic *E. coli* (Romick *et al.*, 1987), *Legionella pneumophila* (Grimont *et al.*, 1985) and *Salmonella typhi* (Rubin *et al.*, 1985) have been developed. These are a few of the many examples available. A variety of applications is possible and probes have been used to detect specific organisms in environmental samples and pathogenic organisms in water systems and foods. Recent studies have shown the value of DNA probes in elucidating the ecology of the rumen (Attwood *et al.*, 1988; Tannock, 1988).

Schleifer *et al.* (1985) employed cloned ribosomal RNA (rRNA) genes from *Pseudomonas aeruginosa* as probes for conserved DNA sequences.

In general, hybridization probes used for microbial identification are highly specific (Table 1) and commonly comprise cloned genes from specific organisms. However, specific hybridization probes usually necessitate culture of the organism of interest to produce the probe and enrichment incubation of the sample to obtain sufficient cells for the probe to be effective.

Giovannoni *et al.* (1988) prepared phylogenetic group-specific probes based on 16S rRNA sequences. The 16S-like rRNAs are large (1500–2000 nucleotides). Some segments in the 16S rRNA do not vary in all organisms and are, therefore, useful as binding sites for oligodeoxynucleotide primers for sequencing protocols. Other sections of the 16S rRNAs are unique to specific organisms and offer targets for hybridization probes (Giovannoni *et al.*, 1988). Oligodeoxynucleotides complementary to the rRNAs were synthesized for the eubacteria, archebacteria and eucaryotes and the method proved effective in direct detection and quantification in environmental samples. Quantification of natural microbial populations was achieved by use of micro-autoradiography to detect *in situ* hybridization. Thus, it is possible to count directly the constituents of microbial populations.

A somewhat different approach was taken by Holben *et al.* (1988) who developed a protocol to obtain purified bacterial DNA from a soil bacterial community. The bacteria were dispersed and separated from soil particles in the presence of polyvinylpyrrolidone to remove humic acid contaminants by absorption to the polymer. The soil bacteria were then collected by centrifugation, lysed and the total bacterial DNA purified from the cell lysate. The isolated, 48 kb DNA was digested and analysed by slot-blot and Southern blot hybridization. With single stranded, ^{32}P-labelled DNA probes, Holben *et al.* (1988) were able to detect and quantitate specific microbial populations in the natural soil communities. The sensitivity of the method was estimated by its ability to detect *Bradyrhizobium japonicum* at concentrations as low as 4.3×10^4 cells per gram dry weight of soil, that is 0.2 pg of hybridizable DNA in 1 μg DNA (Holben *et al.*, 1988).

A similar approach was taken by Somerville *et al.* (1988) to extract DNA directly from the aquatic environment. They used high-capacity micropore filters (Millipore SVGS 01015) to concentrate organisms from litre quantities of water. Cell lysis and proteolysis was performed by addition of the appropriate buffers and enzymes to the inside of the filter-housing and extraction of nucleic acids from the cells on the filter. The efficiency of this method is quite high, yielding up to 10 ng of high molecular weight DNA per million cells when *V. cholerae* is used as the test organism. The extracted DNA can be used as the substrate for filter hybridization with DNA probes or can be purified and further concentrated for other uses.

If recombinant DNA sequences in GEMs are to be detected by DNA probes, these sequences must be present in very high copy number in the sample, otherwise the target sequence concentration will be below the detection threshold of conventional probes. Somerville *et al.* (1988) have developed a method for enhancing the sensitivity based upon the use of specific hybridization probes as primers for chain elongation. Sample DNA is first treated with DNA polymerase I and dideoxynucleotides to eliminate free 3′-hydroxyl residues. Unlabelled probe DNA is then hybridized to target and serves as a primer for extensive chain elongation with labelled deoxynucleotides. In this method, specificity is determined by probe hybridization, while sensitivity is enhanced by polymerization into adjacent genes. The results so far show that sensitivity can be enhanced at least three orders of magnitude over conventional probing methodology.

The presence of extracellular DNA in fresh water and seawater has been known for decades. Recent work by Paul *et al.* (1988) in surveys of offshore and subsurface waters has shown that the lowest dissolved DNA values are found in offshore regions, with concentrations increasing shoreward and in estuarine plumes. Similar results are obtained when quantifying the picoplankton, which increase in samples collected offshore to the estuary. In general, dissolved DNA from marine sources possessed a wide range of molecular sizes (0.12–35.2 kb, De Flaun and Paul, 1986; De Flaun *et al.*, 1987). Seasonal and diel variability in dissolved DNA was observed in a subtropical estuary (Paul *et al.*, 1988).

DETECTION BY NUCLEIC ACID SEQUENCING

Direct extraction of 5S and 16S rRNA from environmental samples for sequencing has been carried out (MacDonnell and Colwell, 1984; Olsen *et al.*, 1986). The methods involved are time-consuming and involve sophisticated laboratory procedures. In addition the data base for sequence comparisons by computer is expanding rapidly but is not yet sufficient for every application. Since access to data bases containing sequences determined to date is on-line, e.g. Genebank, etc., it is possible that direct extraction and sequence analysis may offer a powerful method for detection and monitoring of GEMs in the future (Fox *et al.*, 1980).

MICROBIAL SYSTEMATICS

Fundamental to detection and monitoring of GEMs released into the environment is a reliable, reproducible and information-rich taxonomy and, from that taxonomy, an identification system. Substantial progress has been made in microbial systematics during the past three decades, with improvement achieved from the applications of numerical taxonomy, analytical chemistry and molecular genetics to systematics. Probabilistic identification keys for clinically-important bacteria have reached a relatively advanced stage of development and have also gained widespread acceptance. However, for soil and aquatic systems, the micro-organisms comprising the naturally occurring communities remain very poorly characterized and identification schemes for these are, at best, rudimentary.

The research needs for improved microbial systematics in microbial ecology are most critical for species that comprise the soil, water and atmospheric populations of the natural environment. Although it could be argued that it is possible to monitor, with reasonable assurance, an introduced GEM during the initial stages of introduction, it is not clear that long-term monitoring will be possible without a good knowledge base for the naturally-occurring species. Species interactions, including horizontal and vertical exchange of genetic information, may not yet be measurable with sufficient precision and reproducibility to meet the objections, even of the most reasonable critics of GEM introductions.

The challenge then, is to develop a firm, reliable, information-rich systematics for the autochthonous micro-organisms, many of which perform the key functions of geochemical cycling in nature.

Fortunately, this is an exciting challenge for the microbial systematist. Perhaps for the first time, with the pressure brought to bear on the funding agencies by the proponents of biotechnology, the necessary funding for systematics will become available so that the task can be accomplished.

SUMMARY

During the past several years, the gaps in information available on microbial community structure and function in natural environmental systems have begun to close. Methods are currently available for detection and monitoring GEMs, but the advantages and disadvantages of each must be carefully considered. The selection of a method is case-dependent, that is, it depends on the micro-organism to be introduced, the environment into which it will be introduced and related environmental factors. The combination of fluorescent monoclonal antibody with an epifluorescent microscopy procedure and a target-specific gene probe offers the most sensitive and specific detection method for GEMs, in most situations. It can be stated unequivocally that, at present, every single cell cannot be detected after it has been introduced but the available methods should allow monitoring of population, size, dispersion and persistence. This information is, in many cases, sufficient for introduction of micro-organisms for which data from previous introductions of non-engineered forms are already available.

ACKNOWLEDGEMENTS

The authors gratefully acknowledge the support of Environmental Protection Agency Cooperative Agreement No. *CR81-2246*, National Science Foundation Grant No. *BSR-84-01397*, and University Research Initiative *N 00014-86-K-0696* award from the Office of Naval Research.

REFERENCES

Anon., 1983. *The Bacteriological Examination of Drinking Water Supplies. Reports on Public Health and Medical Subjects No. 71.* London: HMSO.

Anon., 1985. *Standard Methods for the Examination of Water and Wastewater*, 16th edn. Washington DC: American Public Health Association.

Attwood, G. T., Lockington, R. A., Xue, G.-P. and Brooker, J. D., 1988. Use of a unique gene sequence as a probe to enumerate a strain of *Bacteroides ruminicola* introduced into the rumen. *Applied and Environmental Microbiology* **54,** 534–539.

Barkay, T., Fouts, D. L. and Olson, B. H., 1985. Preparation of a DNA gene probe for detection of mercury resistance genes in gram-negative bacterial communities. *Applied and Environmental Microbiology* **49,** 686.

Barry, G. F., 1986. Permanent insertion of foreign genes into the chromosomes of soil bacteria. *Bio/Technology* **4,** 446–449.

Belser, L. W. and Schmidt, E. L., 1978. Diversity in the ammonia-oxidizing nitrifier population of a soil. *Applied and Environmental Microbiology* **36,** 584–588.

Brayton, P. R. and Colwell, R. R., 1987. Fluorescent antibody staining method for enumeration of viable environmental *Vibrio cholerae* O1. *Journal of Microbiological Methods* **6,** 309–314.

Bohlool, B. B. and Schmidt, E. L., 1980. The immunofluoresence approach in microbial ecology. *Advances in Microbial Ecology* **4,** 203.

Colwell, R. R., Brayton, P. R., Grimes, D. J., Roszak, D. B., Huq, S. A. and Palmer, L. M., 1985. Viable but non-culturable *Vibrio cholerae* and related pathogens in the environment: implications for release of genetically engineered microorganisms. *Bio/Technology* **3,** 817–820.

de Flaun, M. F. and Paul, J. H., 1986. Hoechst 33258 staining of DNA in agarose gel electrophoresis. *Journal of Microbiological Methods* **5,** 265–270.

de Flaun, M. F., Paul, J. H. and Jeffrey, W. H., 1987. Distribution and molecular weight of dissolved DNA in subtropical estuarine and oceanic environments. *Marine Biology Progress Series* **38,** 65–73.

Delaney, J. E., McCarthy, J. A. and Grasso, R. J., 1962. Measurement of *E. coli* type I by the membrane filter. *Water and Sewage Works* **109,** 289–294.

Fox, G. E., Stackebrandt, E., Hespell, R. B., Gibson, J., Mamiloff, J., Dyer, T. A., Wolfe, R. S., Balch, W. E., Tanner, R. S., Magrum, L. J., Zablen, L. B., Blakemore, R., Gupta, R., Bonen, L., Lewis, B. J., Stable, D. A., Lurehrsen, K. R., Chen, K. N. and Woese, C. R., 1980. The phylogeny of procaryotes. *Science* **109,** 457.

Gerhardt, P., 1981. *Manual of Methods for General Microbiology.* Washington DC: American Society for Microbiology.

Gibbs, R. A. and Hayes, C. R., 1988. The use of R2A medium and the spread plate method for the enumeration of heterotrophic bacteria in drinking water. *Letters in Applied Microbiology* **6,** 19–22.

Giovannoni, S. J., Delong, E. F., Olsen, G. J. and Pace, N. R., 1988. Phylogenetic group-specific oligodeoxynucleotide probes for identification of single microbial cells. *Journal of Bacteriology* **170,** 720–726.

Grimont, P. A. D., Grimont, F., Desplaces, N. and Tchen, P., 1985. DNA probe specific for *Legionella pneumophila. Journal of Clinical Microbiology* 21, 431–437.

Havelaar, A. H. and During, M., 1988. Evaluation of the Anderson Baird-Parker direct plating method for enumerating *Escherichia coli* in water. *Journal of Applied Bacteriology* **64,** 89–98.

Holben, W. E., Jansson, J. K., Chelm, B. K. and Tiedje, J. M., 1988. DNA probe method for the detection of specific microorganisms in the soil bacterial community. *Applied and Environmental Microbiology* **54,** 703–711.

Howgrave-Graham, A. L. and Steyn, P. L., 1988. Application of the fluorescent-antibody technique for the detection of *Sphaerotilus natans* in activated sludge. *Applied and Environmental Microbiology* **54,** 799–802.

Irby, W. S., Huang, Y. S., Kawanishi, C. Y. and Brooks, W. M., 1985. Immunoblot analysis of exposure polypeptides from some entomorphic microsporidia. *Virology* **143,** 370.

Josserand, A. and Cleyet-Marel, J. C., 1979. Isolation from soils of *Nitrobacter* and evidence for novel serotypes using immunofluorescence. *Microbial Ecology* **5,** 197–205.

Kogure, K., Simidu, U. and Taga, N., 1979. A tentative direct microscopic method for counting living marine bacteria. *Canadian Journal of Microbiology* **25,** 415.

Korolik, V., Coloe, P. J. and Krishnapillai, V., 1988. A specific DNA probe for the identification of *Campylobacter jejuni. Journal of General Microbiology* **134,** 521–529.

Loh, L. C., Hamm, J. J., Kawanishi, C. Y. and Huang, E. S., 1982. Analysis of the *Spodoptera frugiperda* nuclear polyhedrosis virus genome by restriction endonucleases and electron microscopy. *Journal of Virology* **44,** 747.

MacDonell, M. T. and Colwell, R. R., 1984. Identical 5S rRNA nucleotide sequence of *Vibrio cholerae* strains representing temporal, geographical, and ecological diversity. *Applied and Environmental Microbiology* **48,** 199.

McClure, P. J. and Roberts, T. A., 1987. The effect of incubation time and temperature on growth of *Escherichia coli* on gradient plates containing sodium chloride and sodium nitrite. *Journal of Applied Bacteriology* **63,** 401–407.

Olsen, G. J., Lane, D. J., Giovannoni, S. J., Pace, N. R. and Stahl, D. A., 1986. Microbial ecology and evolution: a ribosomal RNA approach. *Annual Reviews of Microbiology* **40,** 337–365.

Paul, J. H., Deflaun, M. F., Jeffrey, W. H. and David, A. W., 1988. Seasonal and diel variability in dissolved DNA and in microbial biomass and activity in a subtropical estuary. *Applied and Environmental Microbiology* **54,** 718–827.

Romick, T. L., Lindsay, J. A. and Busta, F. F., 1987. A visual DNA probe for detection of enterotoxigenic *Escherichia coli* by colony hybridization. *Letters in Applied Microbiology* **5,** 87–90.

Rubin, F. A., Kopecko, D. J., Noon, K. F. and Baron, L. S., 1985. Development of a DNA probe to detect *Salmonella typhi. Journal of Clinical Microbiology* **22,** 600–605.

Rychert, R. C. and Stephenson, G. R., 1981. Atypical *Escherichia coli* in streams. *Applied and Environmental Microbiology* **41,** 1276–1278.

Sakazaki, R. and Donovan, T. J., 1984. Serology and epidemiology of *Vibrio cholerae* and *Vibrio mimicus*. In *Methods in Microbiology*, Bergan, T. (ed.), Vol. 16, Chapt. 11, pp. 271–289. London: Academic Press.

Schleifer, K. H., Ludwig, W., Kraus, J. and Festl, H., 1985. Cloned ribosomal ribonucleic acid genes from *Pseudomonas aeruginosa* as probes for conserved deoxyribonucleic acid sequences. *International Journal of Systematic Bacteriology* **35,** 231–236.

Shimada, T., Sakazaki, R. and Que, M., 1987. A bioserogroup of marine vibrios possessing somatic antigen factors in common with *Vibrio cholerae* O1. *Journal of Applied Bacteriology* **62,** 453–456.

Shimada, T. and Sakazaki, R., 1988. A serogroup of non-O1 *Vibrio cholerae* possessing the Inaba antigen of *Vibrio cholerae* O1. *Journal of Applied Bacteriology* **64,** 141–144.

Somerville, C., Knight, I. T., Straube, W. L. and Colwell, R. R., 1988. Probe-directed, polymerization-enhanced detection of specific gene sequences in the environment. *REGEM 1, Abstract 70.*

Stahl, D. A., Lane, D. J., Olsen, G. J. and Pace, N. D., 1985. Characterization of a Yellowstone hot spring microbial community by 5S RNA sequences. *Applied and Environmental Microbiology* **49,** 1379.

Summers, M. D. and Hoops, P., 1980. Radioimmunoassay analysis of Baculovirus granulins and polyhedrins. *Virology* **103,** 89.

Tannock, G. W., 1988. Molecular gentics: a new tool for investigating the microbial ecology of the gastrointestinal tract? *Microbial Ecology* **15,** 239–256.

Ward, B. B. and Carlucci, A. F., 1985. Marine ammonia- and nitrite-oxidizing bacteria: serological diversity determined by immunofluorescence in culture and in the environment. *Applied and Environmental Microbiology* **50,** 194–201.

Zelibor, J., Tamplin, M. and Colwell, R. R., 1987. A method for measuring bacterial resistance to metals employing epifluorescent microscopy. *Journal of Microbiological Methods* **7,** 143–155.

6 Genetic Transfer in the Natural Environment

STUART B. LEVY[1,2*] and BONNIE M. MARSHALL[1]

Departments of Molecular Biology[1] and of Medicine[2], Tufts University, School of Medicine, Boston, MA 02111, USA

INTRODUCTION

The proposed release of genetically-engineered micro-organisms into an environment generally requires conditions in which the organism and its genetic trait will remain locked together in the host–gene system. The fate of the host organism would effectively dictate that of the introduced genetic trait, if it were maintained in this host. In order to provide an insight into the potential for gene transfer, we have studied such events in different ecosystems. Some information has been obtained by looking at gene transfer in defined ecosystems, such as the intestinal tract of humans and animals. Other data have been obtained by looking at the natural distribution of particular genes in the environments.

The existence of known genetic traits and specific genes in widely dispersed micro organisms that colonize different ecosystems indicates that transfers readily take place. However, there are limits to the kinds of transfer studies that mimic the natural environment and which can be performed in the laboratory. To date, the kinds of environment that have lent themselves to such investigation include soils, fresh water, sewage, plants, the gastro-intestinal (GI) tract and the skin of animals and humans. In these instances genetically-marked donors and potential recipients have been introduced into normal or microcosmal natural environments and gene exchange followed.

The purpose of this account is to provide and summarize pertinent evidence for gene transfer in the environment. The studies cited here are not, however, intended to constitute a comprehensive review of the subject. Rather, they are

---

*Corresponding author.

RELEASE OF GENETICALLY-ENGINEERED MICRO-ORGANISMS ISBN 0-12-677521-4

intended to highlight some of the notable examples, in this rather large field, which convey the extent of diversity of bacterial gene transfer, the ecosystems in which it has been demonstrated and some of the factors that affect the transfer event. The transfer of bacterial genes into plant genomes is the subject of increasing attention but this rather extensive field will not be considered here.

INDIRECT EVIDENCE FOR WIDESPREAD GENE TRANSFER IN THE NATURAL ENVIRONMENT: ANTIBIOTIC RESISTANCE DETERMINANTS

Because of their medical interest and their ease of detection, antibiotic-resistance genes have been the focus of extensive studies. Initial work characterized gene products, e.g. β-lactamases, by substrate analysis or with specific antisera and used these methods to show intergeneric transfer (e.g. Sykes and Richmond, 1970). Later, specific gene sequences and enzymatic assays provided evidence of widespread dissemination of this kind of gene in the natural environment.

For instance, it was shown that the β-lactamase which newly appeared in *Haemophilus influenzae* and *Neisseria gonorrhoeae* was a gene originally found among the Enterobacteriaceae (deGraaff *et al.*, 1976; Elwell *et al.*, 1977) (Fig. 1).

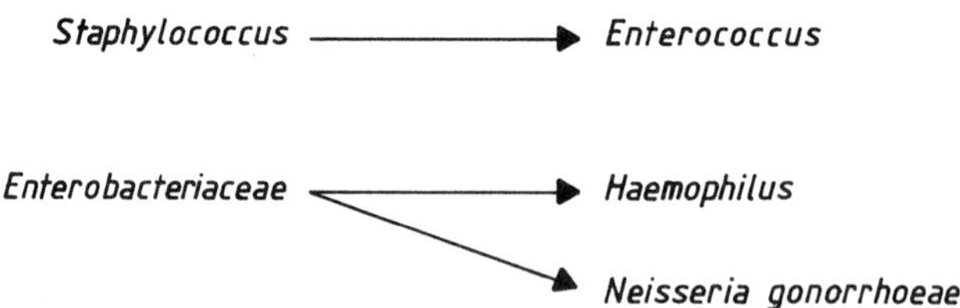

Fig. 1. Intergeneric transfer of β-lactamase genes.

Gene transfer had apparently moved the transposon bearing this determinant into these new hosts. Similarly, the tetracycline resistance determinant borne on Tn*10* was found widely among *Haemophilus* spp. (Marshall *et al.*, 1984) (Fig. 2). In an analogous situation, it was found that the newly-discovered β-lactamase in enterococci was genetically similar to the gene previously considered only in *Staphylococcus* (Murray, 1986) (Fig. 1).

We have been particularly interested in the ecology of tetracycline-resistance determinants (Levy, 1984, 1986). Two different kinds of mechanisms for resistance have been described. In both cases, tetracycline is unaltered. The more common one is an inner membrane efflux system which actively transports the drug out of the resistant cell (McMurry *et al.*, 1980). The second mechanism is

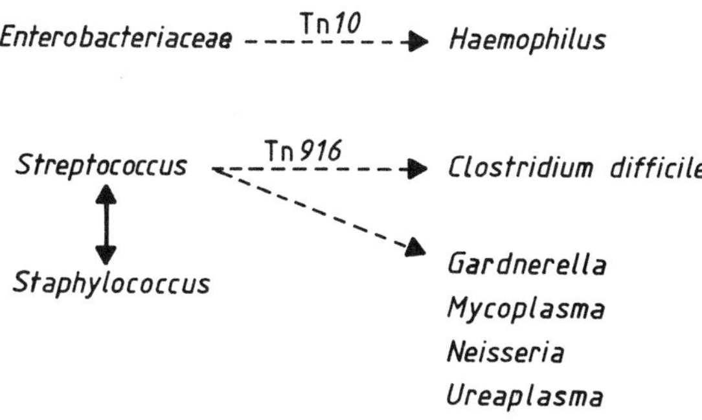

Fig. 2. Intergeneric transfer of tetracycline-resistance genes.

intracellular and involves a cytoplasmic factor which protects the ribosomes from the activity of tetracycline (Burdett, 1986). We have prepared specific DNA probes for five different classes (A–E) of tetracycline resistance determinants in Gram-negative bacteria and used them to determine the natural frequency of particular determinants in different microbial species (Mendez *et al.*, 1980; Marshall *et al.*, 1983, 1986; Levy, 1984).

Classes A, B and C are common in *E. coli, Shigella* and *Salmonella*. Class B was the most common determinant in 225 randomly selected tetracycline-resistant *E. coli* isolated from human and animal faeces. The Class D determinant was found in the fish pathogens *Aeromonas hydrophila, Edwardsiella tarda* and *Pasteurella piscicida* (Aoki and Takahashi, 1987) as well as in *Enterobacter, Salmonella, Shigella* and *Vibrio cholerae*. Class E was the principal determinant for tetracycline resistance in *Aeromonas hydrophila* isolated from cultured channel catfish (DePaola *et al.*, 1988). One of 225 *E. coli* isolates also showed this determinant (Fig. 3).

The wide distribution of tetracycline resistance classes among the Gram-negative organisms is exceeded by those found in Gram-positive organisms. Five genetically-distinguishable determinants have been identified: Classes K, L, M, N and O. The Class L determinant, and presumably the genetically similar Class K on plasmid pT181, encodes an efflux-type mechanism (McMurry *et al.*, 1987). These two classes have, so far, been found only in the Gram-positive organisms.

The broadest distribution has been found for the Class M determinant. Originally identified in *Streptococcus agalactiae* (Burdett *et al.*, 1982), this determinant has been identified in other species of *Streptococcus, Staphylococcus, Mycoplasma, Ureaplasma, Gardnerella, Neisseria* and *Clostridium difficile* (Roberts and Kenny, 1986; Morse *et al.*, 1986; Roberts *et al.*, 1986; Hächler *et al.*, 1987) (Fig. 2). The Class O determinant, which shows over 70% sequence similarity with Class M, was initially identified and characterized in *Campylo-*

GENUS	CLASS A	B	C	D	E
Citrobacter		■			
Enterobacter		■	■	■	
Escherichia	■	■	■		■
Edwardsiella				■	
Klebsiella	■		■		
Proteus	■	■	■		
Salmonella	■	■	■	■	
Shigella	■	■	■	■	
Yersinia		■			
Aeromonas	■			■	■
Haemophilus		■			
Pasteurella				■	
Pseudomonas	■		■		
Vibrio	■	■	■	■	

Fig. 3. Distribution of tetracycline resistance determinants among Gram-negative bacilli. Data from Mendez *et al.* (1980), Levy (1984), Marshall *et al.* (1986), Martinez-Salagar *et al.* (1986), Aoki and Takahashi (1987).

bacter species (Taylor, 1986; Sougakoff *et al.*, 1987). Its similarity to TetM leads us to suspect that further studies will also find it to be widely disseminated.

The finding of Class A, Class B and Class M determinants on transposons explains their widespread appearance on different plasmids and in different bacteria. While direct intergeneric transfer in natural environments has not been demonstrated, such transfer has been documented in the laboratory (Jones *et al.*, 1987; Roberts and Kenny, 1987).

TRANSFER OF GENETIC TRAITS AMONG BACTERIA SHARING THE SAME HUMAN AND ANIMAL ECOSYSTEMS

The human gastrointestinal tract

Most model systems dealing with gene transfer in the GI tract are based on the transfer of antibiotic resistance genes. Transfer of plasmids *in vivo* between *E. coli* in the absence of a selective antibiotic pressure has been demonstrated but the frequency has usually been low and transcipients remained for only short periods of time (Table 1). In general, transfer has been more readily observed and sustained when antibiotics were simultaneously ingested (Anderson *et al.*, 1973a; Burton *et al.*, 1974). Under these circumstances, intergeneric transfer has also been documented (Farrar *et al.*, 1972; Neu *et al.*, 1973). Anderson *et al.* (1973a,b) studied a system which consisted of biochemically-marked wild-type *E. coli* derived from the experimental subjects and two F-like R plasmids, R-1 and R222-1. Large doses (10^{10}–10^{11}) of donors and recipients were ingested by five human subjects, three of whom were given therapeutic doses of tetracycline and ampicillin orally. Even though the test plasmids were readily transmissible *in vitro* ($\sim 10^{-3}$), transfer could not be detected *in vivo* in the absence of antibiotic treatment. However, in subjects receiving antibiotics, transfer was documented not only to the test recipient but also to the resident bacterial flora.

We compared the potential for transfer of a wild-type plasmid and a genetically-engineered ColE1 derivative in the human gut. Human volunteers were given an *E. coli* K-12 strain in which the poorly mobilizable ColE1 plasmid pBR322 co-resided with the Inc FII plasmid pSL222-4 and the Inc P plasmid pLM2, two mobilizing transfer-derepressed plasmids. Only plasmid pSL222-4 was detectably transferred to the indigenous flora in the absence of tetracycline ingestion. However, ingestion of tetracycline increased the number of indigenous organisms that carry the R222-1 plasmid (Marshall *et al.*, 1981).

These findings were investigated by Levine *et al.* (1983) who examined the transfer potential of a similar ColE1 derivative plasmid, pBR325, this time with a host *E. coli* strain originally isolated from the human gut flora. In this case, pBR325 transfer was observed in the gut, but only when the conjugative plasmid F-amp was present in the same *E. coli* host and only in the presence of daily tetracycline ingestion (1000 mg day^{-1}).

The animal gastrointestinal tract

Transfer of genes, including determinants for antibiotic resistance and enterotoxins, has been observed in the gut of numerous species of conventional animals including rats, fowl, pigs, and calves (Table 1). These studies also showed enhanced transfer, including intergeneric transfer, of resistance genes in the presence of antibiotic supplementation.

The presumed transfer of a thermosensitive Inc H2 plasmid in the gastrointestinal tract of cows was shown to have taken place in the external environment. Timoney and Linton (1982) fed calves a donor *E. coli* bearing the Inc H2 plasmid and a nal^R *E. coli* recipient. Transfer of the Inc H2 plasmid did not take place if the cows were muzzled and denied oral contact with their environment. However, transconjugants could be recovered from the faeces of unmuzzled cows or if the muzzles were removed. In addition, transfer could be

Table 1. Transfer *in situ*.

Ecosystem	Mechanism	Organism	Site	Reference
Human	Conjugation	*E. coli* *Shigella sonnei* *Proteus mirabilis*	GI tract	Smith (1969), Farrar *et al.* (1972), Anderson *et al.* (1973a,b), Neu *et al.* (1973), Burton *et al.* (1974), Anderson (1975), Petrocheilou *et al.* (1976), Williams (1977), Marshall *et al.* (1981), Levine *et al.* (1983)
		Staphylococcus spp.	Skin	Naidoo and Lloyd (1984)
Animal	Conjugation	*E. coli* *S. saint-paul* *S. choleraesuis*	GI tract of: calf, chicken, pig, rat, turkey	Walton (1966), Jarolmen and Kemp (1969), Guinée (1970), Smith (1970), Nivas *et al.* (1976), Gyles *et al.* (1978)
		S. typhi	Rabbit bladder	Richter *et al.* (1973)
		Klebsiella aerogenes *P. aeruginosa* *Staphylococcus* spp.	Mouse skin	Roe and Jones (1972), Naidoo and Lloyd (1984)
		Staphylococcus spp.	Dog skin	Naidoo and Lloyd (1984)
	Transduction	*S. aureus*	Mouse kidney	Novick and Morse (1967)
	Transformation	*Streptococcus pneumoniae*	Mouse peritoneum, mouse respiratory tract	Ottolenghi and MacLeod (1963), Conant and Sawyer (1967)

Animal → Human	Not determined	*E. coli*	Calf → human GI	Hirsh and Wiger (1978), Marshall and Petrowski (1988)
		Enterobacteriaceae	Chicken → human GI	Levy *et al.* (1976)
		Streptococcus spp. *E. coli*	Pig → human	Christie and Dunny (1984), Hummel *et al.* (1986)
Soil	Conjugation	*E. coli* *Pseudomonas aeruginosa*	Laboratory models	Weinberg and Stotzky (1972), Devanas and Stotzky (1987), Trevors and Starodub (1987)
	Transformation	*Bacillus subtilis*		Graham and Istock (1979)
	Transduction	*E. coli*		Zeph *et al.* (1987)
Water	Conjugation	*E. coli*	River water	Grabow *et al.* (1975)
	Conjugation	*E. coli, Enterobacter cloacae, Salmonella enteritidis, Proteus mirabilis, Shigella sonnei*	Sewage or treatment facility	Fontaine and Hoadley (1976), Altherr and Kasuwek (1982), Mach and Grimes (1982), Gealt *et al.* (1985), Mancini *et al.* (1987)
	Transduction	*Pseudomonas aeruginosa*	Lake water	Morrison *et al.* (1978), Saye *et al.* (1987)

GI = gastrointestinal tract.

demonstrated in faeces held at 28°C but not at 37°C. Therefore, it appeared that transfer did not take place in the gut environment but rather in the external environment from which the new hosts were subsequently acquired by the animals by ingestion or inhalation.

Transfer in animal ecosystems other than the gastrointestinal tract

A few studies have documented transfer in or on tissues other than the gastrointestinal tract. These include conjugation in rabbit bladder as well as transduction in mouse kidney and transformation of pneumococci in the respiratory tract and in the peritoneum of mice (Table 1).

Roe and Jones (1972) examined the transfer of carbenicillin-resistance in burns on mouse skin. These authors demonstrated conjugal transfer of a multiresistant R factor from *Klebsiella aerogenes* to *Pseudomonas aeruginosa*. Transfer of carbenicillin-resistance was observed within 5 min of application which led these investigators to conclude that bacterial multiplication *in situ* was not necessary for conjugation to take place.

The possible exchange of staphylococcal plasmids between animals and humans prompted Naidoo and Lloyd (1984) to perform mating experiments on the skin of humans, dogs and mice. They observed the transfer of gentamicin-resistance along with other resistance markers from non-pathogenic to pathogenic *Staphylococcus* in the absence of antibiotics. Although the resistance traits were often co-transferred, the determinants were found to reside on separate plasmids. The *in situ* transfer frequency was ten to 10^4-fold greater than that obtained *in vitro*, suggesting that skin facilitated plasmid transfer. Transfer occurred irrespective of the source of the isolates or the species of skin used, i.e. human skin-derived donors could transfer to human or canine skin-derived recipients and transfer could be demonstrated on the skin of humans, dogs and mice regardless of the origin of the parental strains; only the transfer *rates* appeared to differ.

SPREAD OF ANTIBIOTIC RESISTANCE DETERMINANTS AMONG ANIMAL AND HUMAN ECOSYSTEMS

Evidence for the *in vivo* migration of bacteria and their plasmids between human and animal ecosystems has been seen in studies with calves, chickens and pigs.

Hirsh and Wiger (1978) demonstrated the transfer of a marked bovine *E. coli* with a multiple-resistant transmissible plasmid from calves to humans that were in daily contact. The R-factor transferred at low frequency, irrespective of tetracycline consumption and could be found in the enteric flora of the bovines for up to 4 days. In the human subjects it was only recovered in the original *E. coli* host.

We studied the migration of a transferable R plasmid between groups of chickens with and without tetracycline in their feed. Chickens re-colonized with an indigenous gut *E. coli*, now bearing a biochemically-marked plasmid, were introduced into each of two cages that housed uninoculated chickens; one group was given tetracycline-supplemented feed. The test plasmid, specifying tetracycline-resistance and temperature-sensitive chloramphenicol-resistance, moved into the GI tract of the other chickens, but only in the cage in which the chickens were receiving tetracycline. The resistance determinant was also recovered from tetracycline-fed chickens housed in a separate cage located 15 m away from the experimental cage. The plasmid was not found in an identical cage of chickens fed without a tetracycline-supplement. The plasmid was also detected in the gut of one of the animal caretakers. In a second similar experiment, both the plasmid and the *E. coli* host were marked. The plasmid and its host moved among the chickens, not only into the gut of other chickens given tetracycline-supplemented feed but also into the gut flora of one human caretaker not receiving tetracycline (Levy *et al.*, 1976).

By means of DNA : DNA hybridization techniques, Christie and Dunny (1984) demonstrated a link between macrolide–lincosamide–streptogramin B (MLS) resistance of human enterococcus isolates and those from pigs. A probe carrying the MLS determinant detected sequence homology between streptococci from tylosin-fed pigs and MLS-resistant clinical isolates from human infections. Whereas the MLS determinant had previously been localized on the chromosome or on small plasmids, the linked MLS resistance of the antibiotic-fed pigs was found to reside primarily on large conjugative plasmids. Thus, not only did tylosin treatment select for multiple-resistant bacteria but, importantly, it promoted the transfer of the MLS determinant by localizing it on large plasmids of the conjugative type.

Hummel *et al.* (1986) traced a streptothricin-resistance transposon within a single community in which pigs received this antibiotic in their feed. The highest frequency (33%) of the resistance determinant was found in the gut flora of pigs fed the antibiotic and this provided circumstantial evidence that pigs served as the origin of the determinant. The determinant was also found at somewhat lower levels (16–18%) in the gut flora of farm employees (direct contact), family members of employees (indirect contact) and also in outpatients (no contact). The transposon Tn*1825* was located in plasmids of as many as ten different incompatability groups, signifying a widespread dissemination of a single resistance determinant among diverse plasmids.

We recently examined the inter- and intra-species transfer of a wild-type *E. coli* bearing a marked resistance plasmid in a natural farm setting (unpublished data). While transfer of the plasmid to resident bacteria of the ecosystem has not yet been documented, the *E. coli* host which carries the plasmid was seen to move from one bovine ecosystem to diverse other ecosystems, including the gut flora of mice, chickens, pigs and humans. Human carriage lasted as long as 74 days without antibiotic selection. The host and its plasmid were also recovered for up to 26 days in flies.

Gene transfer in soils

So far, few transfer experiments in soils have been able to entirely copy natural conditions; most have approached the problem by using sterile soil ecosystems. *In situ* transfers have been reported to result from conjugation via plasmid transfer, phage transduction and transformation.

Weinberg and Stotzky (1972), who used autoclaved soils, observed the production of multiple-recombinant types by the conjugation of F^+ prototrophic *E. coli* donors with F^- auxotrophic *E. coli* recipients. Transfer was enhanced significantly by the addition of montmorillonite, presumably because it enhanced donor and recipient cell growth, thereby increasing cell-to-cell contact.

Non-conjugative transfer has been observed between *Bacillus subtilis* strains that bear distinct linkage blocks of antibiotic-resistance and auxotrophy genes (Graham and Istock, 1979). *Bacillus* spores were induced to germinate by heat shock. This treatment led to the release of transforming DNA and simultaneously induced competency for transformation. Spores were then incubated in autoclaved potting soil and transformants containing linked blocks of resistance genes and auxotrophy genes were selected. Of the 256 possible gene combinations that might have been obtained, as many as 149 distinct phenotypes were recovered early in the experiment, thus demonstrating the vast potential for genetic exchange. Because of the absence of demonstrable plasmids and transducing phage in *Bacillus*, it was postulated that this exchange occurred by transformation and that cells in close proximity 'sprayed' each other with DNA. The high DNA concentration resulting from heat shock was most probably responsible for the high transformation levels (10^{-4}) that were only ten to 100-fold lower than the optimum levels achieved *in vitro*. By process of natural selection, the number of phenotypes greatly decreased with time until the culture was dominated (79%) by a single phenotype.

Recent studies have reported the potential for conjugation and transduction in natural soils containing an indigenous competitive soil flora (Devanas and Stotzky, 1987; Trevors and Starodub, 1987). Conjugation was observed at levels ten to 57-fold lower than in *in vitro* matings. The authors also demonstrated the positive effects of increasing soil moisture up to 80% of the water-holding capacity and the detrimental effects of anaerobic conditions. Conjugation was unaffected by pH changes within the normal range. Zeph *et al.* (1987) recently reported the apparent recovery of P1 transductants from natural soil to which *E. coli* and phage lysate had been added.

Gene transfer in aquatic environments

In the freshwater environment, diversification of the microbial gene pool has been demonstrated to occur by transduction and conjugation mechanisms. Using cylinders of autoclaved lake water submerged in a lake, Morrison *et al.* (1978) tested transduction of a streptomycin-resistance gene between a susceptible *Pseudomonas* strain and a lysogenic *Pseudomonas* strain containing the generalized transducing phage F116. Under these conditions, successful transduction would require the release of viable phage and concurrent infection of the recipient cells. Interestingly, this transduction event occurred at frequencies equivalent to those obtained when a free transducing phage was present. This demonstrates that a sterile aquatic environment does not inhibit the release of transducing phage.

In a similar experiment, also with *Pseudomonas*, Saye *et al.* (1987) demonstrated transduction of a plasmid but found that the event could only occur between a non-lysogenic donor and a lysogenic recipient. This experiment not only demonstrated that complex transfer events could occur *in vivo* but also that certain conditions were actually required for successful transduction. The donor–recipient ratio was shown to have a significant bearing on transduction and was optimal at 20:1. Therefore, the load of plasmid-containing bacteria released into a freshwater environment becomes an important factor in the determination of gene transfer by transduction. When the sterilized lake water was replaced with natural water, the transduction was clearly reduced but still detectable.

Sewage has been of prime interest as a source for the possible widespread dissemination of genetic elements. Of particular concern has been potential exchange if the novel genetic constructions created by genetic-engineering were discharged into sewage systems. Mach and Grimes (1982), using diffusion chambers containing sterilized sewage, showed transfer of resistance plasmids from clinical isolates of *Shigella, Proteus* and *E. coli* to environmental isolates of *E. coli* and *Shigella sonnei* when the chambers were placed in the primary and secondary settling tanks of a wastewater treatment plant. The frequencies were about 10^{-5}.

Altherr and Kasuweck (1982) also showed transfer of a conjugative plasmid in membrane diffusion chambers placed in raw sewage. A temperature dependency was demonstrated. At 22.5°C, transfer of a 60 Md antibiotic-resistance plasmid occurred at 3.2×10^{-5} but decreased ten-fold at 29.5°C. While transfer could take place in the raw sewage of the degritter tank, it was not observed in the decreased-nutrient, higher-salt environment found 500 m downstream of the release point. Gealt *et al.* (1985) studied the transfer, in sterilized wastewater, of nonconjugative, genetically-engineered plasmids pBR322 and pBR325 from laboratory strains of *E. coli* into natural wastewater strains of *Enterobacter cloacae* and *E. coli.* Transfer occurred only when other *E. coli* were present which contained a transfer-derepressed, mobilizing plasmid. Examination of transconjugants showed not only the presence of pBR322 and pBR325, but also co-integrates resulting from recombinational events, probably mediated by transposons. Recently in a laboratory-scale, waste-treatment facility, Mancini *et al.* (1987) confirmed triparental mating in both laboratory and indigenous strains as mobilizers and recipients. Transfer of a non-conjugative genetically-engineered plasmid could be detected in the bottom and sludge of the primary clarifier but only when at least 10^7 donor cells were present.

SUMMARY

Available evidence indicates that genetic transfer can occur in widely diverse environments involving large numbers of different kinds of bacterial hosts. The conditions of transfer in the environment are not well-understood but appear to be similar to those that affect the survival of the natural bacterial hosts. Besides transmissible plasmids, phages may be vectors of the transfer as well as free DNA via a transformation event.

The background of information on gene transfer suggests that such events occur naturally in the environment without selection but can be propagated by an appropriate selective agent. Subsequently, new hosts for these genes reproduce and augment their members under the protection of the selective agent. In this way genes which are transferred at low frequencies can become prominent in the natural environment through man's use of selective agents whether they be antibiotics, heavy metals or hydrocarbons. Given these conditions, it is important for investigators placing known genetic traits on plasmid vectors or on chromosomes to evaluate the potential for spread of these traits within the natural environment. A trait once introduced into bacteria has a potential to be transferred. However, the frequency can be so small as to not be attained by the numbers of donors and recipients in that environment. Given a low frequency of transfer, it is the fate of the host bacteria which is critical in evaluating the fate of the introduced gene. Moreover, if that gene has no selective advantage, the chance of its ever becoming potentially predominant in the environment would be negligible. It is, therefore, strongly advised that genetic traits should not be linked to common antibiotic resistance genes which are being steadily selected for by the extensive use of antibiotics in man's natural environments.

ACKNOWLEDGEMENTS

This work was supported by grants from the National Institutes of Health, the National Science Foundation and the Center for Environmental Management at Tufts University.

REFERENCES

Altherr, M. R. and Kasuwek, K. L., 1982. *In situ* studies with membrane diffusion chambers of antibiotic resistance transfer in *E. coli*. *Applied and Environmental Microbiology* **44,** 838–843.

Anderson, E. S., 1975. Viability of, and transfer of a plasmid from, *E. coli* K-12 in the human intestine. *Nature* **255,** 502–504.

Anderson, J. D., Gillespie, W. A. and Richmond, M. H., 1973a. Chemotherapy and antibiotic resistance transfer between Enterobacteria in the human gastro-intestinal tract. *Journal of Medical Microbiology* **6,** 461–473.

Anderson, J. D., Ingram, L. C., Richmond, M. H. and Wiedemann, B., 1973b. Studies on the nature of plasmids arising from conjugation in the human gastro-intestinal tract. *Journal of Medical Microbiology* **6,** 475–486.

Aoki, T. and Takahashi, A., 1987. Class D tetracycline resistance determinants of R-plasmids from the fish pathogens *Aeromonas hydrophila, Edwardsiella tarda* and *Pasteurella piscicida. Antimicrobial Agents and Chemotherapy* **31,** 1278–1280.

Burdett, V., 1986. Streptococcal tetracycline resistance mediated at the level of protein synthesis. *Journal of Bacteriology* **165,** 564–569.

Burdett, V., Inamine, J. and Rajagopalan, S., 1982. Heterogeneity of tetracycline resistance determinants in *Streptococcus. Journal of Bacteriology* **149,** 995–1004.

Burton, G. C., Hirsh, D. C., Blenden, D. C. and Zeigler, J. L., 1974. The effects of tetracycline on the establishment of *Escherichia coli* of animal origin, and *in vivo* transfer of antibiotic resistance, in the intestinal tract of man. *Proceedings of the Society for Applied Bacteriology*, 241–253.

Christie, P. J. and Dunny, G. M., 1984. Antibiotic selection pressure resulting in multiple antibiotic resistance and localization of resistance determinants to conjugative plasmids in streptococci. *Journal of Infectious Diseases* **149,** 74–81.

Conant, J. E. and Sawyer, W. D., 1967. Transformation during mixed pneumococcal infection of mice. *Journal of Bacteriology* **93,** 1869–1875.

deGraaff, J., Elwell, L. and Falkow, S., 1976. Molecular nature of two beta-lactamase-specifying plasmids isolated from *Haemophilius influenzae* type B. *Journal of Bacteriology* **126,** 439–446.

DePaola, A., Flynn, P. A., McPherson, R. M. and Levy, S. B., 1988. Phenotypic and genotypic characterization of tetracycline and oxytetracycline resistant *Aeromonas hydrophila* from cultured channel catfish (*Ictalurus punctatus*) and their environments. *Applied and Environmental Microbiology*, in press.

Devanas, M. A. and Stotzky, G., 1987. Survival and conjugal transfer of plasmid RP4 from *Escherichia coli* to *Pseudomonas aeruginosa* in soil: effect of clay minerals and nutrients. *American Society of Microbiology, Abstracts of the Annual Meetings.*

Elwell, L. P., Roberts, M., Mayer, L. W. and Falkow, S., 1977. Plasmid-mediated beta-lactamase production in *Neisseria gonorrhoeae. Antimicrobial Agents and Chemotherapy* **11,** 528–533.

Farrar, W. E., Eidson, M., Guerry, P., Falkow, S., Drusin, L. M. and Roberts, R. B., 1972. Interbacterial transfer of R-factor in the human intestine: *in vivo* acquisition of R-factor-mediated kanamycin resistance by a multi-resistant strain of *Shigella sonnei*. *Journal of Infectious Diseases* **126,** 27–33.

Fontaine, T. D. and Hoadley, A. W., 1976. Transferable drug resistance associated with coliforms isolated from hospital and domestic sewage. *Health Laboratory Science* **13,** 238–245.

Gealt, M. A., Chai, M. D., Alpert, K. B. and Boyer, J. C., 1985. Transfer of plasmids pBR322 and pBR325 in wastewater from laboratory strains of *E. coli* to bacteria indigenous to the waste disposal system. *Applied and Environmental Microbiology* **49,** 836–841.

Grabow, W. O. K., Prozesky, O. W. and Burger, J. S., 1975. Behaviour in a river and dam of coliform bacteria with transferable or non-transferable drug resistance. *Water Research* **9,** 777–782.

Graham, J. B. and Istock, C. A., 1979. Gene exchange and natural selection cause *Bacillus subtilis* to evolve in soil culture. *Science* **204,** 637–639.

Guinée, P. A. M., 1970. Resistance transfer to the resident intestinal *Escherichia coli* of rats. *Journal of Bacteriology* **102,** 291–292.

Gyles, C., Falkow, S. and Rollins, L., 1978. *In vivo* transfer of an *E. coli* enterotoxin plasmid possessing genes for drug resistance. *American Journal of Veterinary Research* **39,** 1438–1441.

Hächler, H., Kayser, F. H. and Berger-Bächi, B., 1987. Homology of a transferable tetracycline resistance determinant of *Clostridium difficile* with *Streptococcus (Enterococcus) faecalis* transposon Tn*916*. *Antimicrobial Agents and Chemotherapy* **31,** 1033–1038.

Hirsh, D. C. and Wiger, N., 1978. The effect of tetracycline upon the spread of bacterial resistance from calves to man. *Journal of Animal Science* **46,** 1437–1446.

Hummel, R., Tschape, H. and Witte, W., 1986. Spread of plasmid-mediated mouse nourseothricin resistance due to antibiotic use in animal husbandry. *Journal of Basic Microbiology* **26,** 461–466.

Jarolmen, H. and Kemp, G., 1969. R-factor transmission *in vivo*. *Journal of Bacteriology* **99,** 487–490.

Jones, J. M., Yost, S. C. and Pattee, P. A., 1987. Transfer of the conjugal tetracycline resistance transposon Tn*916* from *Streptococcus faecalis* to *Staphylococcus aureus* and identification of some insertion sites in the staphylococcal chromosome. *Journal of Bacteriology* **169,** 2121–2131.

Levine, M. M., Kaper, J. B., Lockman, H., Black, R. E., Clements, M. L. and Falkow, S., 1983. Recombinant DNA risk assessment studies in humans: efficacy of poorly mobilizable plasmids in biologic containment. *Journal of Infectious Diseases* **148,** 699–709.

Levy, S. B., 1984. Resistance to the tetracyclines. In *Antimicrobial Drug Resistance*, Byron, L. E. (ed.), pp. 191–239. London: Academic Press.

Levy, S. B., 1986. Ecology of antibiotic resistance determinants. In *Antibiotic Resistance Genes: Ecology, Transfer and Expression*, Levy, S. B. and Novick, R. P. (eds) pp. 17–30. Cold Spring Harbor: Cold Spring Harbor Laboratory.

Levy, S. B., FitzGerald, G. B. and Macone, A. B., 1976. Spread of antibiotic-resistant plasmids from chicken to chicken and from chicken to man. *Nature* **260,** 40–42.

Mach, P. A. and Grimes, D. J., 1982. R-plasmid transfer in a wastewater treatment plant. *Applied and Environmental Microbiology* **44,** 1395–1403.

Mancini, P., Fertels, S., Nave, D. and Gealt, M. A., 1987. Mobilization of plasmid pHSV106 from *E. coli* HB101 in a laboratory scale waste treatment facility. *Applied and Environmental Microbiology* **53,** 665–671.

Marshall, B. M., Schluederberg, S., Tachibana, C. and Levy, S. B., 1981. Survival and transfer in the human gut of poorly mobilizable (pBR322) and of transferable plasmids from the same carrier *E. coli*. *Gene* **14,** 145–154.

Marshall, B., Tachibana, C. and Levy, S. B., 1983. Frequency of tetracycline resistance determinant classes among lactose-fermenting coliforms. *Antimicrobial Agents and Chemotherapy* **24,** 835–840.

Marshall, B., Roberts, M., Smith, A. and Levy, S. B., 1984. Homogeneity of transferable tetracycline-resistance determinants in *Haemophilus* species. *Journal of Infectious Diseases* **149,** 1028–1029.

Marshall, B., Morrissey, S., Flynn, P. and Levy, S. B., 1986. A new tetracycline-resistance determinant, Class E, isolated from Enterobacteriaceae. *Gene* **50,** 111–117.

Marshall, B. and Petrowski, D., 1988. Persistence and transfer of marked wild-type *E. coli* in a farm environment. *Abstracts of the American Society of Microbiology*, in press.

Martinez-Salagar, J. M., Alvarez, G. and Gomez-Eichelmann, M. C., 1986. Frequency of four classes of tetracycline resistance determinants in *Salmonella* and *Shigella* spp. clinical isolates. *Antimicrobial Agents and Chemotherapy* **30,** 630–631.

McMurry, L. M., Petrucci, R. and Levy, S. B., 1980. Active efflux of tetracycline encoded by non-genetically different tetracycline resistance determinants in *E. coli*. *Proceedings of the National Academy of Sciences, USA* **77,** 3974–3977.

McMurry, L. M., Park, B. H., Burdett, V. and Levy, S. B., 1987. Energy-dependent efflux mediated by Class L (Tet L) tetracycline resistance determinant from *Streptococci*. *Antimicrobial Agents and Chemotherapy* **31,** 1648–1650.

Mendez, B., Tachibana, C. and Levy, S. B., 1980. Heterogeneity of tetracycline resistance determinants. *Plasmid* **3,** 99–108.

Morrison, W. D., Miller, R. V. and Sayler, G. S., 1978. Frequency of F116-mediated transduction of *Pseudomonas aeruginosa* in a freshwater environment. *Applied Environmental Microbiology* **36,** 724–730.

Morse, S., Johnson, S. R., Biddle, J. W. and Roberts, M. C., 1986. High-level tetracycline resistance in *Neisseria gonorrhoeae* is result of acquisition of streptococcal *tetM* determinant. *Antimicrobial Agents and Chemotherapy* **30,** 664–670.

Murray, B. E., 1986. Plasmid-mediated penicillinase production in enterococci. In *Antibiotic Resistance Genes: Ecology, Transfer and Expression*, Levy, S. B. and Novick, R. P. (eds), pp. 31–40. Cold Spring Harbor: Cold Spring Harbor Laboratory.

Naidoo, J. and Lloyd, D. H., 1984. Transmission of genes between *Staphylococci* on skin. In *Antimicrobials and Agriculture*, Woodbine, M. (ed.), pp. 285–292. London: Butterworth.

Neu, H. C., Huber, P. J. and Winshell, E. B., 1973. Interbacterial transfer of R-factor in humans. *Antimicrobial Agents and Chemotherapy* **3,** 542–544.

Nivas, S. C., York, M. D. and Pomeroy, B. S., 1976. *In vitro* and *in vivo* transfer of drug resistance for *Salmonella* and *Escherichia coli* strains in turkeys. *American Journal of Veterinary Science* **37,** 433–437.

Novick, R. P. and Morse, S. I., 1967. *In vivo* transmission of drug resistance factors between strains of *Staphylococcus aureus*. *Journal of Experimental Medicine* **125,** 45–59.

Ottolenghi, E. and MacLeod, C. M., 1963. Genetic transformation among living pneumococci in the mouse. *Proceedings of the National Academy of Sciences, USA* **50,** 417–419.

Petrocheilou, V., Grinsted, J. and Richmond, M. H., 1976. R-plasmid transfer *in vivo* in the absence of antibiotic selection pressure. *Antimicrobial Agents and Chemotherapy* **10,** 753–761.

Richter, M. W., Stotzky, G. and Amsterdam, D., 1973. Transfer of R-factor in urine. *American Society of Microbiology, Abstracts of the Annual Meetings*, p. 99.

Roberts, M. C. and Kenny, G. E., 1986. Dissemination of the *tetM* tetracycline resistance determinant to *Ureaplasma urealyticum. Antimicrobial Agents and Chemotherapy* **29,** 350–352.

Roberts, M. C. and Kenny, G. E., 1987. Conjugal transfer of transposon Tn*916* from *Streptococcus faecalis* to *Mycoplasma hominis. Journal of Bacteriology* **169,** 3836–3839.

Roberts, M. C., Hillier, S. L., Hale, J., Holmes, K. K. and Kenny, G. E., 1986. Tetracycline resistance and *tetM* in pathogenic urogenital bacteria *Antimicrobial Agents and Chemotherapy* **30,** 810–812.

Roe, E. and Jones, R. J., 1972. Effects of topical chemoprophylaxis on transferable antibiotic resistance in burns. *Lancet* **1,** 109–111.

Saye, D. J., Ogunseitan, O., Sayler, G. S. and Miller, R. V., 1987. Potential for transduction of plasmids in a natural freshwater environment: effect of plasmid donor concentration and a natural microbial community on transduction in *Pseudomonas aeruginosa. Applied and Environmental Microbiology* **53,** 987–995.

Smith, H. W., 1969. Transfer of antibiotic resistance from animal and human strains of *E. coli* to resistant *E. coli* in the alimentary tract of man. *Lancet* **2,** 1174–1176.

Smith, H. W., 1970. The transfer of antibiotic resistance between strains of Enterobacteria in chickens, calves and pigs. *Journal of Medical Microbiology* **3,** 165–180.

Sougakoff, W., Papadoupoulou, B., Nordmann, P. and Courvalin, P., 1987. Nucleotide sequence and distribution of gene *tetO* encoding tetracycline resistance in *Campylobacter coli. FEMS Microbiology Letters* **44,** 153–159.

Sykes, R. B. and Richmond, M. H., 1970. Intergeneric transfer of a β-lactamase gene between *Ps. aeruginosa* and *E. coli. Nature* **226,** 952–954.

Taylor, D. E., 1986. Plasmid-mediated tetracycline resistance in *Campylobacter jejuni*: expression in *Escherichia coli* and identification of homology with streptococcal Class M determinant. *Journal of Bacteriology* **165,** 1037–1039.

Timoney, J. F. and Linton, A. H., 1982. Experimental ecological studies on H2 plasmids in the intestine and faeces of the calf. *Journal of Applied Bacteriology* **52,** 417–424.

Trevors, J. T. and Starodub, M. E., 1987. R-plasmid transfer in non-sterile agricultural soil. *Systemics in Applied Microbiology* **9,** 312–315.

Walton, J. R., 1966. *In vivo* transfer of infectious drug resistance. *Nature* **211,** 312–313.

Weinberg, S. R. and Stotzky, G., 1972. Conjugation and genetic recombination of *Escherichia coli* in soil. *Soil Biology and Biochemistry* **4,** 171–180.

Williams, P. H., 1977. Plasmid transfer in the human alimentary tract. *FEMS Microbiology Letters* **2,** 91–95.

Zeph, L. R., Onaga, M. A. and Stotzky, G., 1987. Transduction in soil. *American Society of Microbiology, Abstracts of the Annual Meetings.*

7 Fate and Behaviour in an Activated Sludge Microcosm of a Genetically-engineered Micro-organism Designed to Degrade Substituted Aromatic Compounds

D. F. DWYER, F. ROJO and K. N. TIMMIS

Université de Genève, Centre Médicale Universitaire, CH-1211 Genève 4, Switzerland

An enormous variety and amount of xenobiotic pollutants have been and are still being released into our environment. Many of these represent environmental hazards because of their toxicity to man and the world ecosystem as a whole (Keith and Telliard, 1979). The methods currently employed to dispose of these pollutants include the physical containment of affected material, chemical degradation and burning of the compounds. Such methods are often not adequate because they require the collection and concentration of the pollutant which, in many cases, has dispersed into the environment and cannot be recalled. Biodegradation of these compounds by micro-organisms offers an important alternative to these methods (Kobayashi and Rittmann, 1982), especially since *in situ* degradation by introduced micro-organisms is often the only foreseeable possibility for the detoxification of many pollutants. Thus, the biotechnological application of microbial biodegradation of pollutants becomes increasingly attractive as a method for removal of these chemicals from our environment.

Micro-organisms which contain catabolic pathways for the biodegradation of some pollutant compounds can be isolated from the environment. However, because many xenobiotic pollutants are recalcitrant to biodegradation, it is not always possible to isolate by conventional means indigenous micro-organisms that possess the catabolic ability to degrade these compounds. In many cases, however, individual bacteria contain the enzymatic capacity to catabolize some steps in a sequence leading to complete degradation. It may be possible to

RELEASE OF GENETICALLY-ENGINEERED MICRO-ORGANISMS ISBN 0-12-677521-4

maintain these microbes in consortia to catabolize the pollutants (Hovarth, 1972; Grady, 1985), but the interdependent requirements of such associations may make it difficult to apply them to environmental sites, maintain survival of each member and still obtain pollutant-degrading activity. Thus, it is of interest to combine these steps within one individual micro-organism to create a genetically-engineered micro-organism (GEM) with a well-regulated catabolic pathway which can be applied to polluted areas to effect pollutant degradation. Two *in vitro* methods exist to construct such organisms.

IN VITRO CONSTRUCTION OF POLLUTANT-DEGRADING GEMs

Bacteria with complete catabolic pathways for pollutant-degradation may be constructed by means of 'natural genetic exchange' between different micro-organisms. The micro-organisms are often unknown members of environmental inocula grown together in media containing the pollutant compound which select for genetically-rearranged micro-organisms able to use that compound as a growth substrate. They can also be known strains of bacteria which contain specific enzymatic functions. Natural genetic recombination resulted in the expansion of the range of aromatic compounds degraded by *Pseudomonas* sp. strain B13. This bacterium, as isolated, was able to use benzoate, 4-chlorophenol (4CP) and 3-chlorobenzoate (3CB) as growth substrates (Dorn *et al.*, 1974), but was unable to use several other substituted aromatic compounds. This was due to the limited range of substrates oxygenated by 1,2-dioxygenase, the first enzyme in the catabolic pathway for aromatics (Reineke and Knackmuss, 1978). Conjugative transfer of the TOL plasmid which contains the non-specific benzoate 1,2-dioxygenase from *Pseudomonas putida* strain mt-2 to B13 allowed the resultant transconjugants to use 4-chlorobenzoate (4CB) as well as 3- and 4-methylbenozoate (3- and 4-MB) as growth substrates (Reineke and Knackmuss, 1979).

Other examples of natural genetic exchange are worth noting. Plasmid-mediated transfer of the chromosomally-encoded dehalogenase I gene from *Pseudomonas putida* PP3 to other bacteria has resulted in transconjugants with a new ability to degrade halogenated compounds. For example, inclusion of the plasmid R68.45 in *P. putida* PP3 mediated transfer of dehalogenase I gene to *Pseudomonas* sp. strain HB2001. Whereas *Pseudomonas* sp. strain HB2001 was able to use 2-hydroxybutanoic acid as a growth substrate, the resultant transconjugant also used 2-monochlorobutanoic acid (Slater, 1985). A genetic recombinant which uses chlorobenzene has been produced by continuously concentrating and applying the parental bacterial strains onto ceramic beads in the presence of chlorobenzene. The effect was apparently to increase the frequency of genetic transfer since the resultant recombinant strain, *Pseudo-*

monas putida CB1-9 was isolated in less than 1 month of culture (Kröckel and Focht, 1987). This short time span is in contrast to the months often required to obtain similar results from conventional chemostat cultures (Reineke and Knackmuss, 1984) and may therefore be a way of hastening creation of new micro-organisms in cases where the gene transfer frequency is low or where multiple events are necessary to create a particular catabolic sequence.

Genetic engineering techniques can create new catabolic pathways from existing pathways which degrade analogue compounds. These can be expanded to accept degradatively recalcitrant pollutants as substrates by mutating existing enzymes so as to allow reaction with different substrates and/or effectors (Ramos and Timmis, 1987). This ability to expand a degradation pathway (and in the process to understand the evolutionary potential of catabolic enzymes) was first demonstrated by P. H. Clarke and her colleagues in which the substrate specificity of an amidase of *Pseudomonas* and the specificity of its regulator gene were sequentially expanded (Clarke and Slater, 1986). Recently, the catabolic range of *P. putida* carrying the TOL plasmid (pWWO), which specifies a pathway for the degradation of methylbenzoates, was expanded to include 4-ethylbenzoate (4EB). 4EB does not activate the *xylS* protein regulator of the normal catabolic pathway and, therefore, is not used as a substrate by *P. putida* (pWWO). Regulatory mutants of the *xylS* protein were produced that were activated by 4EB (Ramos *et al.*, 1987) and other non-effector substrates (Ramos *et al.*, 1986). A second mutation in the catechol 2,3-dioxygenase of *P. putida* (pWWO) allowed complete degradation of 4EB. This demonstrates how selected mutational changes in existing catabolic enzymes of bacteria can be used to expand the catabolic range of a micro-organism.

An even more versatile strategy for the evolution of new catabolic routes is the patchwork assembly of enzymes from different micro-organisms to form a newly structured catabolic sequence. A GEM was constructed by such a procedure so as to degrade chloro- and methylaromatic compounds exclusively through the *ortho*-cleavage route of degradation (Rojo *et al.*, 1987).

For many micro-organisms which degrade substituted aromatics, combinations of chloro- and methylaromatics lead to dead-end or even toxic intermediate formation (Knackmuss, 1983) due to non productive routing of substituted catechols into non-permissive *ortho*- and *meta*-cleavage routes. The presence of such mixtures in nature has, apparently, not selected for microbes able to degrade these combinations by a single catabolic route. Thus, a GEM was conceived so as to degrade chloro- and methylaromatics through one new pathway with enzymes obtained from three different micro-organisms and reassembled in *Pseudomonas* sp. strain B13 as follows (Fig. 1): (1) cloning and insertion into the B13 chromosome of genes from the TOL plasmid which encode for the enzymes *xylXYZLS* (this allowed B13 to degrade 4CB and to transform methylbenzoates to methylcatechols and then to methyl-2-enelac-

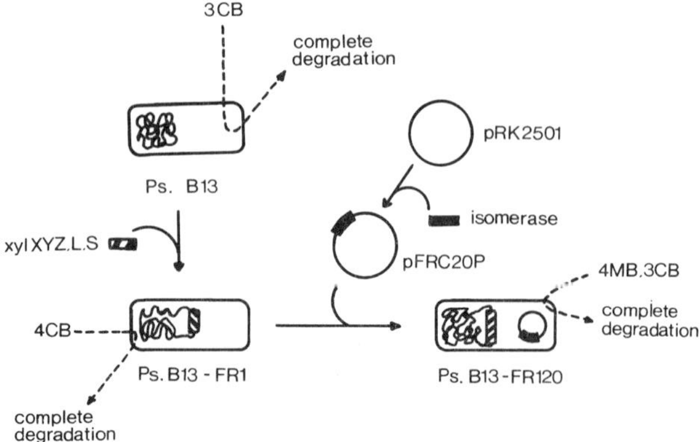

Fig. 1. The constructed pathway for simultaneous degradation of chloro- and methylaromatics in strain FR1(pFRC20P) was based on the modified *ortho*-pathway for 3CB degradation of *Pseudomonas* sp. B13. Incorporation of the TOL plasmid genes that code for toluate 1,2-dioxygenase (*xylXYZ*), dihydroxycyclohexadiene carboxylate dehydrogenase (*xylL*) and the regulator of the operon (*xylS*) expanded the degradation range to include 4CB (strain FR1). Recruitment of 4-methyl-2-enelactone isomerase from *Alcaligenes eutrophus* JPM134 onto plasmid pRK2501 and incorporation of the resultant plasmid, pFRC20P, into strain FR1 expanded the range of the derivative strain, FR1(pFRC20P), to include 4MB, which was then degraded through the same metabolic route as 3CB.

tones (*Pseudomonas* sp. B13 strain FR1), (2) recruitment of a 4-methyl-2-enelactone isomerase from an *Acaligenes* strain to allow for transformation of 4-methyl-2-enelactone to 3-methyl-2-enelactone which is then further degraded by B13 enzymes (*Pseudomonas* sp. B13 strain FR1(pFRC20P)). Except for one enzymatic step, the pathway of the GEM was fully regulated and the component enzymes synthesized only in response to the presence of pathway substrates.

The GEMs described above are representative of those GEMs which can be constructed to catabolize pollutants and which do so under regulated expression of the genetic components which make up the novel catabolic pathway. Although the activities of pollutant-degrading GEMs can be well designed within the laboratory, it is necessary to assess both their risk potential and their ability to function *in situ* before actual environmental introduction. We have used strain FR1(pFRC20P) in experiments designed to test its survival and ability to degrade mixtures of substituted benzoates in microcosms designed to simulate an activated sludge digestor. This microcosm was chosen for its complexity and because such an environment would be appropriate to test the *in situ* ability of strain FR1(pFRC20P) to degrade substituted aromatic compounds since such compounds are often present as chlorinated phenols and benzoates in effluents of treatment systems from wood-processing industries and chemical plants.

DESIGN AND FUNCTION OF MICROCOSM EXPERIMENTS

The diversity of potential GEMs and of the tasks that they will be required to perform in a multitude of different environments makes it difficult to attempt any standardization of procedures with which to analyse the fate and function of GEMs before their actual environmental introduction. Suter (1985) proposed a general assessment scheme which can be applied on a case-by-case basis for such evaluations and which allows for the intricacies of each GEM to be analysed. The approach requires the accumulation of a set of basic data about the GEM to be used in deciding what variables need to be analysed in determining its potential of risk for the environment and to function. This information can then be used to develop test microcosms to allow for further GEM analysis.

Microcosms are fully-contained laboratory ecosystems which simulate an ecological community with its complete ensemble of interacting microbial species. When designing GEMs for the degradation of pollutants, the microcosms should be able to not only assess the fate of the GEM but also its ability to degrade the pollutant in question. The effect of the pollutant on the environment without the presence of the GEM can also be gauged by microcosm analyses. The factors fundamental to GEM risk analysis, survival and function are:

(1) The ability of the GEM to survive in a given environment. There may exist both a low, perhaps non-functioning, population level at which the GEM may survive and also a population size which is desirable to maintain for the GEM to perform the function for which it was designed. In general, micro-organisms which have been well adapted to laboratory life do not survive well when returned to the environment. Thus, it may be important to adjust the GEM, *in vitro*, so as to make it more ecologically fit for both survival and function. Abiotic factors which affect the survival of a GEM include temperature, soil moisture content, pH, salinity, nutrient status and oxygen-levels (Stotzky and Krasovsky, 1981). Biotic factors also influence population sizes since many bacteria appear to survive better in sterilized microcosms (Liang *et al.*, 1982). Biotic antagonists may include bacteriophages, *Bdellovibrio* spp. and protozoan predators.

(2) Stability of genes and potential for *in situ* lateral gene transfer. Important considerations with GEMs are whether the novel genes will be stable *in situ* and whether they will transfer laterally to the indigenous micro-organisms. There are

a number of promising approaches designed to ensure that novel genetic components inserted into a GEM remain within that micro-organism as a stable entity and do not transfer to indigenous micro-organisms (Dwyer and Timmis, 1988). But, this does not obviate the need to assess whether gene transfer occurs under *in situ* conditions. *In situ* gene transfer may be mediated by conjugative transfer (Weinberg and Stotzky, 1972; Gowland and Slater, 1984; Gealt *et al.*, 1985; McPherson and Gealt, 1986; Mancini *et al.*, 1987), transduction by bacteriophage (Saye *et al.*, 1987), and transformation (Graham and Istock, 1979).

(3) Ability of GEM to function. Because a GEM needs to perform outside of the laboratory in which it was constructed, a useful microcosm should give a predictive assessment of the GEM's ability to function *in situ*. Environmental factors often determine how well a micro-organism functions; and microcosms need to be regulated to determine the conditions under which degradation will and will not occur. Potential factors include the matrix on which the GEM and pollutant adsorb, the level of oxygen and the available concentration of the pollutant to be degraded (Pritchard *et al.*, 1987). Because pollutant levels are often too low (ppb) to induce synthesis of native enzymes, alternate regulatory circuits may need to be built into the GEM to ensure enzyme synthesis (Dwyer and Timmis, 1988). When the function of a GEM is disrupted by imbalances in parameters like NADH production, it may be possible to incorporate a second system of metabolism to overcome the disruption (Ingram *et al.*, 1987).

(4) Negative effects of the GEM on the ecosystem. Potential negative effects include alterations in nutrient cycling and the displacement of indigenous micro-organisms because of factors such as competition. Because we are dealing with GEMs designed for degradation of organic compounds, it is important to consider the ability of the GEM to degrade compounds other than that for which it was designed and by which it may alter trophic levels. Other effects include the ability to alter such abiotic factors as pH and the redox potential of the ecosystem.

FATE AND FUNCTION OF FR1(pFRC20P) IN AN ACTIVATED SLUDGE MICROCOSM

The microcosm experiments were designed to assess the fate of both the parental strain B13 and its derivatives, strains FR1 and FR1(pFRC20P), their ability to function as degraders of substituted benzoates and the frequency of lateral transfer of novel genes from the GEMs to micro-organisms indigenous to the microcosm. The microcosm consisted of a 2.5-l container to which was added 900 ml of a sterile synthetic sewage. An inoculum of 100 ml of activated sludge from a municipal sewage treatment plant was added and allowed to equilibrate for 4 h. The sludge was vigorously aerated to simulate activated digestion and to accomplish efficient mixing; a continuous flow system added sterile synthetic sewage at a dilution rate (D) of 0.05 h^{-1}. Substituted benzoates were added to the system through this inflow and liquid samples were taken to determine their concentration in the microcosm.

Survival of the bacteria was measured by direct counts with selective media containing antibiotics and the substituted benzoates as sole substrates. *Pseudomonas* sp. B13 was enumerated on plates containing 3-chlorobenzoate (3CB), FR1 was enumerated on plates containing 4-chlorobenzoate and kanamycin, FR1(pFRC20P) was enumerated on plates containing 4-methylbenzoate and kanamycin; background populations of bacteria from the microcosms did not grow on these plates. The effect on the ecosystem of the added benzoates and of the added bacterial strains was assessed as the change in the total population of indigenous bacteria as enumerated on rich medium and by microscopic observation. The ability of the GEMs to degrade the added benzoates was assessed by analysis of the concentration of the compounds present in the liquid medium of the microcosm.

The parental micro-organism, *Pseudomonas* sp. B13, was able to survive in the microcosm at a level of approximately 10^5 bacteria per ml (Fig. 2). No effect of addition of B13 was observed on the total population level of micro-organisms. 4CB (1mM) was added to the microcosm throughout the 14-day test period; 4CB was not degraded by strain B13, but a population of 4CB-degrading bacteria which were unable to degrade 3CB evolved within the microcosm in response to the presence of 4CB (data not shown). The population appeared on day 10 at 10^4 bacteria per ml and increased by day 14 to 5×10^5 bacteria per ml. The evolution of this bacterial population demonstrated the complexity of the ecosystem maintained within the microcosm and its ability to adapt to the presence of a new substrate.

The population of the derivative GEM, *Pseudomonas* sp. B13 strain FR1, demonstrated a survival pattern almost identical to that of the parental strain B13. Thus, there appeared to be no adverse or promotional affect from genetic manipulation on the survival of the bacterium. FR1 was able to degrade the added 4CB (1 mM) which was not detected in the liquid medium and no population of non-FR1-type 4CB-degraders developed in the system.

The survival of the population of the other derivative GEM, strain FR1(pFRC20P), was similar to that of both the parental strain and FR1 indicating that the presence of the plasmid pFRC20P did not adversely affect its survival ability (Fig. 3). FR1(pFRC20P) was able to degrade mixtures of 3CB and 4MB when added to the microcosm. Because the effect of substituted benzoates might be greater when the organism is already acclimatized to growing on one of the substrates alone (Knackmuss, 1983), the time of addition of the compounds was staggered (Fig. 3). In panel B, 4MB (1 mM) was added

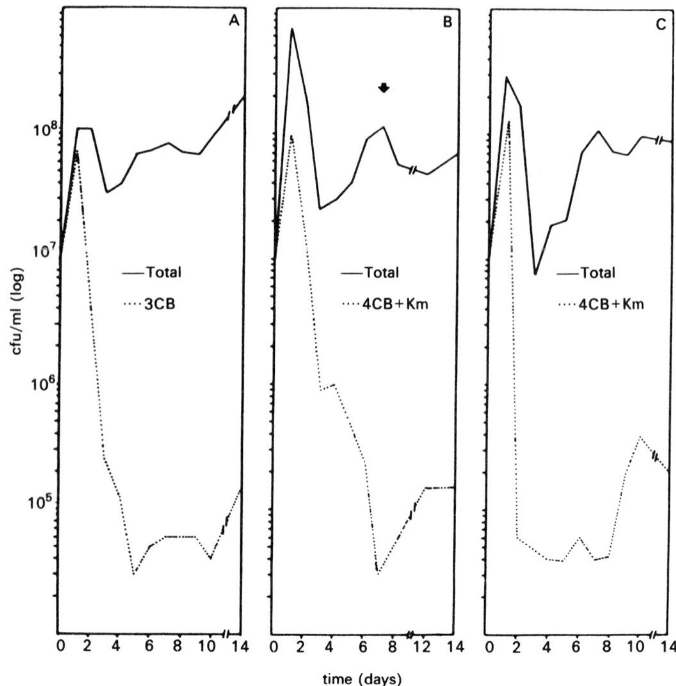

Fig. 2. Survival of bacterial strains added to microcosms. A, Microcosm with *Pseudomonas* sp. B13 added; the synthetic sewage contained 4CB (1 mM). B, Microcosm with *Pseudomonas* sp. B13 strain FR1 added; the synthetic sewage contained 4CB (1 mM) from days 7–14. C, Microcosm with strain FR1 added, but with 4CB (1 mM) added to the synthetic sewage throughout the 14 day experiment.

from the start of the experiment with 3CB (5 mM) added on day 3; in panel C the reverse was done with 3CB (1 mM) added from the start of the experiment and 4MB(5 mM) on day 3. In both cases the initial substrate was degraded, but the substrate added on day 3 was degraded only to about half of the added concentration (2.5 mM). The effect of the simulated shock load of 4MB in panel C was most adverse for the ecosystem; the total population of micro-organisms decreased in number while that of the population of FR1(pFRC20P) increased until it represented fully 50% of the total population of recoverable micro-organisms. An addition of simultaneous shock loads of 3CB and 4MB (5 mM each) on day 3 to an unacclimatized system containing FR1(pFRC20P), severely affected both the total bacterial population and the population of the GEM. In this case (panel D) the GEM population never adequately recovered, while a new population of a 4MB degrader developed which could survive these

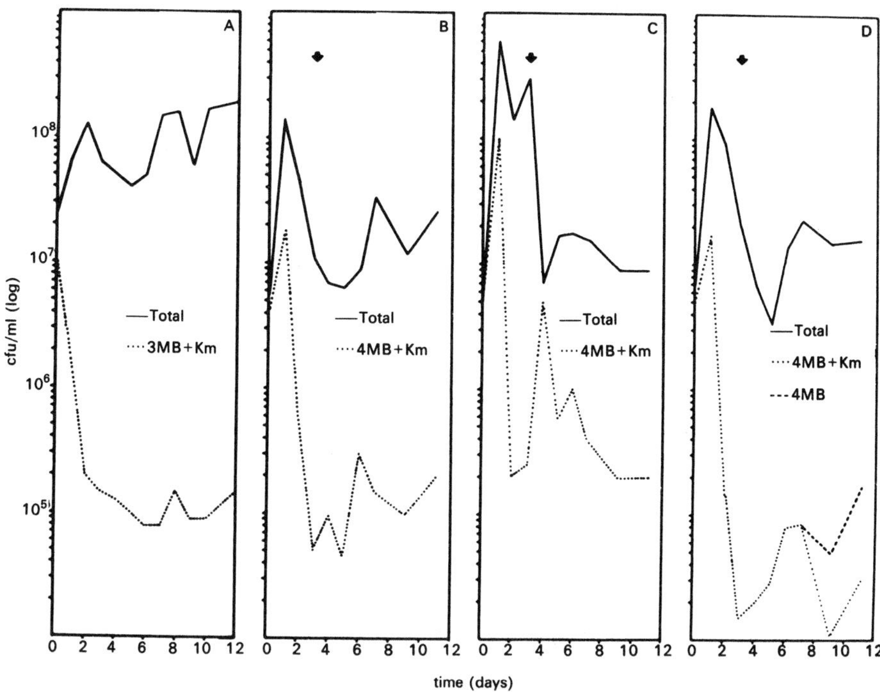

Fig. 3. Survival of *Pseudomonas* sp. B13 strain FR1(pFRC20P) in microcosms. A, Microcosm with no addition of substituted benzoates. B, Microcosm with 4MB (1 mM) added to the synthetic sewage from days 0–12 and 3CB (5 mM) added on days 3–12. C, Microcosm with 3CB (1 mM) added from days 0–12 and 4MB (5 mM) added on days 3–12. D, Microcosm with 3CB (5 mM) and 4MB (5 mM) added together from days 3–12.

conditions. This organism could degrade only 4MB and degradation of 3CB did not occur.

Preliminary evidence suggested that the plasmid pFRC20P was transferred into the indigenous microbial population based on the development of a large population of tetracycline-resistant micro-organisms. This might explain the development of the 4MB-degrading population which arose as a result of the simultaneous addition of 4MB and 3CB to the microcosm (Fig. 3, panel D) and which apparently was able to replace FR1(pFRC20P) as the dominant 4MB-degrading microbe in the microcosm.

The conclusions regarding the fate of the GEMs and their ability to degrade added pollutants that can be derived from this study are as follows:

(1) Both GEMs were able to survive in the microcosms. This demonstrates

the utility of microcosm studies in determining the fate and function of GEMs *in situ*. Because *Pseudomonas* sp. B13 has been cultured for a long time in the laboratory, it was not expected to survive well in the microcosm. Surprisingly, it and the derivative GEMs persisted at a population level of 10^5 bacteria per ml. Pure culture studies had demonstrated an ability of FR1(pFRC20P) to readily degrade simultaneously mixtures of 3CB and 4MB. In the microcosms, however, the GEM did not perform as well as expected, particularly when confronted with a shock load of a 3CB and 4MB mixture. Thus, the microcosm studies are more helpful for making predictions concerning environmental applications of GEMs.

(2) *Pseudomonas* sp. B13 derivative strains FR1 and FR1(pFRC20P) were able to degrade substituted benzoates within the complex ecosystem of the activated sludge microcosm. A good deal of information concerning the degradation pathway for aromatics by *Pseudomonas* sp. B13 was already known. This allowed for the construction of the stable, regulated pathway for the degradation of substituted aromatic compounds in both GEMs (Rojo *et al.*, 1987) and indicates that the construction of similar GEMs for the degradation of environmental pollutants is a promising experimental strategy.

(3) There was not any demonstrable, adverse effect of GEM addition to the microbial population level in the microcosm. Indeed, the GEMs were even able to function in a protective manner for the indigenous populations by buffering them against the adverse effects of addition of substituted benzoates. In microcosms lacking GEM addition, substituted benzoates were not degraded by indigenous microbes and, as intermediates accumulated, a wash-out of the bacterial population in the microcosm occurred (data not shown).

(4) Lateral transfer of new genetic information (*xylXYZLS*) that had been incorporated into the GEM chromosome to indigenous micro-organisms was not detected, whereas transfer of the hybrid mobilizable plasmid pFRC20P carrying the gene for lactone isomerase did apparently occur. In this particular case, transfer may have been beneficial for the community as a whole if it increased the ability of the ecosystem to cope with the presence of toxic pollutants.

As more GEMs are constructed for specific biotechnological applications the diversity and complexity of microcosms used to study their fate and function will increase. The ability of such studies to predict *a priori* the fate of these micro-organisms will help to develop strategies both to decrease the risks associated with introducing GEMs into the environment and to increase and regulate the capacity of GEMs to degrade toxic pollutants.

REFERENCES

Clarke, P. H. and Slater, J. M., 1986. Evolution of enzyme structure and function in *Pseudomonas*. In *The Bacteria*, Sokatch, J. R. (ed.), pp. 71–144. New York: Academic Press.

Dorn, E., Hellwig, M., Reineke, W. and Knackmuss, H.-J., 1974. Isolation and characterization of a 3-chlorobenzoate degrading pseudomonad. *Archives for Microbiology* **99,** 61–70.

Dwyer, D. F. and Timmis, K. N., 1988. Engineering microbes for function and safety in the environment. In *Genetically-Designed Organisms in the Environment*, Mooney, O. and Bernardi G. (eds). In press.

Gowland, P. C. and Slater, J. H., 1984. Transfer and stability of drug resistance plasmids in *Escherichia coli* K12. *Microbial Ecology* **10,** 1–13.

Grady Jr., C. P. L., 1985. Biodegradation: its measurement and microbiological basis. *Biotechnology and Bioengineering* **27,** 660–674.

Graham, J. B. and Istock, C. A., 1979. Gene exchange and natural selection cause *Bacillus subtilis* to evolve in soil culture. *Science* **204,** 637–639.

Hovarth, R. S., 1972. Microbial co-metabolism and the degradation of organic compounds in nature. *Bacteriological Reviews* **36,** 146–155.

Ingram, I. O., Conway, T., Clark, D. P., Sewell, G. W. and Preston, J. F., 1987. Genetic engineering of ethanol production in *Escherichia coli. Applied and Environmental Microbiology* **53,** 2420–2425.

Jain, R. K. and Sayler, G. S., 1986. Problems and potential for *in situ* treatment of environmental pollutants by engineered microorganisms. *Microbiological Sciences* **4,** 59–63.

Keith, L. H. and Telliard, W. A., 1979. Priority pollutants I—a perspective view. *Environmental Science and Technology* **13,** 416–423.

Knackmuss, H.-J., 1983. Xenobiotic degradation in industrial sewage: haloaromatics as target substrates. In *Biotechnology*, Phelps, C. F. and Clarke, P. H. (eds), pp. 173–190. London: The Biochemical Society.

Kobayashi, H. and Rittmann, B. E., 1982. Microbial removal of hazardous organic chemicals. *Environmental Science and Technology* **16,** 170A–181A.

Kröckel, L. and Focht, D. D., 1987. Construction of chlorobenzene-utilizing recombinants by progenitive manifestation of a rare event. *Applied and Environmental Microbiology* **53,** 2470–2475.

Liang, L. N., Sinclair, J. L., Mallory, L. M. and Alexander, M., 1982. Fate in model ecosystems of microbial species of potential use in genetic engineering. *Applied and Environmental Microbiology* **44,** 708–714.

Mancini, P., Fertels, S., Nave, D. and Gealt, M. A., 1987. Mobilization of plasmid pHSV106 from *Escherichia coli* HB101 in a laboratory-scale waste treatment facility. *Applied and Environmental Microbiology* **53,** 665–671.

McPherson, P. and Gealt, M. A., 1986. Isolation of indigenous wastewater bacterial strains capable of mobilizing plasmid pBR325. *Applied and Environmental Microbiology* **51,** 904–909.

Pritchard, P. H., O'Neill, E. J., Spain, C. M. and Ahearn, D. G., 1987. Physical and biological parameters that determine the fate of *p*-chlorophenol in laboratory test systems. *Applied and Environmental Microbiology* **53,** 1833–1838.

Ramos, J. L. and Timmis, K. N., 1987. Experimental evolution of catabolic pathways of bacteria. *Microbiological Sciences* **4,** 228–237.

Ramos, J. L., Stolz, A., Reineke, W. and Timmis, K. N., 1986. Altered effector specificities in regulators of gene expression: TOL plasmid *xylS* mutants and their use to engineer expansion of the range of aromatics degraded by bacteria. *Proceedings of the National Academy of Sciences, USA* **83,** 8467–8471.

Ramos, J. L., Wasserfallen, A., Rose, K. and Timmis, K. N., 1987. Redesigning metabolic routes: manipulation of TOL plasmid pathway for catabolism of alkylbenzoates. *Science* **235,** 593–596.

Ramos, J. L., Gonzalez-Carrero, M. and Timmis, K. N., 1988. Broad-host range expression vectors containing manipulated *meta*-cleavage pathway regulatory elements of the TOL plasmid. *Federation of European Biochemical Societies* **226,** 241–246.

Reineke, W. and Knackmuss, H.-J., 1978. Chemical structure and biodegradability of halogenated aromatic compounds. Substituent effects on 1,2-dioxygenation of benzoic acid. *Biochimica et Biophysica Acta* **542,** 412–423.

Reineke, W. and Knackmuss, H.-J., 1979. Construction of haloaromatic utilizing bacteria. *Nature (London)* **277,** 385–386.

Reineke, W. and Knackmuss, H.-J., 1984. Microbial metabolism of haloaromatics: isolation and properties of a chlorobenzene-degrading bacterium. *Applied and Environmental Microbiology* **47,** 395–402.

Rojo, F., Pieper, D. H., Engesser, K.-H., Knackmuss, H.-J. and Timmis, K. N., 1987. Assemblage of *ortho* cleavage route for simultaneous degradation of chloro- and methylaromatics. *Science* **238,** 1395–1397.

Saye, D. J., Ogunseitan, O., Sayler, G. S. and Miller, R. V., 1987. Potential for transduction of plasmids in a natural freshwater environment: effect of plasmid donor concentration and a natural microbial community on transduction in *Pseudomonas aeruginosa. Applied and Environmental Microbiology* **53,** 987–995.

Slater, J. H., 1985. Gene transfer in microbial communities. In *Engineered Organisms in the Environment: Scientific Issues*, Halvorsan, H. O., Pramer, D. and Rogul, M. (eds), pp. 89–98. Washington DC: American Society for Microbiology.

Stotzky, G. and Krasovsky, V. N., 1981. Ecological factors that affect the survival, establishment, growth and genetic recombination of microbes in natural habitats. In *Molecular Biology, Pathogenicity, and Ecology of Bacterial Plasmids*, Levy, S. B., Clowes, R. C. and Loenig, E. L. (eds), pp. 31–42. New York: Plenum Press.

Suter II, G. W., 1985. Application of environmental risk analysis to engineered organisms. In *Engineered Organisms in the Environment: Scientific Issues*, Halvorsan, H. O., Pramer, D. and Rogul, M. (eds), pp. 211–219. Washington DC: American Society for Microbiology.

Weinberg, S. R. and Stotzky, G., 1972. Conjugation and genetic recombination of *Escherichia coli* in soil. *Soil Biology and Biochemistry* **4,** 171–180.

8 The Use of Fowlpox Virus Vectors to Vaccinate Non-avian Species

JILL TAYLOR[1], RANDALL WEINBERG[1], BERNARD LANGUET[2], PHILIPPE DESMETTRE[2] and ENZO PAOLETTI[1,3]

[1]*Wadsworth Center for Laboratories and Research, New York State Department of Health, Albany, NY 12201, USA*
[2]*Rhone Merieux, 254 Rue Marcel Merieux, Lyon Cedex 07, France*
[3]*Virogenetics Inc., PO Box 2108, Empire State Plaza, Albany, NY 12220, USA*

INTRODUCTION

Poxviruses are a large and diverse group of double-stranded DNA viruses that infect both vertebrate and invertebrate hosts. The family has been subdivided into six genera by the International Committee on Taxonomy of Viruses (Matthews, 1982). The most studied poxvirus is vaccinia virus, the prototypic member of the genus Orthopox. Vaccine strains of this virus were used in the successful World Health Organization sponsored campaign to eradicate smallpox. The success of the campaign was in part due to the physical properties of the virus which is easy to grow, inexpensive to produce, stable under harsh environmental conditions and readily administered by dermal abrasion.

Since the development of vaccinia virus as an expression vector (Panicali and Paoletti, 1982; Mackett *et al.*, 1982) a wide range of foreign DNA sequences which encode antigenic determinants of bacterial, viral and parasitic origin have been inserted at non-essential regions within the viral genome. Inoculation of the recombinants into the appropriate species has resulted in the production of an immune response which, in most cases, is protective against challenge by the live pathogen (Mackett and Smith, 1986). Research is also directed at modifying the virus to generate vaccinia vectors of reduced virulence. For example, Buller *et al.* (1985) reported that disruption of the sequence which encodes the endogenous thymidine kinase resulted in a virus with reduced neurovirulence in a

RELEASE OF GENETICALLY-ENGINEERED MICRO-ORGANISMS ISBN 0-12-677521-4

mouse model system. Vaccinia virus has a broad host range (Fenner, 1985) which allows for the development of the virus as a multi-species vector.

We have been interested in the general development of poxviruses as vectors. Here we describe the development of fowlpox virus (FPV) as a vector capable of expressing foreign antigenic determinants. This vector has obvious implications for use as a vaccination vehicle in the poultry industry. However, unexpected properties of FPV have suggested its usefulness in non-avian species.

FPV is the prototypic species of the Avipox genus of Poxviridae (Matthews, 1982). The group contains eight members, all of which infect avian species. FVP is the most studied member but little is known of its molecular biology. The virus possesses a large double-stranded DNA genome of approximately 240 Kb (Muller *et al.*, 1977), a complex protein composition (Obijeski *et al.*, 1973; Prideaux and Boyle, 1987), and has a high lipid content (Lyles *et al.*, 1976). Little is known of the genomic organization of FPV. Drillien *et al.* (1987) have demonstrated a similarity between the genetic organization of a region of FPV DNA and the conserved Hind III J region of vaccinia virus. However, while the vaccinia virus thymidine kinase gene is encoded within this region, the FPV thymidine kinase is not. The gene has been mapped elsewhere on the vaccinia genome (Boyle and Coupar, 1986; Boyle *et al.*, 1987). Binns *et al.* (1987) have mapped the DNA polymerase of FPV and demonstrated a strong homology at the amino acid level between the FPV enzyme and the vaccinia virus enzyme.

Growth of the virus in tissue culture is restricted to primary or secondary chick embryo fibroblasts (CEF), although growth has also been reported in a quail cell line (Farrow *et al.*, 1975). There are no reports in the literature of a productive FPV infection in tissue culture cells other than of avian origin; however, cytotoxic effects have been noted *in vitro* in a human amniotic cell line (Burnett and Frothingham, 1968). Attenuated strains of FPV have been used as live attenuated vaccines since the 1920s (Tripathy and Hanson, 1975). There have been no reports of transmission of the virus to any non-avian species, including man.

We have asked the question whether inoculation of the virus into non-avian species would result in the expression of FPV encoded viral functions. If so, it was of interest to determine whether a foreign gene would also be expressed by a FPV recombinant virus when inoculated into a non-avian species. In this situation, the level of expression of the foreign antigen might be sufficient to confer protective immunity. Since FPV is restricted in its replication to avian species, inoculation of a FPV recombinant into non-avian species would provide a conditionally replicating viral vector. Using an FPV–rabies glycoprotein recombinant, we will present evidence that the level of expression of the rabies glycoprotein is sufficient to induce a specific and protective immune response in non-avian species in the absence of production of FPV progeny.

EXPERIMENTAL DATA

The recombinant fowlpox viruses were constructed by standard methods. The rabies virus glycoprotein gene was placed under the transcriptional control of vaccinia virus promoters and inserted at a non-essential locus in the FPV genome. Recombinant virus was purified by sequential plaque cloning.

Rabies virus glycoprotein is a viral structural component forming surface projections on the virus particle (Wiktor *et al.*, 1973) which has been shown to be responsible for the induction of virus neutralizing antibody (Cox *et al.*, 1977) and is therefore important in the production of a protective immune response. During rabies virus maturation the glycoprotein becomes localized on the cytoplasmic membrane of infected cells.

To determine whether the rabies glycoprotein produced by recombinant FPV was of the correct size, immunoprecipitation of metabolically labelled antigen was performed. Chick embryo fibroblasts (CEF) cells were infected with the parental virus (FP-1) and the recombinant fowlpox virus (vFP-3) at a high multiplicity of infection (MOI) in the presence of ^{35}S-methionine. For comparative purposes, wild-type vaccinia virus, vTK-79, (Panicali and Paoletti, 1982) and a vaccinia recombinant expressing the rabies glycoprotein were also inoculated onto CEF cells. Infected cells were lysed after 24 h and the rabies glycoprotein precipitated with a specific monoclonal antibody preparation. Figure 1 illustrates that an authentic glycoprotein of 67 Kd was metabolically labelled and was specifically immunoprecipitated from CEF cells infected with either the vaccinia (lane d) or fowlpox virus recombinant (lane a) but not from cells infected with wild-type viruses (lanes b and c). In order to further determine that the rabies glycoprotein was correctly expressed in recombinant FPV infected cells, indirect immunofluorescence studies were performed. Permissive avian (CEF), non-permissive monkey kidney (VERO), or human lung (MRC-5) cells were infected with either the parental fowlpox virus (FP-1) or the recombinant fowlpox virus (vFP-3) at a multiplicity of 10 pfu cell^{-1}. At 24 h post-infection indirect immunofluorescence was performed as described in Taylor *et al.* (1988). Briefly, coverslips were incubated in the presence of a rabies glycoprotein-specific monoclonal antibody preparation, washed and incubated with an FITC goat anti-mouse IgG conjugate. The results of the experiment are shown in Fig. 2. Strong surface fluorescence was detected in all three cell types infected with the recombinant vFP-3 (Fig. 2a, c, e). No such fluorescing material was seen in parental virus-infected cells (Fig. 2b, d, f). A metabolically-labelled glycoprotein of 67 Kd was also immunoprecipitated from lysates of both the non-avian cell lines infected with the recombinant virus (results not shown).

The results presented in Figs 1 and 2 demonstrate that an authentic 67 Kd rabies glycoprotein was expressed by a fowlpox recombinant in both permissive

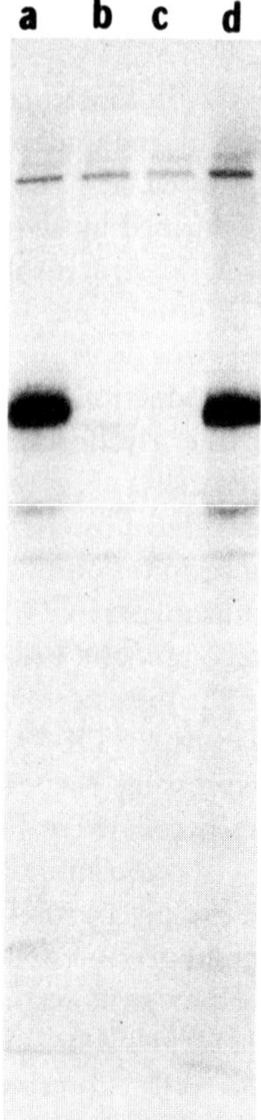

Fig. 1. Immunoprecipitation of the rabies glycoprotein from CEF cells infected with fowlpox and vaccinia-based recombinant viruses: lane a, rabies–fowlpox recombinant; lane b, wild-type fowlpox; lane c, wild-type vaccinia; lane d, rabies–vaccinia recombinant.

(CEF) as well as the non-permissive, non-avian cells (VERO and MRC-5). Furthermore, in all cell types the rabies glycoprotein was demonstrated on the surface membrane of the infected cells.

In order to demonstrate that inoculation of non-avian cells with FPV did not result in a productive infection the following experiments were performed. Six sequential passages of wild-type or recombinant FPV were performed on CEF, VERO or MRC-5 cells. Two growth conditions were chosen for the experiment: (a) 3 days with a MOI of 10 pfu cell^{-1} and (b) 7 days with an MOI of 0.25 pfu cell^{-1}. Three days represented the time during which a CEF monolayer infected with greater than 1 pfu cell^{-1} showed maximal cell degeneration although previous experiments had shown that with this strain of FPV the bulk of progeny virus was produced by 24 h post-infection. The longer growth period (7 days) and low MOI was selected to test the possibility that a putative cellular product was required for productive FPV replication and might be suppressed by inoculation at high MOI or that the growth period of FPV in non-avian cells might be longer than in the permissive CEF cells. After the sixth passage all samples were titrated for viral progeny on permissive CEF cells.

The results are shown in Table 1. As expected, both parental and recombinant fowlpox virus were readily amplified after each passage on CEF cells. However, neither parental nor recombinant fowlpox virus could be amplified on either of the non-avian cells. No infectious virus was detectable after three or four passages. Qualitatively similar results were obtained with either high or low multiplicities of infection.

The levels of residual virus present after the first two passages on VERO or MRC-5 cells was surprisingly high. This was probably due to the inherent thermal stability of fowlpox virus. Additionally, since the virus contains a large amount of lipid it may be difficult to remove all of the input inoculum virus by standard procedures of washing the monolayers of cells after adsorption. In order to determine whether this was indeed residual input virus, the following experiment was performed. Three equivalent monolayers of CEF, VERO or MRC-5 cells were inoculated with parental or recombinant fowlpox virus. After the adsorption period the monolayers were washed well to remove unadsorbed virus. One dish was frozen at this time to measure residual input virus. The remaining two dishes were incubated for 72 h in the absence or presence of 40 μg ml^{-1} cytosine arabinoside (AraC). The drug, a known inhibitor of DNA synthesis, should inhibit any *de novo* viral replication. The results of the experiment are shown in Table 2. Both parental and recombinant fowlpox virus replicated in CEF cells as expected. At 72 h the virus titre increased approximately 100-fold over input virus. No increase in infectious virus titre was detected in the presence of AraC. The levels of measurable infectious fowlpox virus in either VERO or MRC-5 cells were similar in the absence or presence of AraC. The data derived from the experiments in Tables 1 and 2 indicate that productive replication of fowlpox virus is not supported in non-avian cells and support the host-restriction of fowlpox virus recognized in the scientific literature.

Viral gene expression in vaccinia virus is temporarily regulated (Moss, 1985). Certain genes are expressed before DNA replication under the transcriptional control of 'early' promoters, others are expressed only after DNA replication under the control of 'late' promoter sequences and still others are 'constitutively' expressed throughout the replication cycle. Similar observations of temporal expression of FPV proteins have been made (Prideaux and Boyle, 1987). In order to define more precisely where the block to production of fowlpox virus replication in non-avian cells occurs, the following experiment was performed.

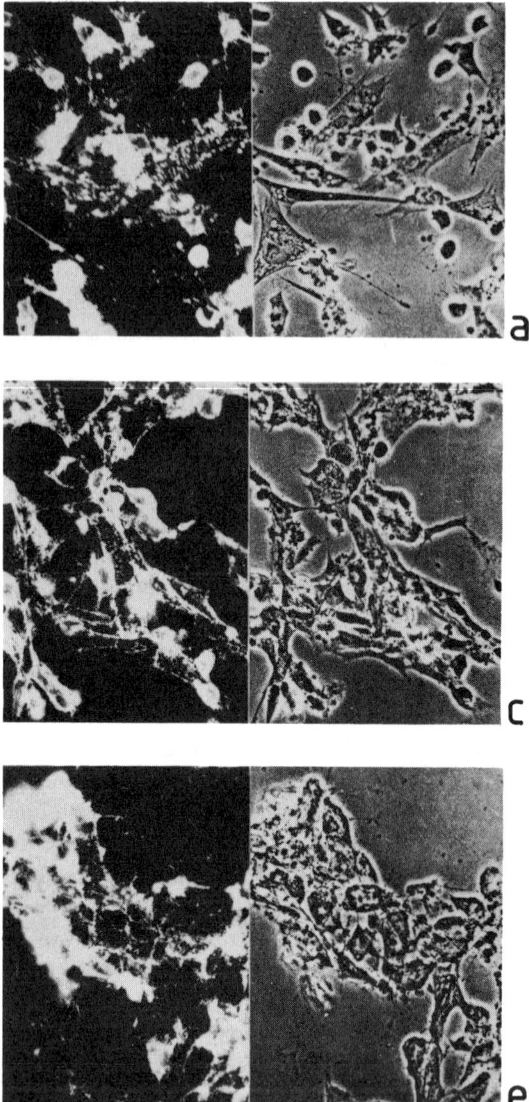

Fig. 2. Detection of surface expression of the rabies glycoprotein on cells infected with a recombinant fowlpox virus. Identical fields of infected cells are shown by phase contrast (a–f) and by fluorescence contrast microscopy immediately to the left: a and b, CEF cells; c and d, VERO cells; e and f, MRC-5 cells: a, c, and e, infected with the rabies–fowlpox recombinant; b, d, and f, cells infected with parental fowlpox.

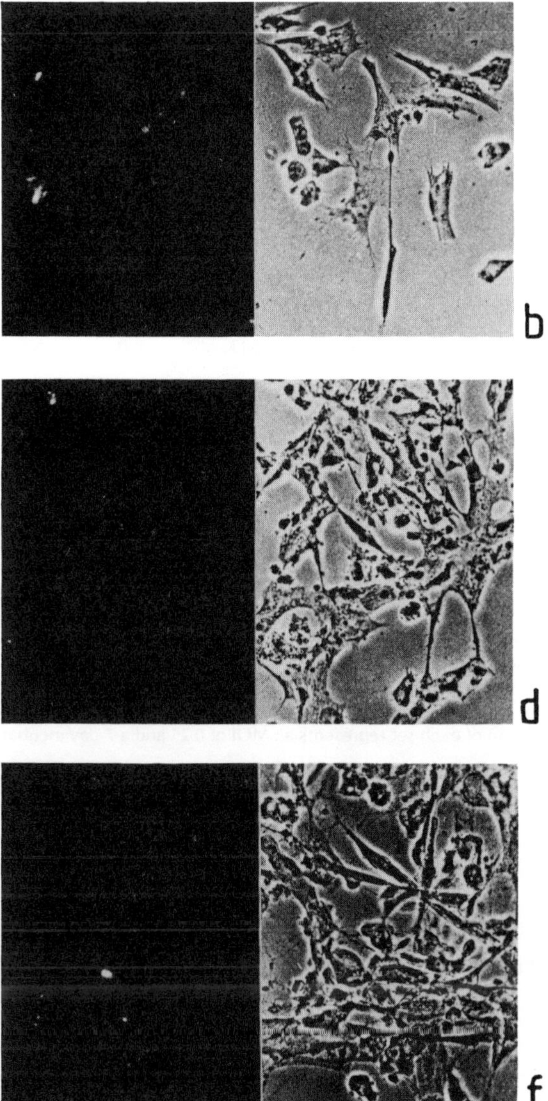
b
d
f

Table 1. Repeated passage of parental (FP-1) and recombinant fowlpox virus (vFP-3) in avian and mammalian cells.

	Cell type					
Inoculum virus	FP-1			vFP-3		
	CEF	VERO	MRC-5	CEF	VERO	MRC-5
Passage 1	6.6* 5.6	4.8 2.7	4.9 3.4	6.6 5.2	5.4 2.4	6.2 3.2
Passage 2	6.7 6.1	2.9 ND†	3.7 ND	6.5 6.0	4.2 ND	5.1 ND
Passage 3	6.4 5.8	1.4 ND	1.0 ND	6.4 5.8	1.7 ND	4.4 ND
Passage 4	6.1 5.9	ND ND	ND ND	6.2 5.8	ND ND	1.0 ND
Passage 5	6.4 5.7	ND ND	ND ND	6.3 5.9	ND ND	ND ND
Passage 6	5.7 5.7	ND ND	ND ND	5.9 5.4	ND ND	ND ND

*Titre expressed as $\log_{10}$ pfu ml^{-1}.
†Not detectable.
The upper figure of each set represents a multiplicity of infection (MOI) of 10 and 3 day incubation period. The lower figure of each set represents an MOI of 0.25 and a 7 day incubation period between passages.

Table 2. Effect of cytosine arabinoside on growth of wild-type or recombinant fowlpox virus.

Virus	Time post-inoculation (h)	CEF	VERO	MRC-5
Wild-type	0	4.5*	4.9	4.9
	72	6.1	4.1	5.2
	72†	4.0	4.3	4.9
Recombinant	0	5.0	4.6	4.9
	72	7.0	4.4	4.7
	72†	3.6	4.0	4.9

*Titre expressed as $\log_{10}$ pfu ml^{-1}.
†Culture incubated in the presence of 40 $\mu g\ ml^{-1}$ cytosine arabinoside.

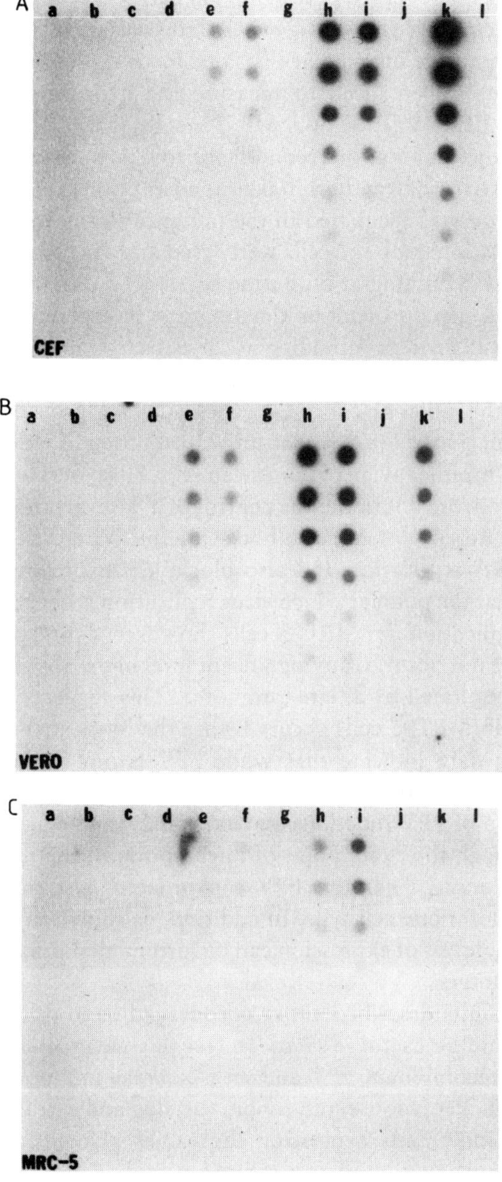

Fig. 3. Expression of the rabies glycoprotein in fowlpox virus recombinants under the transcriptional control of different vaccinia promoters: A, Viruses grown in CEF cells; B, Viruses grown in VERO cells; C, Viruses grown in MRC-5 cells. Lanes a, b and c, wild-type fowlpox; lanes d, e and f, 'early' promoter-regulated; lanes g, h and i, 'constitutive' promoter-regulated; lanes j, k and l, 'late' promoter-regulated. The first lane in each series represents a 0 h sample, the second a 72 h sample and the third a 72 h sample in the presence of 40 μg ml^{-1} AraC. Vertical rows represent serial two-fold dilutions beginning at 1 : 4.

The rabies glycoprotein gene was fused to an 'early' vaccinia virus promoter, a 'constitutive' vaccinia virus promoter or a 'late' vaccinia virus promoter. The gene and promoter were inserted at the same locus in the FPV genome. These recombinant viruses were used to measure the expression levels of rabies glycoprotein on CEF, VERO, and MRC-5 cells. The three cell lines were inoculated with the parental or recombinant fowlpox viruses. One dish from each cell type was frozen after the initial virus adsorption period. A separate dish from each cell type was incubated in the presence (40 μg ml^{-1}) or absence of AraC. At 72 h post-infection the cells were lysed and the lysate applied in serial two-fold dilutions beginning at a dilution of 1:4 to a nitrocellulose filter. The amount of rabies glycoprotein in the lysate was determined with specific antiserum and I-125 Protein A binding as the detection method. The results of the experiment are shown in Fig. 3.

Expression of the rabies glycoprotein gene in the fowlpox virus recombinants utilizing either an 'early' or a 'constitutive' promoter is not affected by the presence of AraC in either avian or non-avian cells. However, when transcription of the rabies glycoprotein is under the control of a 'late' promoter, expression is inhibited by the presence of AraC in both CEF and VERO cells. The fact that AraC blocks DNA replication and also blocks the expression of the foreign gene, indicates that the point at which virus replication is aborted in VERO cells is after DNA replication. In MRC-5 cells, however, expression of the rabies glycoprotein does not occur to any significant level in the absence of AraC when transcription is regulated by a 'late' promoter. This suggests that the block in FPV replication in MRC-5 cells occurs before the block in VERO cells.

The foregoing data indicate that while FPV is not capable of causing a productive infection in non-avian cells, the virus is nevertheless able to initiate genetic expression of FPV functions and express foreign genes engineered into the vector. Although the exact point of interruption of the replication cycle is not known and may be variant, FPV can proceed past the point of DNA replication in at least one cell type. In addition, as shown in Fig. 3, and other unpublished data, levels of expression can be further manipulated by utilization of different promoters.

The *in vitro* results described above encouraged us to ask whether an FPV recombinant would be useful *in vivo*. To test this we inoculated a number of species with the recombinant FPV and measured the immune response. Seven animal species (chicken, mouse, rat, rabbit, cat, dog and cattle) were inoculated with an FPV recombinant expressing the rabies glycoprotein. All animals responded by the production of anti-rabies antibody detectable by one to two weeks post-inoculation. An example of the immune response is shown in Fig. 4. Cattle were inoculated by three different routes; intradermal, intramuscular, and subcutaneous. All responded with similar antibody titres. Increased antibody

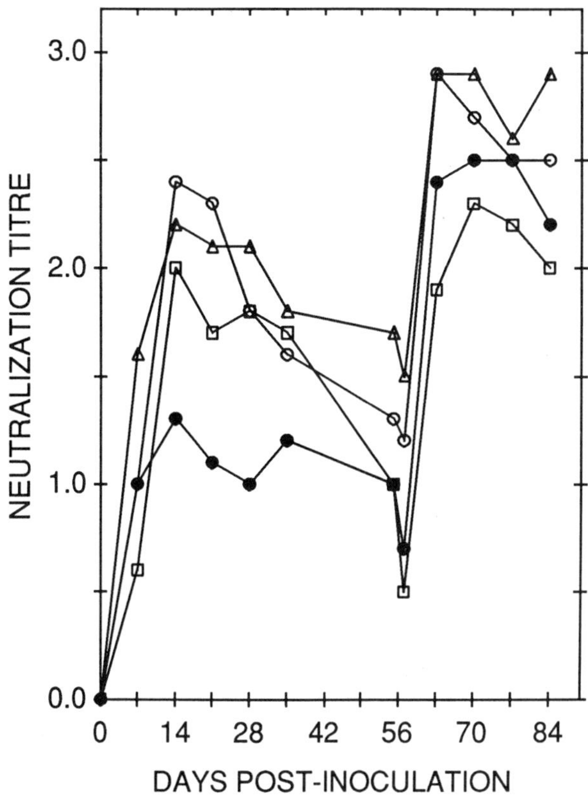

Fig. 4. Antibody response in cattle inoculated with fowlpox virus recombinants expressing the rabies glycoprotein: △, 1 × 10^8 of vFP-3 inoculated subcutaneously; □, 2 × 10^7 of vFP-3 inoculated intradermally; ○, 1 × 10^8 of vFP-3 inoculated by the intramuscular route; ●, 2 × 10^7 of vFP-2 inoculated intradermally. Neutralization titres are expressed as $\log_{10}$ of the highest serum dilution giving greater than 50% reduction in the number of fluorescing wells in an RFFI test performed as described in Smith *et al.* (1973).

levels were obtained after a second inoculation of the recombinant at 55 days post-inoculation.

In order to determine whether the levels of anti-rabies antibody elicited by inoculation of the FPV–rabies recombinant were sufficient to protect against a live rabies virus challenge, the following experiment was performed. Two dogs and two cats were immunized with a single subcutaneous dose of 1 × 10^8 $TCID_{50}$ of the FPV–rabies recombinant. At 94 days post-inoculation the animals, as well as equivalent age and weight-matched non-vaccinated controls, were challenged with live rabies virus. The dogs were inoculated in the temporal muscle with two doses of 0.5 ml of a salivary gland homogenate of the NY strain of rabies virus corresponding to 10 000 mouse LD_{50}. The cats were similarly challenged by inoculation in the neck muscle with 40 000 mouse LD_{50} of the same rabies virus strain. The results are shown in Table 3. All non-vaccinated animals died 12–16 days post-challenge. All animals vaccinated with the FPV–rabies recombinant survived challenge. All animals were observed for 21 days beyond the death of the last control animal and remained without symptoms and healthy.

Table 3. Protection of cats and dogs immunized with a rabies–fowlpox virus recombinant against a live rabies virus challenge.

Species		Vaccination	Titre at 94 days post-inoculation	Survival/ day of death
Cat	7015	vFP-3*	1.5†	+
	7016	vFP-3	1.3	+
	8271	c‡	0	d/13
	T10	c	0	d/12
	T41	c	0	d/13
	T42	c	0	d/12
Dog	426	vFP-3	1.2	+
	427	vFP-3	1.9	+
	55	c	0	d/15
	8240	c	0	d/16

*Both cats and dogs vaccinated with the rabies–fowlpox virus recombinant (vFP-3) received 1×10^8 $TCID_{50}$ by the subcutaneous route.
†Titre expressed as $\log_{10}$ highest serum dilution giving greater than 50% reduction in the number of fluorescing wells in an RFFI test.
‡Non-vaccinated control animals.
d = Animal died.
+ = Survived challenge.

CONCLUSION

FPV possesses the physical characteristics desirable in a live vaccine candidate, i.e. genetic and environmental stability, ease of growth and inexpensive to produce. To this list can now be added the fact that the virus will stably incorporate and express foreign genes. This opens the way for the development of multivalent vaccines for use in the poultry industry based on FPV as a delivery vehicle. An additional and exciting prospect is also presented. FPV recombinants have now been shown to express foreign genes in non-avian species in the absence of the production of infectious virus. With the rabies virus glycoprotein gene under the transcriptional control of a vaccinia virus promoter, we have demonstrated that a protein of the correct size is synthesized and localized on the surface of avian or non-avian cells infected with the recombinant. On inoculation into animals the level of expression is sufficient to induce an immune response which is protective against a live rabies virus challenge. That the immunological response is due to the *de novo* expression of the foreign gene and not to glycoprotein adventitiously introduced with the inoculum has been shown by other experiments. Rabbits inoculated with chemically inactivated fowlpox virus–rabies recombinant produced an immune response to FPV antigens but not to the rabies glycoprotein (Taylor *et al.*, 1988).

Experiments to date have failed to produce any evidence for productive replication in non-avian cells. Initial results indicate that in at least one cell line the virus is able to proceed past the point of DNA replication. However, the precise nature of the block to the production of infectious virus progeny is not at present known and may vary with each cell type. Historical evidence also supports the fact that FPV will not replicate in non-avian cells since transmissions of the virus to other species have not been reported. After years of extensive use of an attenuated live fowlpox virus vaccine in the poultry industry, infections in humans have not been documented.

When selecting the appropriate vaccination strategy for a given disease, the safety features of subunit and killed vaccines must be weighed against the increased efficacy and longevity of response achieved by live vaccines. Live vector-based vaccines combine the advantages of both approaches. Foreign antigenic determinants are presented to the immune system as if in a replicating state in a relatively innocuous background and without actual exposure of the vaccine to the pathogens of interest. As a further safety measure, the use of a live vector in which replication is limited to the originally vaccinated individual would provide advantages, especially in disease situations where contact transmission of the recombinant vector presents particular problems. Such conditions exist in the unique requirements of wildlife and open-range herd vaccination where animal movement and possible inter-species contacts cannot be controlled.

The potential use of FPV recombinants as vaccine vectors for non-avian species presents a number of advantages. The live nature of the infection presents the foreign immunogen in such a way that both the humoral and cell mediated components of the immune system can be elicited. The non-productive aspect of the vector allows the incorporation of an additional safety feature into the procedure since the possibility of transmission of the recombinant to non-vaccinated individuals and to other species would be greatly reduced. The use of FPV as a vector may also be indicated in immunocompromised individuals where use of a productively replicating vector would be undesirable.

REFERENCES

Binns, M. M., Stenzler, L., Tomley, F. M., Campbell, J. and Bournsell, M. E. G., 1987. Identification by a random sequencing strategy of the fowlpox virus DNA polymerase gene, its nucleotide sequence and comparison with other viral DNA polymerases. *Nucleic Acids Research* **15,** 6563–6573.

Boyle, D. B. and Coupar, B. E. H., 1986. Identification and cloning of the fowlpox virus thymidine kinase gene using vaccinia virus. *Journal of General Virology* **67,** 1591–1600.

Boyle, D. B., Coupar, B. E. H., Gibbs, A. J., Seigman, L. J. and Both, G. W., 1987. Fowlpox virus thymidine kinase: nucleotide sequence and relationships to other thymidine kinases. *Virology* **156,** 355–365.

Buller, R. M. L., Smith, G. L., Cremer, K., Notkins, A. L. and Moss, B., 1985. Decreased virulence of recombinant vaccinia virus expression vectors is associated with a thymidine-kinase negative phenotype. *Nature (London)* **317,** 813–815.

Burnett, J. W. and Frothingham, T. E., 1968. The cytotoxic effect of fowlpox virus on primary human amniotic cell cultures. *Archiv für die Gesamte Virus Forschung* **24,** 137–147.

Cox, J. H., Dietzschold, B. and Schneider, L. G., 1977. Rabies virus glycoprotein. II. Biological and serological characterization. *Infection and Immunity* **16,** 754–759.

Drillien, R., Spehner, D., Villeval, D. and Lecocq, J. P., 1987. Similar genetic organization between a region of fowlpox virus DNA and the vaccine virus Hind III J fragment despite divergent location of the thymidine kinase gene. *Virology* **160,** 203–209.

Farrow, W. M., Schmitt, M. W. and Groupe, V., 1975. Responses of isolator derived Japanese quail cell cultures to selected animal viruses. *Journal of Clinical Microbiology* **2,** 419–424.

Fenner, F., 1985. Poxviruses. In *Virology*, Fields, B. N. (ed.), pp. 661–684. New York: Raven Press.

Lyles, D. S., Randall, C. C., Gafford, L. G. and White, H. B. Jr., 1976. Cellular fatty acids during fowlpox virus infection of three different host systems. *Virology* **70,** 227–229.

Mackett, M. and Smith, G. L., 1986. Vaccinia virus expression vectors. *Journal of Virology* **67,** 2067–2082.

Mackett, M., Smith, G. L. and Moss, B., 1982. Vaccinia virus: a selectable eukaryotic cloning and expression vector. *Proceedings of the National Academy of Sciences USA* **79,** 7415–7419.

Matthews, R., 1982. Classification and nomenclature of viruses. *Intervirology* **17,** 42–46.

Moss, B., 1985. Replication of Poxviruses. In *Virology*, Fields, B. N. (ed.), pp. 685–703. New York: Raven Press.

Muller, H. K., Wittek, R., Schaffner, W., Schumperli, D., Menna, A. and Wyler, R., 1977. Comparison of five poxvirus genomes by analysis with restriction endonucleases Hind III, Bam I and EcoRI. *Journal of General Virology* **38,** 135–147.

Obijeski, J. F., Palmer, E. L., Gafford, L. G. and Randall, C. C., 1973. Polyacrylamide gel electrophoresis of fowlpox and vaccinia virus proteins. *Virology* **51,** 512–516.

Panicali, D. and Paoletti, E., 1982. Construction of poxviruses as cloning vectors: Insertion of the thymidine kinase gene from herpes simplex virus into the DNA of infectious vaccinia virus. *Proceedings of the National Academy of Sciences USA* **79,** 4927–4931.

Prideaux, C. T. and Boyle, D. B., 1987. Fowlpox virus polypeptides: sequential appearance and virion associated polypeptides. *Archives of Virology* **96,** 185–199.

Smith, J. S., Yager, P. A. and Baer, G. M., 1973. A rapid tissue culture test for determining rabies neutralizing antibody. In *Laboratory Techniques in Rabies*, Kaplan, M. M. and Koprowski, H. (eds), 3rd edn, pp. 354–357. Geneva: World Health Organization.

Taylor, J., Weinberg, R., Languet, B., Desmettre, P. and Paoletti, E., 1988. A recombinant fowlpox virus induces protective immunity in non-avian species. Submitted for publication.

Tripathy, D. N. and Hanson, L. E., 1975. Immunity to fowlpox. *American Journal of Veterinary Research* **36,** 541–544.

Wiktor, T. J., Gyorgy, E., Schlumberger, H. D., Sokol, F., and Koprowski, H., 1973. Antigenic properties of rabies virus components. *Journal of Immunology* **110,** 269–276.

9 Progress Towards Vaccines Against Bacterial Oral and Enteric Pathogens

ALF A. LINDBERG

Karolinska Institute, Department of Clinical Bacteriology, Huddinge University Hospital, S-141 86 Huddinge, Sweden

INTRODUCTION

Vaccines were originally developed by trial and error. This approach led to the successful development of effective toxoid vaccines against bacterial disease like diphtheria and tetanus. For other bacterial infections the efficacy of developed vaccines has been less spectacular. Thus, diseases like tuberculosis, pertussis, leprosy and enteric infections still remain major health problems in developing countries. Bacterial vaccine development, successful or not, was based on a poor foundation of scientific understanding. Neither the chemical nature of the immunogens in inactivated vaccines, nor the molecular basis for the attenuation of living bacterial vaccines was known and understood. The nature of the immune response required to give protective immunity was also little understood. Furthermore, there was no means of knowing how to elicit appropriate immune reponses with an attenuated vaccine.

At present, successful bacterial vaccines are directed against invasive organisms where circulating antibodies either neutralize toxins, like those of diphtheria and tetanus toxins that enter the circulation, or function as opsonins that facilitate phagocytosis of invading pathogens like *Neisseria meningitidis, Streptococcus pneumoniae* and *Haemophilus influenzae* which gain access to tissues and the circulation.

Attempts to elicit protective immunity at mucosal surfaces or to stimulate efficient cell-mediated immune responses, have been less successful. This is due not only to an incomplete understanding of how to regulate the immune system but also to a less than satisfactory understanding of virulence mechanisms and how they may be neutralized and the nature of the protective epitopes.

RELEASE OF GENETICALLY-ENGINEERED MICRO-ORGANISMS ISBN 0–12–677521–4

During the last decade, however, the use of recombinant DNA techniques has considerably facilitated identification and characterization of attenuation mechanisms, has targeted manipulation of virulence factors of importance for vaccine development and, finally, has resulted in the construction of a number of promising candidate vaccines. Some of these achievements will be discussed with a focus on vaccines for bacterial enteric pathogens.

LIVE VACCINES FOR MUCOSAL IMMUNE DEFENCE AGAINST NON-INVASIVE PATHOGENS

Adherence of pathogens to mucosal surfaces is accepted as a major mechanism for promoting colonization of the gastrointestinal, genito-urinary and respiratory tracts (Evans *et al.*, 1984). The adherence is frequently specific for both the bacterial species and the host tissue and is mediated by the expression of bacterial surface structures termed adhesins. The presence of predominantly secretory IgA (sIgA) antibodies at mucosal surfaces is considered to be the primary line of host defence against bacterial infections (McCaughan and Basten, 1983). Therefore, elicitation of specific sIgA responses could potentially be important in preventing or altering the natural course of infectious diseases. Bacteria such as enterotoxigenic *Escherichia coli* (ETEC) and *Vibrio cholerae* are non-penetrating and hence the elicited host immune response should be mucosal.

Enterotoxigenic *Escherichia coli* (ETEC) vaccines

A recent study has estimated the worldwide annual incidence of ETEC diarrhoea to be almost 400 million cases, with 700 000 deaths in children under 5 years (Institute of Medicine, 1986). ETEC is also an important cause of travellers' diarrhoea. Although oral rehydration therapy can reduce mortality significantly, it has no effect on morbidity and hence, the development of vaccine(s) is of great importance. Because of the recent recognition of ETEC as a pathogen the development of vaccine(s) against this disease has received attention only in the last decade.

At least four different groups of *E. coli* strains are associated with diarrhoeal disease; ETEC is, perhaps, the most common, particularly in the developing world. In addition, enteropathogenic *E. coli* (EPEC), enteroinvasive *E. coli* (EIEC) and enterohaemorrhagic *E. coli* (EHEC) cause diarrhoeal disease. Of the four types only ETEC strains are exclusively extracellular and produce diarrhoea by colonizing the small intestinal mucosa, followed by elaboration of one or more toxins: heat-labile (LT) and heat-stable (ST) toxins. Among ETEC a series of different adhesins has been identified: colonization factor antigen 1

(CAFI), CFAII (CSI-6), etc. These adhesins are associated with a wide range of serovars based on different lipopolysaccharide (LPS = O antigen), capsular and flagellar antigens among *E. coli*. A vaccine for ETEC should protect against disease caused by strains of differing serovars, which produce enterotoxin (LT, ST or both) and different colonization factors.

An *E. coli* strain E1392/75 2A (O6 : H16) lacked the genes for LT and ST when probed in DNA hybidization (Levine *et al.*, 1984). When fed to volunteers by 24 h post-ingestion the upper small intestine was colonized in all individuals who were given 6×10^{10} bacteria. All individuals receiving this dose showed a significant rise in intestinal sIgA against the CFAII adhesin. Twelve volunteers who ingested a single dose (5×10^{10} bacteria) of the live oral vaccine strain E 1392/75 2A and six unimmunized control volunteers were challenged with 5×10^8 enterotoxigenic (LT^+, ST^+) *E. coli* of a different serovar (O139 : H28) but with the identical CFA II adhesins (CS 1 and 3) as the vaccine strain. Diarrhoea occurred in six of six controls but in only three of 12 vaccinees ($P < 0.005$). The challenge strain was recovered from the duodenal fluid of five of six controls but from only one of 12 vaccinees ($P < 0.004$). These results show that a protective mucosal immunity can be elicited by a non-invasive colonizing strain and that immune mechanism(s) prevent the adherence of the pathogen to the mucosal surface. Although limited, the study suggests that protection is independent of both the *E. coli* serovar and the toxins produced. Oral immunization of man with purified fimbrial antigen was also ineffective, probably as a consequence of adverse effects of the gastric acid on the fimbrial protein.

Mucosal antitoxic immunity should either complement or be an alternative to anti-adhesin immunity. The number of epitopes of LT and ST is small compared with the multitude of colonization factors. Evidence for a lasting protective anti-enterotoxic immunity, however, is lacking both from volunteer studies (Levine *et al.*, 1979) and the common occurrence of multiple ETEC infections in developing countries (Black *et al.*, 1981). Of the two toxins only LT stimulates a neutralizing immune response. In the recent cholera vaccine trial in Bangladesh, where inactivated *V. cholerae* bacteria plus the toxin B subunit was given orally, a high-grade but short-lasting protection against *E. coli* ETEC diarrhoea was detected: initially for 2 months, 75%; 1 year, 37%; and after 2 years, 10% (Clements *et al.*, 1988b). The antitoxic immunity primarily reduced the severity of the diarrhoea. The observed short-lived protection was not particularly encouraging, especially when seen in the context that the B toxin subunit had to be given in a large dose (1 mg) which makes the vaccine expensive.

Another experimental approach has been to incorporate cloned sequences of the LT-B subunit into an attenuated but invasive carrier strain which functions as an antigen-delivery vehicle (Clements and El-Morshidy, 1984; Maskell *et al.*,

Table 1. Vaccines and vaccine candidates.

Immunizing agent	Genotype	Phenotype	Efficacy	Safety	Reference
1. Strains with deleted virulence factors					
Escherichia coli E1392/75 2A	CFA/II$^+$, spontaneous loss of LT and ST genes	Colonizing upper small intestine	Yes, single dose, sIgA to CFA/II	Mild diarrhoea in 11% vaccinees	Levine *et al.* (1984)
Vibrio cholerae CVD 103 Hg-R	CT A$^-$B$^+$, Hg^{2+}-resistance	Colonizing small intestine	Yes, single dose	Yes	(M. Levine)
2. Mutant attenuated strains					
Salmonella typhi Ty21a	*galE* plus unidentified mutations	Invasive, limited replication	Yes, multiple doses	Yes	Germanier and Furer (1975)
Salmonella typhi 541Ty	*aroA, purA*	Invasive, limited replication, Vi-positive	Yes, cell-mediated but not humoral immune response	Yes	Stocker (1988)
Shigella flexneri Sfl114	*aroD*	Invasive, limited replication	Yes	Yes	Lindberg *et al.* (1988)
3. Interspecies hybrid strains					
Salmonella typhi SE12	*S. typhi* Ty21a, pJC217 coding for LT-B	Invasive, limited replication, expresses LT-B	Elicits toxin neutralizing ab in mice	Not tested	Clements and El-Morshidy (1984)

Salmonella typhimurium SL3261 (pBRD026)	*aroA*, pBRD026 coding for L^--B	Invasive, limited replication, expresses LT-B	Elicits toxin neutralizing ab in mice	Not tested	Maskell *et al.* (1987)
Salmonella typhi Ty21aS	*S. typhi* Ty21a, pCSlt1 coding for CFA/I and ST	Invasive, limited replication, expresses CFA/I and ST	Not tested	Not tested	Yamamoto *et al.* (1985)
Salmonella typhimurium SR11	*cya*, *crp*, pStSR100 negative, pYA179	Invasive, avirulent expresses *Streptococcus sobrinus* surface protein antigen	Not tested	Yes, in mice	Curtiss *et al.* (1988)
Escherichia coli EC104	*E. coli* K12 with *S. flexnin* 2a *his* and *pro* genes and 140 Mdal invasion plasmid	Invasive, nonreplicating expresses *S. flexnin* 2a **O** antigen and outer membrane proteins	Yes, in monkeys	No, insufficient attenuation	Formal *et al.* (1984)
Salmonella typhi 5076-1C	*S. typhi* Ty21a with *S. sonnei* 120 Mdal plasmid	Invasive, limited replication expresses *S. sonnei* **O** antigen	Inconsistent	Yes	Black *et al.* (1987)
Shigella flexneri M22-18X16	*S. flexneri* 2a with *E. coli* K12 *xyl-rha* genes	Invasive, nonreplicating	Yes	No, insufficient attenuation	Formal and Levine (1984)

1987) (Table 1). *Salmonella typhi* Ty21a with plasmid pJC217 produced a LT-B subunit which was structurally and immunologically indistinguishable from that produced by *E. coli* strains harbouring the same plasmid. When guinea-pigs were fed the pJC217-containing strain *S. typhi* SE12 at a dose of 3×10^9 cfu and mice injected intraperitoneally with a dose of 2.5×10^7 cfu no adverse effects were seen. The mice produced significant levels of anti-LT antibody responses. The *S. typhi* SE12 vaccine candidate has not, so far, been tested in human volunteer studies. Likewise, both CFA I adhesin and ST are expressed in the *S typhi* Ty21a vaccine strain when the plasmid pCSI of *E. coli* ETEC strain H 10407 was transferred as its non-conjugative Tn*5* derivative pCSIt1 (Yamamoto *et al.*, 1985).

Recombinant DNA techniques have been used to fuse genes for LT and ST to produce chimaeric LT–ST molecules (Kupersztoch, 1986; Sanchez *et al.*, 1986). The chimaeric proteins reacted with antibodies specific for either LT or ST which showed that LT and ST epitopes were intact after fusion. Thus, recombinant DNA techniques have made it possible to construct bacterial strains that express either or both of the enterotoxins and also colonization factors. So far, studies of the immunogenicity, safety or efficacy of these carrier vaccines in human beings have not been reported. To develop a safe and efficacious ETEC vaccine a great deal of additional work remains to be done.

Vibrio cholerae vaccines

Of the ten species of the genera *Vibrio* and *Aeromonas* which are aetiological agents in diarrhoeal disease, *Vibrio cholerae* O group 1 is the best known. Cholera is caused by strains producing the cholera toxin (CT) which inhibits absorption of sodium and causes secretion of chloride and bicarbonate ions. The major objective is to develop an effective cholera vaccine that is safe, produces lasting protection against both illness and, if possible, asymptomatic infection after one or two oral doses. The current killed, whole-cell, parenteral vaccines are of limited usefulness; vaccine efficacy is limited and protection lasts for only 3–6 months. *Vibrio cholerae* is a non-invasive organism, and the goal is to produce a local intestinal immune response which prevents the adhesion of the bacteria to mucosal cells and the toxin-mediated triggering of electrolyte and fluid losses. In recent years, two approaches have been taken to produce more efficacious cholera vaccines, both based on oral immunization.

In 1985 a large-scale field trial was initiated in Bangladesh where children aged 2–15 years and women over 15 years were given three doses of one of two inactivated vaccines: (a) 2×10^{11} killed whole cells plus 1.0 mg of cholera toxin B subunit or (b) 1×10^{11} killed whole cells only (Clements *et al.*, 1988a). The placebo control group received 1×10^{11} *E. coli* K12 bacteria. More than 62 000 individuals took three complete oral doses of the vaccines given at monthly intervals. The protection levels with the whole-cell vaccine plus B subunit vaccine after 1 year was 63% and after 2 years was 57%. For the whole-cell vaccine alone, the corresponding figures were 58 and 57%, respectively. These results show that the addition of the B subunit only marginally increased the efficacy during the first year after vaccination. The efficacy of the inactivated whole-cell plus CT-B-subunit-vaccine appears to be too low to make it the cholera vaccine of choice in the future.

A second approach has been to construct attenuated *V. cholerae* strains to be used as live vaccines (Table 1). Genes encoding the enterotoxin were deleted with recombinant DNA techniques (Mekalanos *et al.*, 1983; Kaper *et al.*, 1984). Restriction enzyme fragments encoding CT were deleted *in vitro* from cloned chromosomal DNA and the resulting mutations introduced into the chromosome of a *V. cholerae* strain of proven immunogenicity. Various candidate vaccine strains, which did not produce cholera toxin tested in human volunteers, had problems associated with reactogenicity, i.e. up to 50% of volunteers had an associated diarrhoea. Since no CT is produced other explanations for the reactogenicity have been suggested such as the production of another toxin (Betley *et al.*, 1986) or possibly an effect connected with "the mere colonization of the small intestine" (Taylor *et al.*, 1987). The production of the pilus is co-regulated with the production of cholera toxin, and hence it has been named toxin-coregulated pilus (TCP). Mutations in the regulatory gene results in reduced production of cholera toxin and a colonization defect. It should be noted that the inactivated whole-cell cholera vaccine tested in Bangladesh probably contained little, if any, TCP antigen because the cholera strain and growth conditions used were not optimal for pilus production. The recent discovery of the TCP will certainly influence both our understanding of the colonization of *V. cholerae* and influence the composition of live and inactivated cholera vaccines.

Vibrio cholerae vaccine candidate strain CVD 102 is a CT A^-B^+ derivative of Inaba 569B. It caused mild diarrhoea in 12% of volunteers but also vigorous serum vibriocidal and antitoxin titres in 96 and 94%, respectively, of the volunteers after a single oral dose of 1×10^8 live bacteria. A recently engineered derivative of CVD 102, labelled CVD 103-HgR because of the introduction of mercury-resistance genes into the *Hly* locus, caused no diarrhoea in 18 volunteers (M. M. Levine, personal communication). The reason(s) for the additional attenuation after introduction of Hg-resistance, i.e. less reactogenicity, remains to be explained. Since CVD 103-HgR elicits serum vibriocidal and antitoxic immune responses and is protective in challenge studies of volunteers with live wild-type *V. cholerae*, it is now the most advanced of the live oral *V. cholerae* vaccine candidates.

Streptococcus mutans vaccines

Several approaches have been taken to develop a vaccine against human dental caries where *Streptococcus mutans* is the principal causative agent. Induction of a *S. mutans* antigen-specific IgA response in saliva resulted in a reduction in the level of *S. mutans* organisms in dental plaque (Taubman and Smith, 1987; Katz *et al.*, 1987). Oral adminstration of *S. mutans* whole-cell or cell-wall antigen, or purified glucosyltransferase and elicitation of IgA antibodies correlated with a reduction on the incidence of caries activity.

Oral immunization with capsules containing killed *S. mutans* whole cells, or with soluble antigens, has been achieved. The latter are less effective than the particulate forms in inducing secretory responses. Live vaccines, however, are in turn superior to killed vaccines in eliciting secretory responses. Identification of surface protein antigen A (Spa A) and a glucosyl-transferase (GtfA) as immunogens eliciting protective immune response, led to the cloning of genes coding for these two proteins. Recombinant plasmids specifying SpaA (from *S. mutans* and *S. sobrinus*) and GtfA (from *S. mutans*) have been used to transfect *Salmonella typhimurium* carrier strains (Curtiss *et al.*, 1988) (Table 1). The most recent carrier has been avirulent because of deletion mutations in genes coding for adenylate cyclase (*cya*) and the cyclic AMP receptor protein (*crp*) and elimination of a 100 kb virulence plasmid (pStSR100). Oral immunization of mice with these hybrid strains resulted in the appearance of salivary IgA responses after approximately 1 week. Immunization of both 4- and 8-week-old mice had no adverse effect on weight gain. Studies to investigate the level and duration of the sIgA and serum IgG responses against the streptococcal antigens and the efficacy in reducing dental caries have not yet been reported.

LIVE VACCINES FOR MUCOSAL AND INTRACELLULAR IMMUNITY AGAINST INVASIVE PATHOGENS

Pathogens such as *Salmonella* spp. and *Shigella* colonize the intestine and then penetrate into the epithelial cells of the mucosa. Multiplication intracellularly and intercellular spread is then a dominant feature in the pathophysiology of these enteric infections. *Shigella* organisms do not spread beyond the *lamina propria*, except in malnourished and immunocompromised individuals, whereas a bacteraemic stage is common in typhoid and paratyphoid fevers. An effective immune defence should be targeted so that local as well as cell-mediated immune responses are elicited.

Salmonella typhi vaccines

Typhoid fever persists with a high incidence in countries where there is neither a safe water supply nor adequate sewage disposal (Hornick, 1985). *S. typhi* induces disease only in human beings and is disseminated by carriers by the faecal–oral route. The measures needed to interrupt the infectious cycle are apparent but are unlikely to be available to all in the near future. Hence, an effective vaccine against typhoid fever would help to decrease the incidence of this disease.

Typhoid vaccines have been available for a long time. Inactivated vaccines, either acetone-treated or heat-killed and formalin-fixed, have been used for parenteral administration. These vaccines give good protection to children and adults living in endemic areas (Hornick, 1985). They have never been universally accepted and used in endemic areas because of the need to administer a large number of doses and because of a high rate of reactivity among recipients. In recent years two different approaches for the development of more efficient typhoid vaccines have been used: (a) live vaccines that elicit local (antibody and cell-mediated) and humoral immunity and (b) a polysaccharide vaccine that elicits humoral immunity.

Attenuated live *Salmonella typhi* vaccines for oral administration

Salmonella typhi strain Ty21a is a spontaneous mutant which, because of genetic lesions, is unable to form the O-antigenic polysaccharide side-chains of its lipopolysaccharide (= endotoxin) and the Vi antigen (Germanier and Furer, 1975). Administration of the vaccine to more than 32 000 Egyptian children in a total of three doses without adverse reaction, resulted in a 95% protection after 3 years (Wahdan *et al.*, 1982). Subsequent field trials in Chile showed a protection level of 50%. The high protection level in Egypt was obtained with a liquid formulation, which it is impractical to distribute. The low protection level in Chile was obtained with gelatin capsules containing lyophilized vaccine and sodium bicarbonate. In a recent field trial in Santiago, Chile, involving 109 000 school children who were given an enteric-coated formulation, which does not require bicarbonate, the protection rate was 67% for at least 3 years (Levine *et al.*, 1987a). This vaccine was administered in three doses within 1 week. The protective efficacy of the liquid compared with the enteric-coated formulation of the *S. typhi* Ty21a vaccine is, at present, being studied in a large field trial in Indonesia. Preliminary and unpublished data after 1 year of surveillance indicate that both vaccine formulations elicit protective immunity in which the liquid formulation is marginally better than the enteric-coated capsule formulation.

Another development has been to convert a virulent *S. typhi* strain to non-virulence by introduction of one or more mutations of known molecular character, which cause loss of virulence by known mechanisms. In this way non-reverting mutations that cause a requirement for certain metabolites (auxotrophy) have been transduced to virulent strains of *S. typhi* (Stocker, 1988) (Table 1).

Strain 541Ty has deletions at *aroA* and *purA* which cause a requirement for the aromatic metabolites *p*-aminobenzoic acid and 2,3-dihydroxybenzoic acid and for adenine. None of 37 volunteers who were given a liquid formulation of up to 10^{10} bacteria of *S. typhi* 541Ty, or its Vi antigen-negative variant 543Ty, showed any adverse effects. All excreted the strain for 2 days and all gave evidence of a cellular immune response measured as peripheral blood lymphocyte blast transformation when stimulated with *S. typhi* antigens. However, a significant rise in antibody titres was not seen. It is surmised that this was a consequence of the fact that strains 541Ty and 543Ty had defects that cause both aromatic and adenine dependence. Animal experiments have shown that adenine-requiring bacteria, and even more so bacteria requiring both adenine and aromatic metabolites, survive less well in mouse tissues than those requiring only the aromatic compounds. Therefore *S. typhi* strains with deletions in genes for different steps in the aromatic biosynthesis pathway only have been constructed and are ready to be tested.

Vi antigen polysaccharide vaccine

Most *S. typhi* strains produce a capsular polysaccharide, a homopolymer of *N*- and O-acetylated *N*-acetyl-D-galactose manaronic acid, termed Vi antigen. The idea of using Vi antigen as a vaccine was first tested in the early 1950s, but it did not protect volunteers. A re-examination of the protective role of Vi antigen, based on newer methods for the production of endotoxin-free preparations, showed it to be safe and immunogenic when tested in volunteers. A field trial in South Africa involving more than 11 000 school children showed a vaccine efficacy of 64% after a 21-month surveillance period after a single intramuscular injection of 25 μg of Vi polysaccharide (Klugman *et al.*, 1987). The protective efficacy is similar to that observed with the enteric-coated, live *S. typhi* Ty21a vaccine in trials in Chile and Indonesia. However, the duration of the single dose Vi polysaccharide remains to be determined.

The protective immunity seen after intramuscular immunization with the Vi polysaccharide is most probably an effect of circulating antibodies preventing the bacteraemic stage of typhoid fever. Oral immunization with a live oral Vi antigen-positive strain should be expected to elicit both humoral and cell-mediated immune responses directed against the Vi antigen but also against bacterial outer membrane components such as the O antigen and proteins. This immune defence may reduce both the colonization and intracellular multiplication stages of the typhoid, as well as its bacteraemic stage. It would be desirable to conduct a field trial where the protective efficacy of a live oral Vi positive *S. typhi* vaccine is compared with that of the Vi polysaccharide vaccine for intramuscular injection.

Salmonella vaccines other than *Salmonella typhi*

Most data support the concept that the important immune response elicited by live oral *Salmonella* vaccines is cell-mediated (Lindberg and Robertsson, 1983; Eisenstein *et al.*, 1984). Animal and human volunteer studies have shown that a specific cellular immune response is directed against the O antigenic polysaccharide (Lindberg and Robertsson, 1983; Levine *et al.*, 1987b). Studies with the Ty21a vaccine demonstrated that the O antigen was necessary for Ty21a to be immunogenic and protective (Gilman *et al.*, 1987).

In *Salmonella* as in other Enterobacteriaceae there are numerous O antigens in each genus. If O antigens immunity is critical for protective efficacy then a large number of candidate vaccine strains are constructed. In many *Salmonella* strains studied, all the specific genetic information needed for building the O repeat unit (the O polysaccharide chain is a linear polymer of the repeat unit) resides in the *rfb* gene-cluster near the *his* (histidine biosynthesis) operon. Strains which expressed both O antigen 4 (of *S. typhimurium* = serogroup B) and O antigen 9 (of *S. enteritidis* = serogroup D) were constructed (Johnson *et al.*, unpublished). Individual bacteria made lipopolysaccharide with both O4 and O9 reactivity. The immunogenicity and protective efficacy of such *Salmonella* O4/O9 hybrid strains is currently being studied in experimental animals.

Shigella vaccines

The major objective is to produce vaccine(s) which will protect against illness caused by *Shigella dysenteriae* type 1 and prevalent serovars to *Shigella flexneri*. A recent estimate of the annual worldwide disease burden of bacillary dysentery indicated that there are more than 100 million cases and approximately 600 000 deaths (Institute of Medicine, 1986). Since oral rehydration has no effect on the outcome of bacillary dysentery other means must be sought. Dysentery is a disease of man and dissemination is by the faecal–oral route. Vaccination is one possible intervention strategy and has been tested since the 1940s. Parenteral immunization with heat-inactivated *Shigella* vaccines resulted in elicitation of high titres of circulating antibodies. However, protection was not recorded after experimental challenge or natural infection in field trials (Formal and Levine, 1984). Subcutaneous immunization with live, virulent *S. flexneri* 2a which gave high serum titres did not elicit a protective immune response in monkeys (Formal and Levine, 1984). On the other hand, natural infections result in immunity to subsequent infection with the same serovar. This suggests that the natural infection elicits a protective local immune response in the intestine. Consequently, effective vaccines for these infections will probably be live avirulent bacteria given orally.

Over the last decade there have been three different approaches in the development of *Shigella* vaccines.

Mutant attenuated *S. flexneri* strains

These have been tested since the early 1960s. The *S. flexneri* 2a strain 24570 had unidentified mutation(s) which reverted to the parental wild-type virulent organism and was deemed too unstable for human use (Formal and Levine, 1984). *S. flexneri* 2a strain TS32 Istrati is a non-invasive strain which has been extensively tested in Roumania and China; it is safe and protective against homologous and heterologous *Shigella* species. The vaccine must be given in multiple doses annually which limits its usefulness. A third type of avirulent mutant are the streptomycin-dependent *S. flexneri* and *S. sonnei* vaccine strains. Although *S. flexneri* serovar-specific protection was observed in large-scale Yugoslav field trials and vaccines are not in current use because of the requirement for multiple doses and annual boosters, a reversion to prototrophy was observed (Formal and Levine, 1984).

Attenuated *Shigella* hybrids

These have been obtained in two ways (Table 1). Conjugative transfer of the *xyl-rha* region from *E. coli* K12 into *S. flexneri* gave a strain which was insufficiently attenuated, since it caused disease in approximately one-third of human volunteers (Formal and Levine, 1984).

A more recent approach has been to attenuate *S. flexneri* by making it auxotrophic and dependent on aromatic metabolites not available in mammalian tissues. An *aroD* gene of *E. coli* K12 strain NK5131, inactivated by the insertion of the Tn*10* transposon, was transduced with phage P1 into a virulent *S. flexneri* serovar Y strain designated Sfl1 (Lindberg *et al.*, 1988). One of the transductant strains Sfl114 was found to invade HeLa cells *in vitro* and to cause plaque formation in HeLa monolayers. It maintains intracellular multiplication *in vitro*, though with a reduced efficiency when compared with the wild-type Sfl1 parent strain but it was also unable to cause kerato-conjunctivitis in guinea-pigs. When the strain was given orally to *Macaca fascicularis* monkeys at a dose of 1×10^{11} bacteria it was well-tolerated, excreted for 1–4 days, and found to elicit a local intestinal sIgA and serum IgA, IgM and IgG responses. Monkeys challenged with $100 \times ID_{50}$ doses (1×10^{11} bacteria) of the virulent parent strain Sfl1 were completely protected from development of diarrhoea, none of nine monkeys had diarrhoea compared with four of five unimmunized control monkeys ($P < 0.001$).

All monkeys were subjected to colonoscopy and biopsies were taken from the proximal colon to the rectum before immunization and 3 days after immunization and challenge. Histopathological examination showed surface damage with epithelial erosions and patchy mucosal erosions from the ascending colon down to the rectum in unimmunized and Sfl114-challenged monkeys.

Moderate interstitial inflammation with engagement of the mucosal crypts was noted. Secondary epithelial hyperplasia was seen in many areas of the surface and at the neck of the crypts. Monkeys immunized with *S. flexneri* strain Sfl114 lacked gross morphological signs of infection and inflammation. Biopsies after three doses showed a mild epithelial hyperplasia and inflammatory changes in the crypts. Colonoscopy after the challenge infection showed normal mucosa. Histological examination of the biopsies revealed no surface erosion. In the interstitial tissue, however, moderate inflammation was observed. Only a few crypts were inflamed and the reactive epithelial hyperplasia was mild and focal. The histological results are interpreted to mean that vaccination with strain Sfl114 elicits a local immune defence which prevents damage to the intestinal mucosa. The defence is caused not only by sIgA which prevents invasion but probably also by cellular immune mechanisms, since *S. flexneri* is found intracellularly in low numbers after challenge.

The *S. flexneri* vaccine candidate strain Sfl114 was recently fed to 49 healthy volunteers in a safety and immunogenicity test. At a dose of 1×10^9 bacteria given orally in a bicarbonate buffer it was well-tolerated, excreted from 92% of the volunteers up to 5 days and it elicited an intestinal sIgA response specific for the *S. flexneri* serovar Y O antigen. The duration of the immune response is at present being studied.

Shigella hybrid vaccines based on *E. coli* or *S. typhi*

The approaches have been based either on the construction of invasive vaccines based on an avirulent *E. coli* recipient (Formal *et al.*, 1984), or provision of the *S. typhi* Ty21a vaccine strain with presumed protective antigens from *Shigella* (Black *et al.*, 1987) (Table 1). These vaccines were found to elicit protective immunity in monkeys. They are also effective in human volunteers. Both vaccine types are at present being modified in order to reduce the frequency of side-effects (*E. coli*–*S. flexneri* hybrid), or to enhance immunogenicity (*S. typhi*–*S. sonnei* hybrid).

FUTURE DEVELOPMENT

There is no doubt that by the use of modern genetic techniques significant progress has been made in the development of better defined and effective vaccines. Even more rapid and spectacular progress can be expected in coming years, particularly in studies of strains functioning as carriers of heterologous antigens. However, there are important areas where progress has been less spectacular. We still do not have the means to produce an inexpensive formulation of a live vaccine such that the strain can pass through the acid environment of the stomach unharmed and then start multiplying in the small intestine. Furthermore, we are unaware of how best to stimulate a local intestinal immunity which results in a long-lasting protective efficacy against enteric pathogens. Perhaps such a goal is unattainable and the aim of stimulating a long-lasting protection with only one immunization may remain merely a vision. It is clear that the technical and conceptual advances in the construction of vaccine candidates at present outstrip our capacity on two other fronts. Our capacity to comprehend the immunobiology of infectious disease and our capacity to overcome the political and economic impediments to the use of vaccines.

ACKNOWLEDGEMENTS

Work from the author's laboratory was supported by the Swedish Medical Research Council (grant No. 16x-656) and the Swedish Agency for Research Cooperation with Developing Countries (SAREC). The skilled secretarial assistance by Ms E. Belardo is gratefully acknowledged.

REFERENCES

Betley, M. J., Miller, V. L. and Mekalanos, J. J., 1986. Genetics of bacterial enterotoxins. *Annual Review of Microbiology* **40,** 577–596.

Black, R. E., Levine, M. M., Clements, M. L., Losonsky, G., Herrington, D., Berman, S. and Formal, S. B., 1987. Prevention of Shigellosis by a *Salmonella typhi–Shigella sonnei* bivalent vaccine. *Journal of Infectious Diseases* **155,** 1260–1265.

Black, R. E., Merson, M. H., Rowe, B., Taylor, D., Alim, A. R. M. A., Gross, R. J. and Sack, D. A., 1981. Enterotoxigenic *Escherichia coli* diarrhoea: acquired immunity and transmission in an endemic area. *Bulletin of the World Health Organization* **59,** 263–273.

Clements, J. D. and El-Morshidy, S., 1984. Construction of a potential live oral bivalent vaccine for typhoid fever and cholera–*Escherichia coli*-related diarrheas. *Infection and Immunity* **46,** 564–569.

Clements, J. D., Harris, J. R., Sack, D. A., Chakraborty, J., Khan, M. R., Stanton, B. F., Ali, M., Huda, N., Yunus, M., Kay, B. A., Khan, M. U., Rao, M. R., Svennerholm, A. M. and Holmgren, J., 1988. Field trial of oral cholera vaccines in Bangladesh: results of one year of follow-up. *Journal of Infectious Diseases*, in press.

Clements, J. D., Sack, D. A., Harris, J. R., Chakraborty, J., Neogy, P. K., Stanton, B. F., Huda, N., Khan, M. U., Kay, B. A., Khan, M. R., Yunus, Md., Rao, M. R., Svennerholm, A. M. and Holmgren, J., 1988. Efficacy of oral cholera toxin B subunit as a protective immunogen against diarrhea associated with heat-labile toxin-producing enterotoxigenic *Escherichia coli*: results of a large scale field-trial. Submitted.

Curtiss III, R., Goldschmidt, R. M., Fletchall, N. B. and Kelly, S. M., 1988. Avirulent *Salmonella typhimurium Δ1cya Δcrp* oral vaccine strains expressing a streptococcal colonization and virulence antigen. *Vaccine* **6,** 155–160.

Eisenstein, T. K., Killar, L. M., Stocker, B. A. D. and Sultzer, B. M., 1984. Cellular immunity induced by avirulent *Salmonella* in LPS-defective C3H/HeJ mice. *Journal of Immunology* **133,** 958–961.

Evans, D. G., Evans, D. J., Sack, D. A. and Clegg, S., 1984. Enterotoxigenic *Escherichia coli* pathogenic for man: biological and immunological aspects of fimbrial colonization factor antigens. In *Attachment of Organisms to the Gut Mucosa*, Boedeker, E. C. (ed.), Vol. 1, pp. 63–78. Boca Raton: CRC Press.

Formal, S. B. and Levine, M. M., 1984. Shigellosis. In *Bacterial Vaccines*. Germanier, R. (ed.), pp. 167–186. New York: Academic Press.

Formal, S. B., Hale, T. H., Kapfer, C., Cogan, J. P., Snoy, P. J., Chung, R., Wingfield, M. E., Elisberg, B. L. and Baron, L. S., 1984. Oral vaccination of monkeys with an invasive *Escherichia coli* K-12 hybrid expressing *Shigella flexneri* 2a somatic antigen. *Infection and Immunity* **46,** 465–469.

Germanier, R. and Furer, E., 1975. Isolation and characterization of *galE* mutant Ty21a of *Salmonella typhi*: a candidate strain for a live oral typhoid vaccine. *Journal of Infectious Diseases* **131,** 553-558.

Gilman, R. H., Hornick, R. B., Woodward, W. E., Du Pont, H. L., Snyder, M. J., Levine, M. M. and Libonati, J. P., 1987. Immunity in typhoid fever: evaluation of TY21a—an epimeraseless mutant of *S. typhi* as a live oral vaccine. *Journal of Infectious Diseases* **136,** 717–723.

Hornick, R. B., 1985. Selective primary health care: strategies for control of disease in the developing world. XX. Typhoid fever. *Review of Infectious Diseases* **7,** 536–546.

Institute of Medicine, 1986. *New Vaccine Development: Establishing Priorities*. Vol. II. *Diseases of Importance in Developing Countries*. Washington DC: National Academy Press.

Johnson, B. N., Stocker, B. A. D., Weintraub, A. and Lindberg, A. A., 1988. Construction and characterization of Salmonella strains with both antigen 04 (of group B) and antigen O9 (of group D). Submitted.

Kaper, J. B., Lockman, H., Baldini, M. M. and Levine, M. M., 1984. Recombinant nontoxigenic *Vibrio cholerae* strains as attenuated cholera vaccine candidates. *Nature* **308,** 655–656.

Katz, J., Michalek, S. M., Curtiss III, R., Harmon, C., Richardsson, G. and Mestecky, J. 1987. Novel oral vaccines: the effectiveness of cloned gene products on inducing secretory immune responses. *Advances in Experimental Medicine and Biology* **216B,** 1741–1747.

Klugman, K. P., Gilbertson, I. T., Koornhof, H. J., Robbins, J. B., Schneerson, R., Schultz, D., Cadoz, M., Armand, J. and Vaccination Advisory Committee, 1987. Protective activity of Vi capsular polysaccharide vaccine against typhoid fever. *Lancet* **ii,** 1165–1169.

Kupersztoch, Y. M. Fusion between the ST enterotoxin and the B subunit of the heat-labile toxin of *Escherichia coli*, 1986. *Abstracts of the 22nd Conference on Cholera.* US–Japan Cooperative Medical Sciences Program, Toyama, Japan, pp. 51–52.

Levine, M. M., Nalin, D. R., Hoover, D. L., Bergquist, E. J., Hornick, R. B. and Young, C. R., 1979. Immunity to enterotoxigenic *Escherichia coli. Infection and Immunity* **23,** 729–734.

Levine, M. M., Black, R. E., Clements, M. L., Young, C. R., Cheney, C. P., Schad, P., Collins, H. and Boedeker, E. C., 1984. Prevention of enterotoxigenic *Escherichia coli* diarrheal infection in man by vaccines that stimulate anti-adhesion (anti-pili) immunity. In *Attachment of Organisms to the Gut Mucosa*, Boedeker, E. C. (ed.), Vol. II, pp. 223–244. Boca Raton: CRC Press.

Levine, M. M., Ferreccio, C., Black, R. E., Germanier, R. and Chilean Typhoid Committee, 1987a. Large-scale field trial of Ty21a live oral typhoid vaccine in enteric-coated capsule formulation. *Lancet* **i,** 1049–1052.

Levine, M. M., Herrington, D., Murphy, J. R., Morris, J. G., Losonsky, G., Tall, B., Lindberg, A. A., Svenson, S., Baqar, S., Edwards, M. F. and Stocker, B. A. D., 1987b. Safety, infectivity, immunogenicity and in vivo stability of two attenuated auxotrophic mutant strains of *Salmonella typhi* 541Ty and 543Ty, as live oral vaccines in humans. *Journal of Clinical Investigation* **79,** 888–902.

Lindberg, A. A. and Robertsson, J. Å., 1983. *Salmonella typhimurium* infection in calves: cell-mediated and humoral immune reactions before and after challenge with live virulent bacteria in calves given live or inactivated vaccine. *Infection and Immunity* **41,** 751–757.

Lindberg, A. A., Kärnell, A., Stocker, B. A. D., Katakura, S., Sweiha, H. and Reinholt, F. P., 1988. Development of an auxotrophic oral live *Shigella flexneri* vaccine. *Vaccine* **6,** 146–150.

McCaughan, G. and Basten, A., 1983. Immune system of the gastrointestinal tract. *International Review of Physiology* **28,** 131–153.

Maskell, D. J., Sweeney, K. J., O'Callaghan, D., Hormaeche, C. E., Liew, F. Y. and Dougan, G., 1987. *Salmonella typhimurium aroA* mutants as causes of the *Escherichia coli* heat-labile enterotoxin B subunit to the murine secretory and systemic immune systems. *Microbial Pathogenicity* **2,** 211–221.

Mekalanos, J. J., Swartz, D. J., Pearson, G. D. N., Harford, N., Groyne, F. and de Wilde, M., 1983. Cholera toxin genes: nucleotide sequence, deletion analysis and vaccine development. *Nature* **306,** 551–557.

Sanchez, J., Uhlin, B. E., Grundström, T., Holmgren, J. A., Hirst, T. R., 1986. Immunoactive chimeric ST-LT enterotoxins of *Escherichia coli* generated by *in vitro* gene fusion. *FEBS Letters* **208,** 194–200.

Stocker, B. A. D., 1988. Auxotrophic *Salmonella typhi* as live vaccine. *Vaccine* **6,** 141–145.

Taubman, M. A. and Smith, D. J., 1987. A mucosal approach to immunoprophylaxis of dental infections. *Advances in Experimental Medicine and Biology* **216B,** 1721–1730.

Taylor, R. K., Miller, V. L., Furlong, D. B. and Mekalanos, J. J., 1987. Use of *phoA* gene fusions to identify a pilus colonization factor coordinately regulated with cholera toxin. *Proceedings of the National Academy of Sciences, USA* **84,** 2833–2837.

Taylor, R., Shaw, C., Peterson, K., Spears, P. amd Mekalanos, J. J., 1988. Safe, live *Vibrio cholerae* vaccines? *Vaccine* **6,** 151–154.

Wahdan, M. H., Serie, C., Cerisier, Y., Sallam, S. and Germanier, R., 1982. A controlled field trial of live *Salmonella typhi* strain Ty21a oral vaccine against typhoid: three year results. *Journal of Infectious Diseases* **145,** 292–296.

Yamamoto, T., Tamura, Y. and Yokota, T., 1985. Enteroadhesion fimbriae and enterotoxin of *Escherichia coli*: genetic transfer to a streptomycin-resistant mutant of the *galE* oral-route live-vaccine *Salmonella typhi* Ty21a. *Infection and Immunity* **50,** 925–928.

10 Field Tests of Recombinant Ice$^-$ *Pseudomonas syringae* for Biological Frost Control in Potato

STEVEN E. LINDOW and N. J. PANOPOULOS

Department of Plant Pathology, University of California, Berkeley, CA 94720, USA

This review will not attempt to examine all ecological, epidemiological and genetic features of *Pseudomonas syringae*, the first recombinant micro-organism approved for release into the environment. Instead, pertinent features of its natural habitats, processes such as competition which determine its population size on plants, appropriate genetic methodologies utilized in its modification and the design of field studies in which mutant strains of this species have been studied for their potential for control of frost injury to potato will be reviewed. The results of these initial field trials will be summarized with emphasis on measurements of the fate and dispersal of the recombinant bacteria utilized in these studies. Results of major scientific objectives and the efficacy of Ice$^-$ bacterial strains for biological frost control are still being compiled and will be presented elsewhere. General features of the Ice$^-$ bacteria used and the physical design of the experimental site pertinent to the study of other organisms in uncontained environments will be addressed.

BACTERIAL ICE NUCLEI AND PLANT FROST INJURY

Five bacterial species have the ability to catalyse ice formation in water at temperatures near 0°C. Many pathovars of *P. syringae*, and some strains of *P. fluorescens*, *P. viridiflava*, *Erwinia herbicola* and *Xanthomonas campestris* pv. *translucens* are catalysts for ice formation at temperatures warmer than −5°C (Gross *et al.*, 1984; Maki *et al.*, 1974; Maki and Willoughby, 1978; Paulin and Luisetti, 1978; Hirano *et al.*, 1978; Lindow *et al.*, 1978b,c, 1982; Yankofsky *et al.*, 1981; Ashworth *et al.*, 1985; Lim *et al.*, 1987). Other bacterial species and

RELEASE OF GENETICALLY-ENGINEERED MICRO-ORGANISMS ISBN 0-12-677521-4

micro-organisms have no ice nucleation activity at temperatures warmer than −15°C. Most organic or inorganic compounds are very inefficient catalysts for ice formation at temperatures warmer than −10°C (Schnell and Vali, 1976). Thus, these bacteria constitute the most abundant and efficient ice nucleating agents in nature. In the absence of such heterogeneous ice nuclei, pure water can supercool to −40°C before the ice–water phase transition occurs (Hobbs, 1974). Ice nucleation-active (Ice$^+$) bacteria occur in several habitats including water (Maki and Willoughby, 1978), plant surfaces (Lindow *et al.*, 1978c; Lindow, 1982, 1983a,b, 1985b; Makino, 1982; Hirano and Upper, 1983), and in the atmosphere above plants (Lindemann *et al.*, 1982). Ice$^+$ bacteria occur on most plant species that have been examined (Lindow *et al.*, 1978c). The population size of Ice$^+$ bacteria on different plant species ranges from less than 100 to over 10^7 cells per gram fresh weight.

Frost injury is a serious abiotic disease of many important agricultural and native plants. Losses in agricultural production to frost-sensitive plants in the US due to frost injury are more than one billion dollars annually (White and Haas, 1975). Frost-sensitive plants are characterized by the inability to tolerate ice formation (Cary and Mayland, 1970; Burke *et al.*, 1976). Ice formation is not restricted to the intercellular spaces of frost-sensitive plants as it is in frost-hardy plant species, and such intracellular ice formation is damaging. Thus, frost-sensitive plant species have no tolerance of ice formation and must avoid ice formation to avoid freezing injury. Water in many plant species can supercool to −5° to −6°C in the absence of bacterial ice nuclei before ice formation is likely to occur (Arny *et al.*, 1976; Burke *et al.*, 1976; Lindow *et al.*, 1978a, 1982; Lindow, 1982). While non-bacterial ice nuclei may occur in some plant species, particularly perennial woody plants (Ashworth *et al.*, 1985), the numbers of such nuclei and their importance in the freezing behaviour of such plants is undetermined. Frost injury to many plants is proportional to the logarithm of the population size of Ice$^+$ bacteria or the numbers of bacterial ice nuclei on plants at the time of freezing (Lindow *et al.*, 1978a; Lindow, 1982). Treatments that reduce the numbers of heterogeneous ice nuclei on plant tissues are promising in the avoidance of plant freezing injury.

BIOLOGICAL METHODS OF PLANT FROST INJURY CONTROL

Determination of the population dynamics of Ice$^+$ bacteria on plants has indicated methods by which these species can be manipulated and plant frost injury reduced. The population size of Ice$^+$ bacteria on many plants in nature undergoes large seasonal variations (Lindow *et al.*, 1978c; Lindow, 1982; Hirano and Upper, 1983; Gross *et al.*, 1983; Lindow and Connell, 1984). The total population size of bacteria, as well as that of Ice$^+$ bacterial species, is generally lowest on young vegetative tissues of plants. Only about 0.01–40% of the total bacteria on plant surfaces are Ice$^+$ strains that are involved directly in plant freezing injury. Thus, non-ice nucleation-active and Ice$^+$ bacteria coexist on many plants in natural habitats and competition between such bacterial species on leaves appears likely. However, the degree of natural competition provided by indigenous non-ice nucleation-active micro-organisms is generally inadequate to prevent the development of large populations of Ice$^+$ bacteria on many plants. Competition has been enhanced by direct application of antagonistic bacteria. Antagonistic non-ice nucleation-active bacteria have been established on plants by applications to leaves or to seeds of plants before they are colonized by Ice$^+$ bacteria (Lindow, 1982, 1983a, 1985a,b, 1987; Lindow *et al.*, 1983a,b). Many bacterial antagonists colonize emerging plant tissues for 1–4 months after single leaf or seed applications. Population sizes of Ice$^+$ bacteria have been reduced from ten to 500-fold on plants treated with antagonistic bacteria when compared with untreated plants under field conditions. Frost injury to treated plants was reduced from 20 to 95% compared with untreated plants in natural fields frosts (Lindow, 1982, 1985a; Lindow *et al.*, 1983a,b). Significant reductions of epiphytic population sizes of Ice$^+$ bacteria were achieved only when non-ice nucleation-active bacteria were applied to plants before subsequent colonization by Ice$^+$ bacteria (Lindow, 1982; Lindow *et al.*, 1983a). Although not all bacteria are highly antagonistic to Ice$^+$ bacteria on leaves, all identified antagonists act by preventing increases in population size of Ice$^+$ bacteria and not by displacement of established populations (Lindow *et al.*, 1983a).

ROLE OF COMPETITION IN ANTAGONISM ON LEAF SURFACES

Competition for limiting environmental resources appears to account for most antagonism observed between bacteria on leaves. While many naturally-occurring non-ice nucleation active bacteria that are antagonistic to Ice$^+$ bacteria on plants produce antiobiotic-like compounds in culture, antibiosis does not appear to be necessary for biological control on plant surfaces. Only about half of a collection of naturally-occurring non-ice nucleation-active bacteria that were antagonistic to *P. syringae* on leaf surfaces produced antibiotics against this species in culture (Lindow, 1985a, 1988a). Antibiosis$^-$ mutants of strains which were inhibitory to *P. syringae* in culture did not differ from their respective parental strains in controlling *P. syringae* populations on leaf structures under greenhouse and field conditions (Lindow, 1985a). Ice$^-$ mutants of *P. syringae* induced by chemical mutagenesis prevented the growth of near-isogenic Ice$^+$ parental strains or other Ice$^+$ *P. syringae* strains on plants under greenhouse and field conditions (Lindow, 1985a, 1987). The maximum epiphytic bacteria population size on most plants is between 10^7 and 10^8 cells per gram fresh weight. While the total population size of *P. syringae* strains on plants treated with Ice$^-$ mutants of this species approached this number, the proportion of Ice$^+$ strains among the population decreased with increasing population size of Ice$^-$ *P. syringae* strains (Lindow, 1987). Thus a limited number of habitable sites, nutrients on leaves, or other resources may limit the population size of Ice$^+$ bacteria.

While competition appears sufficient to account for the exclusion of Ice$^+$ bacteria by Ice$^-$ mutants of *P. syringae*, the specificity of competitive exclusion is unknown. Non-ice nucleation-active bacterial strains differed greatly in their ability to exclude *P. syringae* from leaf surfaces (Lindow, 1985a). However, the specificity of competition among *P. syringae* strains on leaves is unknown. A single Ice$^-$ *P. syringae* strain applied to leaves may exclude all other Ice$^+$ *P. syringae* strains only if there is a high degree of ecological similarity between such strains. Natural epiphytic populations of *P. syringae* are comprised of a diversity of ecotypes (Gross *et al.*, 1984). Thus, more information on the specificity of competition on leaf surfaces is needed to determine whether Ice$^-$ and Ice$^+$ bacterial species coexist on the same plant. Effective biological control of plant frost injury could be achieved only if coexistence of such dissimilar strains does not occur.

The process of competitive exclusion of Ice$^+$ *P. syringae* strains by Ice$^-$ mutants on leaves was examined by testing the hypothesis that a similarity in ecological habitat preference, and thus in genotype, is required to optimize pre-emptive competitive exclusion of *P. syringae* from leaf surfaces by members of this and related species (Lindow, 1988b). Isogenic Ice$^-$ derivatives of several different Ice$^+$ *P. syringae* strains were developed for such competition studies by site-directed mutagenesis *in vitro*. DNA sequences conferring ice nucleation activity in *P. syringae* have been isolated by cosmid cloning procedures and partially characterized (Orser *et al.*, 1983, 1985). The *iceC* gene from *P. syringae* strain Cit7 is approximately 4200 base pairs in length and confers the production of a single large protein. Ice nucleation activity and the *iceC* gene product are both localized primarily in the outer membrane of *P. syringae* and in *Escherichia coli* strains carrying this cloned gene. Ice nucleating sites in these bacteria have been hypothesized to be an oligomeric structure composed wholly or partially of aggregates of the *iceC* gene product (Govindarajan and Lindow, 1988). Large deletions internal to the *iceC* gene have been made with the restriction enzyme *Sal*I *in vitro* and result in an Ice$^-$ phenotype in *E. coli* (Orser *et al.*, 1985). Reciprocal exchange of the modified *iceC* gene for the complementary functional *iceC* gene in the chromosome of Ice$^+$ *P. syringae* strain Cit7 and a homologous gene in strain TLP2 was made by recombinational marker exchange (Ruvkun and Ausubel, 1981). Additional genetic material was not added to these strains by this process of homogenization, nor should modifications occur at any other genetic locus. However, such strains are classified as recombinant because of the use in their construction of the *in vitro*-generated deletion and recombinant plasmids.

Recombinant Ice$^-$ deletion mutants of *P. syringae* strains Cit7 and TLP2 do not cause frost damage to plants and prevent the colonization of inoculated plants by both isogenic and non-isogenic Ice$^+$ *P. syringae* strains (Lindow, 1985c, 1988b). Frost damage to plants treated with Ice$^-$ deletion mutants of strains Cit7 or TLP2 (Cit7del1b and TLP2dell, respectively) does not differ significantly from untreated control plants at all temperatures above −7°C (Lindow, 1985c). Pairwise comparisons of different Ice$^+$ and Ice$^-$ strains have indicated that while considerable competition occurred between all Ice$^+$ and Ice$^-$ bacterial strains, there were differences in magnitude of such competition (Lindow, 1988b). A major goal of field tests of Ice$^-$ *P. syringae* strains was to demonstrate the specificity (differences in magnitude) of competition among *P. syringae* strains under field conditions. The hypothesis was made that the great diversity of Ice$^+$ *P. syringae* strains that would impinge on potato plants colonized by a given Ice$^-$ strain in the field would lead to enrichment for strains which differed from that Ice$^-$ strain in growth rate, or in resources or sites for which they competed. Field studies were, therefore, conducted to compare the competitive exclusion of isogenic Ice$^+$ parental strains of Ice$^-$ *P. syringae* strains Cit7del1b and TLP2dell on potato plants colonized by these Ice$^-$ strains with the exclusion of natural populations of Ice$^+$ *P. syringae*. Thus, at the University of California Tulelake Field Station in northern California, plants were treated with these Ice$^-$ *P. syringae* strains, selected as vigorous colonizers of potato, with and without challenge inoculations of appropriate Ice$^+$ parental strains.

CHARACTERISTICS OF Ice$^-$ *P. SYRINGAE* STRAINS DETERMINED IN THE LABORATORY THAT PREDICTED THE FATE AND EFFECTS OF THESE STRAINS UNDER FIELD CONDITIONS

Extensive studies were made of the biochemical, biological and ecological properties both of Ice$^+$ parental strains Cit7 and TLP2, and Ice$^-$ derivative strains to anticipate their survival, dispersal and activity in natural habitats. Since extensive studies of the natural abundance and population dynamics of Ice$^+$ *P. syringae* strains, including strains Cit7 and TLP2, had been made at Tulelake and in other areas of California (Lindow, 1982), most studies were designed to identify even small differences between Ice$^-$ strains and Ice$^+$ parental strains. No significant difference in the behaviour of Ice$^-$ strains, as compared with Ice$^+$ parental strains, was anticipated because of the nature of the *ice* gene product which was eliminated. Nonetheless, parental Ice$^+$ and Ice$^-$ strains were examined under the conditions of a number of simulated natural environments and their growth, death or activities in these environments compared to reveal even subtle differences in behaviour. Since the exact behaviour of many micro-organisms may not be predictable based on individual laboratory measurements, we emphasized the laboratory assessment of the *comparative* behaviour of our mutant strains since parental strains could be tested in realistic settings in the field site in which Ice$^-$ strains would be released.

Ice$^-$ *P. syringae* strains TLP2del1 and Cit7del1b grew at rates in culture and on plants that were not measurably different from those of the respective parental strains (Lindow, 1985c). An indirect, but more sensitive, test of differences in growth rates on plants was made by co-inoculation of equal numbers of isogenic Ice$^+$ and Ice$^-$ *P. syringae* strains and measurement of relative population size of each strain as a function of time after inoculation. No significant difference in growth rate between Ice$^+$ and Ice$^-$ strains occurred, as inferred from the observation that there was no change in the proportion of these strains during growth representing about ten generations.

Many different native and cultivated plant species were examined as potential hosts for the epiphytic growth of Ice$^+$ and Ice$^-$ *P. syringae* strains. Ice$^+$ strains TLP2 and Cit7 grew on most of 67 plant species onto which they were inoculated and incubated under favourable (moist) conditions and maintained a significant population size upon stressing such plants with high light intensities and low relative humidities (Lindow, 1985c). The population size achieved by Ice$^+$ parental strains differed greatly between plant species examined but few plant species supported populations as high as did potato. Ice$^-$ *P. syringae* strains Cit7del1b and TLP2del1 did not differ from their respective parental strains in their growth or survival on these 67 plant species. Thus, Ice$^-$ *P. syringae* strains were not anticipated to have a restricted range of plants on which they might survive. However, Ice$^-$ strains clearly did not differ from endemic Ice$^+$ *P. syringae* strains in their relative colonization of cultivated and native plants in the Tulelake, California region, with which these strains might come into contact. *P. syringae* is a taxon containing a diversity of strains (Dye *et al.*, 1980). Many strains classified as 'pathovars' of *P. syringae* are pathogenic to one or more plant species. Strains of this species can be isolated from symptomatic tissues and are placed within the taxon *P. syringae* by virture of key biochemical characteristics and within pathovars of *P. syringae* based on a unique host range (Dye *et al.*, 1980). In contrast, strains such as TLP2 and Cit7, which were isolated as epiphytes from healthy plant tissues, were not pathogenic to any of 75 plant species into which they were inoculated. Thus, while many plant pathogens can be designated as *P. syringae*, many strains with biochemical characteristics consistent with their designation as *P. syringae* are not necessarily pathogenic. While Ice$^+$ strains Cit7 and TLP2 differed from each other in certain biochemical characteristics such as carbon source utilization, both clearly fit within the description of *P. syringae*. Strains Cit7del1b and

TLP2del1 did not differ from their parental strains in any of 178 biochemical characteristics tested (Lindow, 1985c).

Isogenic Ice$^+$ and Ice$^-$ *P. syringae* strains did not differ from each other in their survival of freezing stress. While an alteration of the Ice phenotype was not expected to significantly affect the growth of bacteria on plants or their survival in various habitats, this phenotype might have been selected because of its utility to the cell in coping with freezing stress. Isogenic Ice$^+$ and Ice$^-$ *P. syringae* strains were, therefore, subjected to repeated freezing and thawing cycles both in an aqueous environment as well as on leaves. No significant difference between isogenic Ice$^+$ and Ice$^-$ *P. syringae* strains was observed in their survival of these freezing processes (Lindow, 1985c).

Soil was shown not to be a reservoir for Ice$^-$ *P. syringae* strains. The population size of Ice$^+$ as well as Ice$^-$ *P. syringae* strains decreased rapidly in laboratory studies when incorporated into organic soils collected at Tulelake, California (Lindow, 1985c). Both Ice$^+$ and Ice$^-$ *P. syringae* strains were undetectable 18 days or more after incorporation into such soils. While these strains survived longer when incorporated into the soil in the presence of plant material, such strains were also undetectable upon the decay of the incorporated leaf matter.

Isogenic Ice$^+$ and Ice$^-$ *P. syringae* strains did not differ in their competition with each other or with other *P. syringae* strains. Ice$^+$ *P. syringae* strains exhibited little growth on plants co-inoculated with a much larger number of Ice$^-$ Cit7del1b or TLP2del1 cells (Lindow, 1985c, 1988b). However, Ice$^-$ *P. syringae* strains also exhibited little growth when co-inoculated in plants with much larger numbers of cells of different Ice$^+$ *P. syringae* strains. Thus, the competition exhibited between Ice$^+$ and Ice$^-$ bacteria on plants was principally dependent on the relative population size and not the genotype of the bacterial strains utilized. It was expected that Ice$^-$ *P. syringae* strains would succumb to the competition if dispersed in small numbers to plants already containing a relatively large number of Ice$^+$ *P. syringae* strains.

The dispersal of Ice$^+$ *P. syringae* strains during and subsequent to spray inoculation onto plants was evaluated in field trials at Tulelake, California, to simulate such applications of Ice$^-$ strains. Numbers of viable Ice$^+$ *P. syringae* strains, aerosolized during spray application, that were deposited at ground level decreased exponentially with distance away from the edge of treated plots (Lindow *et al.*, 1988). While large numbers of viable cells of sprayed strains were detected within 1–3 m from the edge of inoculated plots, few cells were detected more than 10 m away. The survival of aerosolized inoculum on plants beyond the sprayed plot was lower than that of inoculum directly applied to treated plants (Lindow *et al.*, 1988).

The results of laboratory and field studies performed before the release of recombinant Ice$^-$ *P. syringae* strains indicated that rather small numbers of cells would be dispersed 10 m or more away from the edge of the plot, were capable of growing on plants not already supporting large numbers of bacteria including Ice$^+$ bacteria, but were unlikely to grow on plants already colonized by Ice$^+$ *P. syringae* strains. Such strains were not expected to be phytopathogenic or to survive in soil. Since Ice$^-$ strains were indistinguishable in all measured characteristics from their parental Ice$^+$ strains, which had been shown to exhibit population sizes and changes typical of Ice$^+$ *P. syringae* populations as a whole, they were expected to mimic endemic strains under field conditions.

DESIGN FEATURES THAT MITIGATED DISPERSAL AND SURVIVAL OF RECOMBINANT Ice$^-$ *P. SYRINGAE* STRAINS IN A FIELD TRIAL

The physical setting of the experimental site at which *P. syringae* strains were tested was designed to minimize dispersal of these strains to unintended habitats. An experimental plot of potato, planted in a randomized complete block design, was restricted to an area of 33 × 58 m (Fig. 1). Ice$^-$ strains Cit7del1b and TLP2del1 were applied to potato seed pieces at planting on 29 April and 12 August at a concentration of 3×10^9 cells ml^{-1}, and were also sprayed onto foliage of plants after emergence at a concentration of 3×10^8 cells ml^{-1} on 28 May and 3 September 1987. A zone of soil kept free of all plants, that extended at least 30 m north, east and west of the plot and 20 m south of the plot, was established to minimize the number of bacteria dispersed as aerosols during spray inoculation that would contact surrounding vegetation (Fig. 1). The experimental site was located 40 m or more from the nearest surface water sources and more than 1 km from the nearest inhabited dwelling. A physical restraint was erected around the plot at a distance of 20 m to discourage unnecessary movement of people and equipment in and out of the treated plot area. Disposable shoe covers were utilized when entry into the plot for sampling or treating purposes was necessary. Disposable clothing was also worn during plot treatment. Machinery such as planting and harvesting equipment was sterilized after use in the plot with steam or 70% ethanol. Thus, every attempt was made to minimize the movement of Ice$^-$ *P. syringae* strains outside of the plot by passive means. Spray application of Ice$^-$ *P. syringae* strains was conducted only under extremely calm wind conditions (less than 0.2 m s^{-1}). Spraying was done with a carbon dioxide pressurized hand-held sprayer with an orifice which generated large droplets (average droplet diameter greater than 100 microns) to minimize transportable aerosols.

Ice$^-$ *P. syringae* strains were largely eradicated at the termination of the experiment (after a killing frost on 15 September 1987) by removal of plant material from the field site. Potato plants inside of the experimental plot were

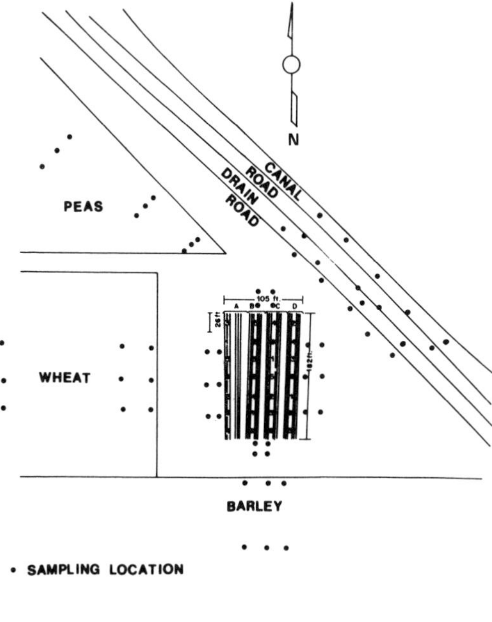

Fig. 1. Location of experimental potato plot in relation to surrounding vegetation and water. Numbers within the experimental plot represent each replication of seven different treatments arranged in a randomized complete block design. Sampling locations (●) are also shown.

uprooted, air-dried on the surface of the soil, and then incinerated with propane burners. Mature progeny tubers and undecayed potato seed pieces were harvested from the soil with a mechanical harvester and steam-sterilized. All other remaining plant material in the plot was collected from the plot area by hand and steam-sterilized. Since Ice$^-$ *P. syringae* strains do not survive in the soil in the absence of plant material, their numbers could be greatly reduced by removing such habitable sites.

DISPERSAL AND SURVIVAL OF RECOMBINANT Ice$^-$ *P. SYRINGAE* STRAINS IN AND AROUND A FIELD EXPERIMENTAL SITE

Frequent estimates of the population size of Ice$^-$ *P. syringae* strains TLP2del1 and Cit7dellb in various habitats in and around the field site where these strains were tested were performed in compliance with the requirements of an experimental use permit issued by the US Environmental Protection Agency to conduct this experiment. Population sizes of these two strains were obtained by enumeration of colonies arising on a highly selective culture medium developed for this use. SSM-rif medium, highly selective for *P. syringae* and containing 100 μg ml^{-1} rifampicin, to which strains Cit7del1b and TLP2del1 were resistant, efficiently recovered these strains and, with few exceptions, no other bacterial strains from environmental samples. Bacteria were removed from environmental samples by sonication in 0.1 M phosphate buffer containing 0.1% Bacto peptone and appropriate dilutions were plated onto selective media and non-selective media such as King's medium B (King *et al.*, 1954). When appropriate, the identity of colonies resembling strains Cit7del1b or TLP2del1 on selective media was verified by dot-blot hybridization with a ^{32}P-labelled oligonucleotide probe specific to these two strains. Since the deletion within the *ice* gene of these two strains involved digestion with *Sal*I, re-ligation of the distal non-contiguous DNA sequences resulted in the formation of a 'novel joint' centred at a *Sal*I site. The DNA sequence in the region of this junction was determined and a 21-base pair oligonucleotide homologous to this region was prepared, which, under proper stringency conditions, hybridized only with DNA from these two *P. syringae* strains. Weekly or bi-weekly samples of soil were taken from 28 locations within the experimental plot, and at four locations 5 and 10 m away from the potato plot. Samples of leaves were taken from 28 locations within the experimental plot, from the nearest vegetation, from vegetation 50 m from the edge of the potato plot, and from vegetation 100 m north, east, south and west from the edge of the plot. A total of eight samples of water was analysed from two canals which ran north of the experimental site (Fig. 1). Bacteria were enumerated on the roots of potato plants collected from 12 locations and from insects collected at locations within the plot and from insects collected 100 m north, east, south and west of the plot. The dispersal of bacteria outside of the experimental plot was assessed by measuring deposition of aerosolized inoculum onto open Petri dishes containing SSM-rif medium placed in an array at distances of 1, 3, 5, 10, 20 and 30 m away from the plot in all directions.

While large numbers of aerosolized cells of Ice$^-$ *P. syringae* strains were detected in the buffer zone near sprayed plants, the numbers of cells deposited decreased greatly with distance away from the edge of the experimental plot (Fig. 2). Considerable variation in the numbers of Ice$^-$ *P. syringae* strains deposited even near the experimental plot was related to their proximity to sprayed plants (Figs 1 and 2). While a constant exponential decrease in the number of Ice$^+$ *P. syringae* strains deposited with distance during spray events in slightly more turbulent atmospheric conditions was observed (Lindow *et al.*, 1988), a non-constant exponential decrease in the numbers of Ice$^-$ *P. syringae* strains deposited in this spray event occurred (Fig. 3). This inoculation was performed at night at an air temperature of about 4°C and with very calm winds. The dilution of inoculum by air turbulence and death of inoculum due to

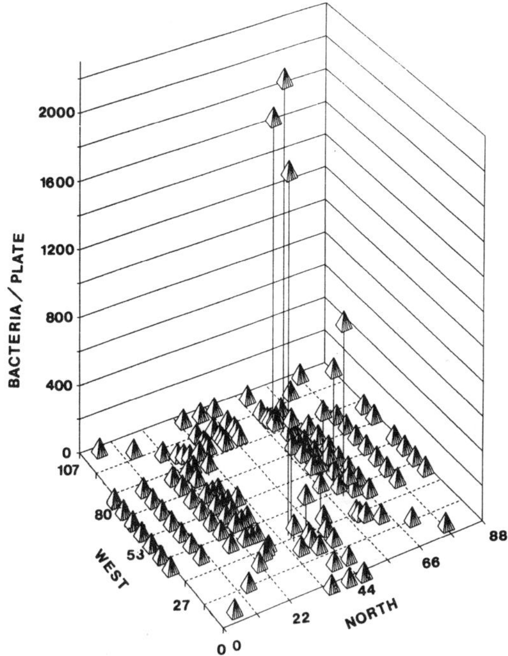

Fig. 2. The recovery of Ice⁻ *P. syringae* strains on plates of SSM-rif medium placed at various locations around a sprayed potato plot during a spray application on 3 September 1987. The rectangular area at the centre of the figure represents the plot itself, where no sampling was undertaken. The dimensions of the plot in relation to sampling locations are shown in meters. Sample plates were exposed to the atmosphere from 10 min before to 20 min after cessation of spraying *P. syringae* strains Cit7del1b and TLP2del1. Each point represents the number of colonies recovered per SSM-rif plate.

desiccation may not have been as great as in earlier trials. The numbers of Ice⁻ *P. syringae* strains deposited at distances greater than 10–20 m, however, was very low.

Ice⁻ *P. syringae* strains were not detected at any time on vegetation surrounding the experimental site (Fig. 4). Total bacterial population sizes on vegetation such as wheat outside of the experimental plot were high (greater than 10^6–10^7 cells per gram fresh weight) and large numbers (over 10^4–10^5 cells per gram fresh weight) of Ice⁺ *P. syringae* strains occurred on plants at the time of spray application of Ice⁻ *P. syringae* strains to potato in May. Thus, wheat plants were heavily colonized by bacteria, including Ice⁺ bacteria, that might compete with any immigrant Ice⁻ *P. syringae* strains.

No Ice⁻ *P. syringae* strains were detected at any time in surface waters surrounding the experimental site (Fig. 5). Total bacterial population size in

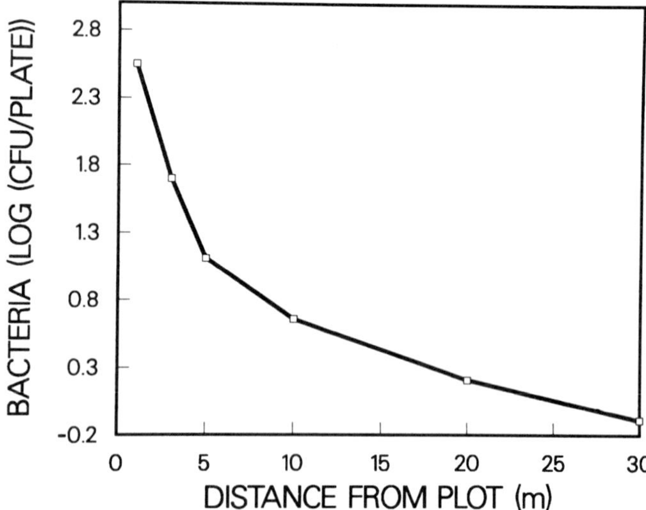

Fig. 3. Logarithm of mean colony-forming units recovered on SSM-rif plates as a function of increasing distance away from the edge of a potato plot sprayed with Ice$^-$ *P. syringae* strains Cit7del1b and TLP2del1 on 3 September 1987. The logarithm of mean colony-forming units from plates exposed at each of the six distances shown on the abscissa from data presented in Fig. 2 is shown.

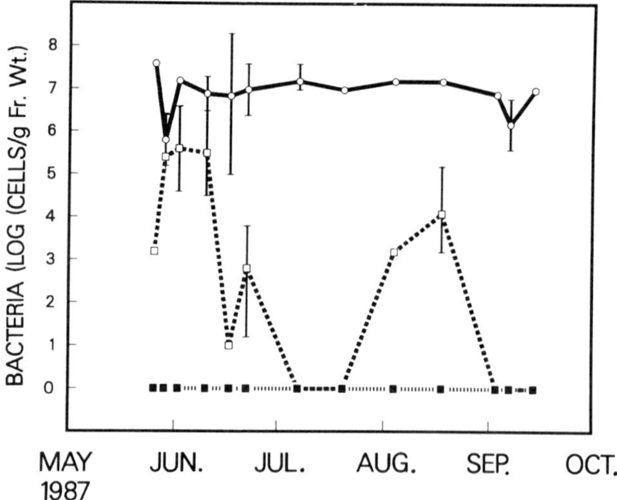

Fig. 4. Total bacterial population size (○), and population size of Ice$^+$ bacteria detected on King's medium B (□), and Ice$^-$ rifampicin-resistant *P. syringae* strains (■) detected on wheat located 20 m south of an experimental plot of potato at various times after spray inoculation of the plot with Ice$^-$ *P. syringae* strains. The vertical bars represent the standard error of the mean of log population size at each sampling time.

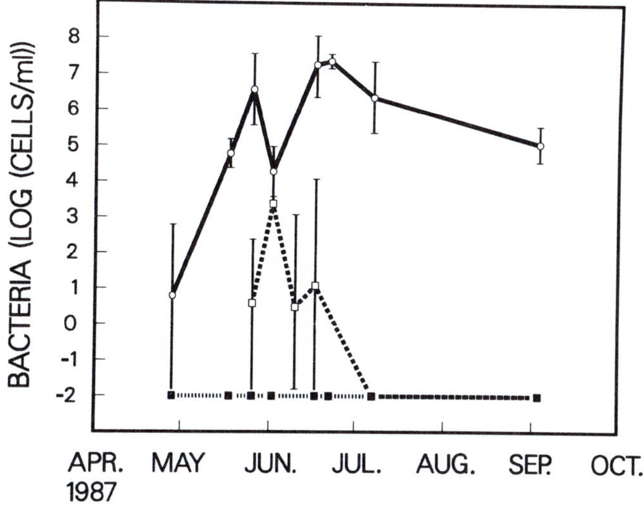

Fig. 5. Total bacterial population size (○) and population size of Ice$^+$ bacteria (□) and rifampicin-resistant Ice$^-$ *P. syringae* strains (■) in water collected from a canal about 50 m north of an experimental potato plot at Tulelake, California, before and after treatment of plot with Ice$^-$ strains. The vertical bars represent the standard error of the mean of log population size at each sampling time.

these waters was quite high. No surface water run-off to these water bodies occurred during the growing season and, because of their great distance from the experimental site, deposition of aerosolized inoculum from the sprayed plot was very low.

Ice$^-$ *P. syringae* strains did not survive well in soil. A considerable inoculum of strains Cit7del1b and TLP2del1 was applied to soil during spray inoculation and when inoculated tubers were planted in soil. However, maximum population sizes of these strains in soil were less than about 100 cells per gram of soil even shortly after spray inoculation to the surface of this dry soil (Fig. 6). Strains Cit7del1b and TLP2del1 were not detected in soil sampled at any time from one week after spray inoculation until the termination of the experiment in October.

While aerosolized inoculum did not colonize vegetation surrounding the experimental plot, strains Cit7del1b and TLP2del1 maintained a high population size (greater than 10^4–10^6 cells per gram fresh weight) for about 4 weeks on inoculated potato plants inside the plot (Fig. 7). Potato plants were only approximately 5–10 cm tall at the time of spray inoculation, but had grown to 30 cm or more in height after 4 weeks. These strains had a small population size on many new leaves of large plants which were not spray inoculated with these strains 4 or more weeks earlier.

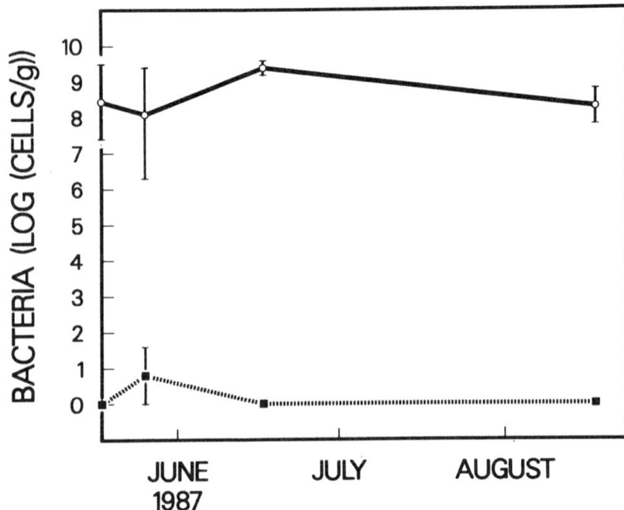

Fig. 6. Total bacterial population size (○) and population size of rifampicin-resistant *P. syringae* strains in soil (■) within an experimental plot of potato at various times after spray treatment of plants with Ice⁻ *P. syringae* strain Cit7del1b on 28 May 1987. The vertical bars represent the standard error of the mean of log population size at each sampling time.

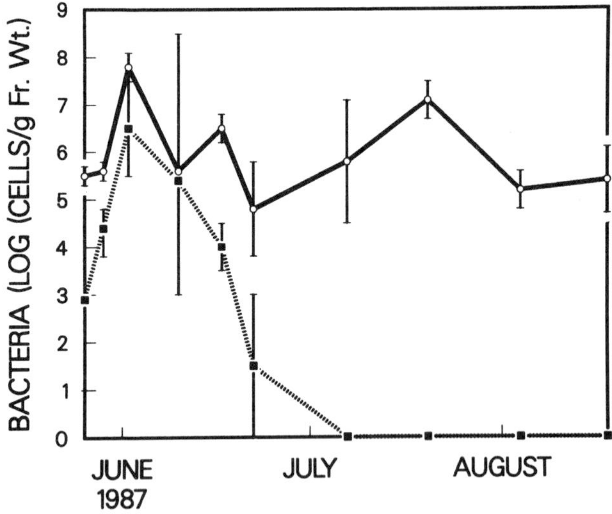

Fig. 7. Total bacterial population size (○) and population size of Ice$^-$ *P. syringae* strain TLP2del1 (■) on plants treated with this strain on 28 May 1987, at various times during the 1987 growing season. The vertical bars represent the standard error of the mean of log population size at each sampling time.

CONCLUSIONS

The dispersal and survival of Ice$^-$ *P. syringae* strains in various habitats during and after experimentation in an uncontained field site was closely predicted by the results of experiments done in contained environments with these and non-recombinant Ice$^+$ parental strains. Studies of the mechanisms and magnitude of competition of Ice$^+$ and Ice$^-$ bacteria were of primary importance in the design of this experimental site. It appears that competition was largely responsible for the lack of survival of these strains after their dispersal to surrounding vegetation. The identification of soil as a poor site of survival of these strains also indicated a convenient method for the reduction of the numbers of these strains in the experimental site without the need for other mitigation practices. Thus, minimal isolation of the plot, with a buffer zone, was apparently effective at facilitating competitive exclusion of Ice$^-$ *P. syringae* strains after their dispersal from the plot by ensuring their small relative population sizes in new habitats. The design of this experiment was aided greatly by the comparative studies of recombinant Ice$^-$ strains and their Ice$^+$ parental types both in contained and uncontained experiments. The results of studies of population dynamics and dispersal of parental strains in similar uncontained experimental settings were very helpful in predicting the eventual fate and effects of Ice$^-$ *P. syringae* strains. Whenever possible such comparative studies should be considered before releases of recombinant micro-organisms. While valuable information can be gained on the survival and effects of recombinant micro-organisms by careful monitoring of experimental sites, care must be taken that only the most appropriate measures are made so that the scientific objectives of such experiments are not superseded by monitoring activities.

ACKNOWLEDGEMENTS

We thank G. Andersen, G. Lim, C. Pierce, J. Watkins, D. Kirby and J. Smalley for valuable technical assistance in the field-monitoring studies reported. Some of the work was supported by Cooperative Agreement CR813419 from the US Environmental Protection Agency.

REFERENCES

Arny, D. C., Lindow, S. E. and Upper, G. D., 1976. Frost sensitivity of *Zea mays* increased by application of *Pseudomonas syringae*. *Nature (London)* **262,** 282–284.

Ashworth, E. N., Anderson, J. A. and Davis, G. A., 1985. Properties of ice nuclei associated with peach trees. *Journal of the American Society for Horticultural Science* **110,** 287–291.

Burke, J. J., Gusta, L. A., Quamme, H. A., Weiser, C. J. and Li, P. H., 1976. Freezing and injury to plants. *Annual Review of Plant Physiology* **27,** 507–528.

Cary, J. W. and Mayland, H. F., 1970. Factors influencing freezing of supercooled water in tender plants. *Agronomy Journal* **62,** 715–729.

Dye, D. W., Bradbury, J. F., Goto, M., Hayward, A. C., Elliot, R. A. and Schroth, M. N., 1980. International standards for naming pathovars of phytopathogenic bacteria and a list of pathovar names and pathotype strains. *Review of Plant Pathology* **59,** 153–168.

Govindarajan, A. G. and Lindow, S. E., 1988. Size of bacterial ice nucleation sites measured *in situ* by gamma radiation inactivation analysis. *Proceedings of the National Academy of Sciences USA*, in press.

Gross, D. C., Cody, Y. S., Proebsting, E. L., Radamaker, G. and Spotts, R. A., 1983. Distribution, population dynamics, and characteristics of ice nucleation active bacteria in deciduous fruit tree orchards. *Applied and Environmental Microbiology* **46,** 1370–1379.

Gross, D. C., Cody, Y. S., Proebsting, E. L., Radamaker, G. and Spotts, R. A., 1984. Ecotypes and pathogenicity of ice nucleation active *Pseudomonas syringae* isolated from deciduous fruit tree orchards. *Phytopathology* **74,** 241–248.

Hirano, S. S. and Upper, C. D., 1983. Ecology and epidemiology of foliar plant pathogens. *Annual Review of Phytopathology* **21,** 243–269.

Hirano, S. S., Maher, E. A., Kelman, A. and Upper, C. D., 1978. Ice nucleation activity of fluorescent plant pathogenic pseudomonads. In *Proceedings of the 4th International Conference on Plant Pathogenic Bacteria, Angers, France, Vol. 2,* Station de Pathologie Vegetale et Phytobacteriologie (ed.), pp. 717–725. Beaucouze, France: Institut National de la Recherche Agronomique.

Hobbs, P. V., 1974. *Ice Physics.* Oxford: Clarendon Press.

King, E. O., Ward, M. K. and Raney, D. E., 1954. Two simple media for the demonstration of pycocyanin and fluorescein. *Journal of Laboratory and Clinical Medicine* **44,** 301–307.

Lim, H. K., Orser, C., Lindow, S. E. and Sands, D. C., 1987. *Xanthomonas campestris* pv. *translucens* strains active in ice nucleation. *Plant Disease* **71,** 994–997.

Lindemann, J., Constantinidou, H. A., Barchet, W. R. and Upper, C. D., 1982. Plants as sources of airborne bacteria, including ice nucleation active bacteria. *Applied and Environmental Microbiology* **44,** 1059–1063.

Lindow, S. E., 1982. Population dynamics of epiphytic ice nucleation active bacteria on frost sensitive plants and frost control by means of antagonistic bacteria. In *Plant Cold Hardiness and Freezing Stress*, Li, P. H. and Sakai, A. (eds), pp. 394–416. New York: Academic Press.

Lindow, S. E., 1983a. Methods of preventing frost injury through control of epiphytic ice nucleation bacteria. *Plant Disease* **67,** 327–333.

Lindow, S. E., 1983b. The role of bacterial ice nucleation in frost injury to plants. *Annual Review of Phytopathology* **21,** 363–384.

Lindow, S. E., 1985a. Integrated control and role of antibiotics in biological control of fireblight and frost injury. In *Biological Control on the Phylloplane*, Windels, C. and Lindow, S. E. (eds), pp. 83–115. Minneapolis: American Phytopathological Society Press.

Lindow, S. E., 1985b. Strategies and practice of biological control of ice nucleation active bacteria on plants. In *Microbiology of the Phyllosphere*, Fokkema, N. (ed), pp. 293–311. London: Cambridge University Press.

Lindow, S. E., 1985c. Ecology of *Pseudomonas syringae* relevant to the field use of Ice$^-$ deletion mutants constructed *in vitro* for plant frost control. In *Engineered Organisms in the Environment*; *Scientific Issues*, Halverson, H. O., Pramer, D. and Rogul, M. (eds), pp. 23–25. Washington DC: American Society of Microbiology.

Lindow, S. E., 1987. Competitive exclusion of epiphytic bacteria by Ice$^-$ mutants of *Pseudomonas syringae. Applied and Environmental Microbiology* **53,** 2520–2527.

Lindow, S. E., 1988a. Lack of correlation of antibiosis in antagonism of ice nucleation active bacteria on leaf surfaces by non-ice nucleation active bacteria. *Phytopathology* **76,** in press.

Lindow, S. E., 1988b. Construction of isogenic Ice$^-$ strains of *Pseudomonas syringae* for evaluation of specificity of competition on leaf surfaces. In *Microbial Ecology* Megusar, F. (ed.), in press.

Lindow, S. E. and Connell, J. H., 1984. Reduction of frost injury to almond by control of ice nucleation active bacteria. *Journal of the American Society for Horticultural Science* **109,** 48–53.

Lindow, S. E., Arny, D. C., Barchet, W. R. and Upper, C. D., 1978a. The role of bacterial ice nuclei in frost injury to sensitive plants. In *Plant Cold Hardiness and Freezing Stress*, Li, P. H. and Sakai, A. (eds), pp. 249–261. New York: Academic Press.

Lindow, S. E., Arny, D. C. and Upper, C. D., 1978b. *Erwinia herbicola*: an active ice nucleus incites frost damage to maize. *Phytopathology* **68,** 523–527.

Lindow, S. E., Arny, D. C. and Upper, C. D., 1978c. Distribution of ice nucleation active bacteria on plants in nature. *Applied and Environmental Microbiology* **36,** 831-838.

Lindow, S. E., Arny, D. C. and Upper, C. D., 1982. Bacterial ice nucleation: a factor in frost injury to plants. *Plant Physiology* **70,** 1084–1089.

Lindow, S. E., Arny, D. C. and Upper, C. D., 1983a. Biological control of frost injury. I. An isolate of *Erwinia herbicola* antagonistic to ice nucleation active bacteria. *Phytopathology* **73,** 1097–1102.

Lindow, S. E., Arny, D. C. and Upper, C. D., 1983b. Biological control of frost injury. II. Establishment and effects of an antagonistic *Erwinia herbicola* isolate on corn in the field. *Phytopathology* **73,** 1102–1106.

Lindow, S. E., Knudsen, G. R., Seidler, R. J., Walter, M. V., Lambou, V. W., Amy, P. S., Schmedding, D., Prince, V. and Hern, S., 1988. Aerial dispersal and epiphytic survival of *Pseudomonas syringae* during a pre-test for the release of genetically engineered strains into the environment. *Applied and Environmental Microbiology*, in press.

Maki, L. R. and Willoughby, J. K., 1978. Bacteria as biogenic sources of freezing nuclei. *Journal of Applied Meteorology* **17,** 1049–1053.

Maki, L. R., Galyon, E. L., Chang-Chien, M. and Caldwell, D. R., 1974. Ice nucleation induced by *Pseudomonas syringae. Applied Microbiology* **28,** 456–459.

Makino, T., 1982. Micropipette method: a new technique for detecting ice nucleation activity of bacteria and its application. *Annals of the Phytopathology Society of Japan* **48,** 452–459.

Orser, C. S., Staskawicz , B. J., Loper, J., Panopoulos, N. J., Dahlbeck, D., Lindow, S. E. and Schroth, M. M., 1983. Cloning of genes involved in bacterial ice nucleation and fluorescent pigment-siderophore production. In *Molecular Genetics of the Bacteria–Plant Interaction*, Puhler, A. (ed.), pp. 353–361. Berlin: Springer-Verlag.

Orser, C., Staskawicz, B. J., Panopoulos, N. J., Dahlbeck, D. and Lindow, S. E., 1985. Cloning and expression of bacterial ice nucleation genes in *Escherichia coli. Journal of Bacteriology* **164,** 359–366.

Paulin, J. P. and Luisetti, J., 1978. Ice nucleation activity among phytopathogenic bacteria. In *Proceedings of the 4th International Conference on Plant Pathogenic Bacteria, Angers, France, Vol. 2*, Station de Pathologie Végétale et Phytobactériologie (ed.), pp. 725–733. Beaucouze, France: Institut National de la Recherche Agronomique.

Ruvkun, G. B. and Ausubel, F. M., 1981. A general method for site-directed mutagenesis in prokaryotes. *Nature (London)* **289,** 85–88.

Schnell, R. C. and Vali, G., 1976. Biogenic ice nuclei, Part I. Terrestrial and marine sources. *Journal of Atmospheric Science* **33,** 1554–1564.

White, G. F. and Haas, J. E., 1975. In *Assessment of Research on Natural Hazards*, pp. 304–312. Cambridge, Mass.: The MIT Press.

Yankofsky, S. A., Levin, Z. and Moshe, A., 1981. Association with citrus of ice nucleating bacteria and their possible role as causative agents of frost damage. *Current Microbiology* **5,** 213–217.

11 Inventory of Natural Rhizobacterial Populations from Different Crop Plants

H. JOOS, B. LAMBERT, F. LEYNS, A. DE ROECK and J. SWINGS

Plant Genetic Systems NV, Plateaustraat 22, 3-9000 Ghent, Belgium

Several micro-organisms have been shown to have a potential as bio-pesticides and/or bio-fertilizers for a number of different crop plants. (Howell and Stipanovic, 1979; Gardner *et al.*, 1984; Kloepper *et al.*, 1988; Nelson, 1988).

The success of these micro-organisms as a commercial product will depend on their ability to compete *in planta* with the indigenous microflora, on their versatility to survive and to express the genes essential for bio-control or plant growth promotion and on the ability to produce these micro-organisms in such a form that they can readily be used on one or several crop plants. One of the major drawbacks for the rational development of microbial products for agronomic applications is the lack of basic knowledge on the molecular processes that drive plant–micro-organism interactions and on the ecological relations in the plant niches where the microbial products must be active. This explains why most of the selection procedures for potential microbial biological agents have essentially been empirical processes based on one plant–pathogen model within the context of a limited set of controlled conditions.

The goal of the presented study was two-fold: (a) to obtain qualitative and quantitative data on dominant bacterial populations associated with roots from different crop plant and (b) to build up a rationale for the selection of potential bacterial pesticides and/or fertilizers.

The methodology used in this study can be summarized in four essential steps and is explained in more detail by Lambert *et al.* (1987):

(1) Isolation of dominant rhizobacteria from various crop plants, developmental stages of the plant and field situations.

(2) Characterization of single bacterial isolates by electrophoresis of total cellular proteins. This results in protein fingerprints (FP) characteristic for

RELEASE OF GENETICALLY-ENGINEERED MICRO-ORGANISMS ISBN 0-12-677521-4

subspecies of different bacterial taxons, a method that has previously been applied for taxonomic purposes by Kersters and DeLey (1980). The method allows rapid characterization of thousands of strains.

(3) Classification of the bacterial isolates in fingerprint types (FP-type = class of identical fingerprints).

(4) Qualitative and quantitative comparison of natural rhizobacterial populations from different crop plants, fields and developmental stages within one crop, based on the FP-type classification.

With this methodology, single bacteria isolates originating from different crops have been analysed (Table 1). The qualitative comparison of the populations between different crops revealed very little overlap, indicating that dominant rhizobacterial populations are crop-specific.

Table 1. The number of strains characterized via fingerprint typing and the original plant crops.

Crop	No. of Plants	No. of Isolates
Sugarbeet	1550	6780
Corn	503	1508
Soybean	450	1139
Sunflower	450	1119
Barley	36	175
Witloof chicory	22	662
Total	3011	11 383

In Table 2, the bacterial species are listed which are represented in the most frequently occurring fingerprint types (FP-types that harbour the highest number of single isolates). One species, which has not previously been described in the rhizosphere, *Xanthomonas maltophilia*, is abundantly present.

Quantitative comparisons of populations originating from different fields, growth stages and plants within the sugarbeet crop resulted in the following conclusions:

(1) The rhizobacterial flora is very dynamic during the growth season. Young seedlings are characterized by a relatively high density colonization with a limited number of strains. Later in the season, these strains are replaced by a new, more diverse population of bacteria which are present in relatively lower densities.

(2) The rhizobacterial flora differs both qualitatively and quantitatively in different soil types.

(3) A minority of fingerprint types is represented very frequently in the

Table 2. Bacterial species represented in the most frequently occurring fingerprint types.

Pseudomonas fluorescens
Xanthomonas maltophilia
Achromobacter sp.
Pseudomonas paucimobilis
Alcaligenes paradoxus

rhizosphere and is found at different locations; the majority of fingerprint types occurs only occasionally at a limited number of locations.

In parallel with the fingerprint typing, the majority of the bacterial isolates has been tested for antifungal activity towards a number of phytopathogenic fungi which are agronomically relevant for the analysed crop. In addition, indirect screening for antifungal strains has been selected directly in fungal spore plates. This analysis showed that 46% of the plants harbour antifungal rhizobacteria. Table 3 lists the antifungal bacterial species that have been isolated from the rhizosphere.

The information recovered from this study will in the future be used to design more rational selection procedures for plant growth promoting and/or pathogen controlling rhizobacteria. This study clearly illustrates that natural or genetically-engineered rhizobacteria applied as root inoculant are confronted with an extremely dynamic and complex ecosystem. This is most probably one of the explanations for the variable units obtained in a number of trials with bacterial strains as root inoculants.

The methodology applied for this study is, in principle, applicable to any ecosystem that harbours a dynamic and complex microflora. One laboratory is currently using these methods to study bacterial populations in silage fermentations.

Table 3. Antifungal rhizobacteria recovered from the rhizosphere of different crop plants.

Pseudomonas fluorescens
Pseudomonas paucimobilis
Pseudomonas cepacia
Pseudomonas aureofaciens
Xanthomonas maltophilia
Achromobacter sp.
Serratia liquefaciens
Serratia plymuthica
Erwinia herbicola
Bacillus subtilis
Other *Bacillus* sp.
Actinomycetes

REFERENCES

Gardner, J. M., Chandler, J. L. and Feldman, A. W., 1984. Growth promotion and inhibition by antibiotic-producing fluorescent pseudomonads on citrus roots. *Plant Soil* **77,** 103–113.

Howell, C. R. and Stipanovic, R. D., 1979. Control of *Rhizoctonia solari* on cotton seedlings with *Pseudomonas fluorescens* and with an antibiotic produced by the bacterium. *Phytopathology* **69,** 480–482.

Kersters, K. and De Ley, J., 1980. Classification and identification of bacteria by electrophoresis of their proteins. In *Microbiological Classification and Identification*, Goodfellow, M. and Board, R. G. (eds), pp. 273–293. Society for Applied Bacteriology Symposium Series, No. 8. London: Academic Press.

Kloepper, J. W., Flune, D. J., Scher, F. M., Singleton, C., Tipping, B., Laliberte, M., Frauley, K., Kutchaw, T., Simonson, C., Litshitz, Zalaska, I. and Lec, L., 1988. Plant growth promoting rhizobacteria on Canola (Rapeseed). *Plant Disease* **72,** 42–46.

Lambert, B., Leyns, F., Van Rooyen, L., Gossele, F., Papon, Y. and Swings, F., 1987. Rhizobacteria of maize and their antifungal activities. *Applied and Environmental Microbiology* **53,** 1866–1871.

Nelson, E. B., 1988. Biological control of Pythium seed rot and preemergence damping-off of cotton with *Enterobacter cloacae* and *Erwinia herbicola* applied on feed treatments. *Plant Disease* **72,** 140–142.

12 Field Trials of Genetically-engineered Baculovirus Insecticides

D. H. L. BISHOP, P. F. ENTWISTLE, I. R. CAMERON, C. J. ALLEN and R. D. POSSEE

NERC Institute of Virology, Mansfield Road, Oxford OX1 3SR, UK

INTRODUCTION AND OBJECTIVES

Since the last century, naturally-occurring baculoviruses have been used to control insect pests. They have found particular favour because, unlike many chemical insecticides, they affect only a few species of insect (the so-called 'permissive hosts'). These viruses have no effect on other types of insect, or other invertebrates, or plants, or vertebrates; nor do they pollute the environment (groundwater, streams, lakes, etc.). Members of the Baculoviridae family are unique in that they only infect arthropods. No member of this family of viruses infects any other form of life.

In 1973, a joint WHO/FAO meeting on insect viruses (WHO/FAO, 1973) endorsed the potential use of baculoviruses as field control agents. The report noted that, in addition to their specific host ranges, baculoviruses exhibited good storage properties, were safe to handle, relatively easy to produce and widely distributed in nature among insects. In fact, several hundred baculoviruses have now been isolated and described, and more than a dozen have been employed commercially to control insect pests. Reports of some of the baculoviruses which have been used for control purposes are available (Podgewaite, 1985; Entwistle and Evans, 1985). Their environmental safety is a matter of record and a major factor when considering their further development by genetic engineering. This record of safety has been confirmed by experience gained over many decades involving the use of naturally-occurring baculovirus insecticides in agriculture and forestry. Studies of normal epizootics of baculovirus disease have also shown that baculoviruses are safe and limited in their effects to certain insect species.

RELEASE OF GENETICALLY-ENGINEERED MICRO-ORGANISMS ISBN 0–12–677521–4

The objective of the genetic engineering of baculoviruses is to improve their speed of action. This is desirable since, during the normal infection process of a permissive host, a baculovirus undergoes several cycles of replication. These cycles take time—up to several days, or weeks, depending on the virus, the host and the environmental conditions (such as temperature). By contrast, most chemical insecticides act quickly, killing the target insect and often other, beneficial insects in a matter of hours. By means of genetic engineering procedures it will be possible to minimize the time taken by a viral insecticide to exert an affect on a target species by incorporating into the viral genome genes for other, quickly acting, products (e.g. bacterial, or scorpion insect-specific toxins, or insect hormone genes, etc.). Since the use in the environment of a genetically-engineered organism or virus represents a new era of scientific endeavour, it is important to develop the subject systematically and cautiously, taking into consideration the consequences, and particularly the risks, of any prospective introduction. It would be irresponsible to do otherwise.

The programme of research and development of genetically-engineered baculovirus insecticides that is underway at the Natural Environment Research Council's (NERC) Institute of Virology in Oxford, UK, includes risk assessment and environmental impact analyses. The objectives are to assess the consequences of a deliberate release into the environment of a genetically-engineered viral insecticide, specifically a baculovirus insecticide. Since 1986, when field trials were initiated, two forms of a genetically-altered baculovirus insecticide have been developed and released in a field facility. The facility that was used is in open arable land at the Oxford University Field Station at Wytham, Oxfordshire. The first release, in 1986, involved a genetically-marked nuclear polyhedrosis virus of *Autographa californica* (AcNPV) (Bishop, 1986), the second, undertaken in 1987, involved a genetically-marked *and* crippled AcNPV from which the protective polyhedrin gene had been removed. This virus has been called a 'self-destructive' virus because it was rapidly degraded in the environment (*vide infra*). Two more releases are planned for 1988, one also involving the crippled virus, the other, a crippled virus containing a 'junk' gene, *viz*: the bacterial β-galactosidase gene. The latter virus is to be used to determine the levels of expression of a foreign gene (β-galactosidase) by the virus in infected moth caterpillars, a permissive host for this virus. This paper will describe only the field trials that were undertaken in 1986 and 1987.

An additional point should be made concerning the suitability of using genetically-engineered baculoviruses as a model for risk assessment analyses. Baculoviruses, like all viruses, are obligate parasites. On their own, they are inert in the environment, unable to replicate. Baculoviruses have to be ingested by a *permissive* host and gain entry to a cell before they can replicate. In this regard baculoviruses differ from most bacteria, or other free-living micro-organisms. The inability of viruses to replicate *per se*, therefore, restricts the opportunity for virus to escape and multiply *ad libitum*. Also, the field trials that we have conducted were undertaken at a time of year, the autumn, when the natural permissive insect host was not present in the environment; infected caterpillars were supplied from the laboratory. In addition, the field facility was designed to restrict egress of infected hosts and limit entry to the site of other insects, rodents, moles, or larger animals. By such physical and temporal restraints any risk to the environment due to the introduction of a genetically-engineered virus, or its spread, was minimized. Finally, after each study the field site was disinfected and demonstrated to be free from virus before new insect species emerged in the following spring.

BACKGROUND INFORMATION

Baculoviruses (Matthews, 1982) have genomes consisting of circular, covalently closed, double-stranded DNA of between 100 and 150 kb pairs. The viral DNA molecule is packaged into a rod-shaped nucleocapsid which is surrounded by a lipoprotein envelope. In some viruses the enveloped particles are occluded by protein to form a polyhedron, or granule. Based on morphological considerations baculoviruses have been divided into three subgroups.

Subgroup A comprise the nuclear polyhedrosis viruses (NPVs). In the NPVs, groups of enveloped viruses are embedded in a proteinaceous polyhedral inclusion body that consists almost entirely of a single type of protein, called the polyhedrin protein which has a size of 29 kDa. Such occluded virus particles may contain singly or multiply enveloped nucleocapsids. Each form is genetically determined. The multiple-enveloped NPVs only infect members of the Lepidoptera, the single-enveloped NPVs infect either members of the Lepidoptera, or Hymenoptera.

Subgroup B baculoviruses comprise the granulosis viruses, with a host-range confined to lepidopteran species. The occlusion bodies, or granules, of these viruses usually package a single virus particle containing one nucleocapsid. The major occlusion body protein, while of similar size to the polyhedrin of the NPVs, is called the granulin protein.

Subgroup C baculoviruses consist of a single-enveloped nucleocapsid that is in a non-occluded form.

Over the last 15 years, staff of the Institute of Virology in Oxford have been developing four baculoviruses for commercial use: baculoviruses of pine sawfly (Hymenoptera, *Neodiprion sertifer*, NsNPV) (Entwistle *et al.*, 1985), pine beauty moth (Lepidoptera, *Panolis flammea*, PfNPV) (Entwistle and Evans, 1987; Doyle and Entwistle, 1988), the rhinoceros beetle (Coleoptera, *Oryctes rhinoceros*) (Lomer, 1986) and the brown-tail moth *Euproctis chrysorrhoea* (Kelly *et al.*, 1988). In addition, field trials have been undertaken with the baculoviruses of *Autographa californica* (AcNPV), against a complex of lepidopterous pests of brassicas in the West Indies (Small, 1984; personal communication). Other publications have reported the use of these naturally-occurring, and other, baculoviruses for pest control, so that this subject will not be considered further here (Hunter *et al.*, 1984; Entwistle and Evans, 1985).

The principal target of a baculovirus, such as an NPV, is the larval (caterpillar) stage of the host. Viruses are ingested together with the food. The polyhedrin protein is removed when the virus reaches the alkaline environment of the midgut of the permissive host. A schematic infection cycle of an occluded baculovirus is shown in Fig. 1. Infectious virus particles infect the columnar cells in the epithelium of the caterpillar midgut. Virus DNA enters the cell nucleus and replication ensues. In Lepidoptera (e.g. for AcNPV, PfNPV) non-occluded virus particles are released from these cells to spread the infection via the haemocoel to cells in other tissues of the larva. Late in infection viral inclusion bodies (polyhedra) are formed. An electron micrograph of an AcNPV infection of *Trichoplusia ni* tissues is shown in Fig. 2. Both non-occluded and occluded viruses are visible in the cell nucleus that has been sectioned. The virulence and extent of the infection induces death. Up to 10^9 progeny polyhedra may be present in the corpse of an infected insect. In Hymenoptera (e.g. for NsNPV), the replication is confined to the insect midgut cells with the formation of mature polyhedra late in infection. There is severe damage to the gut wall, resulting in death of the insect.

Infected insects commonly exhibit behavioural abnormalities which may be of benefit to the subsequent spread of virus to other larvae. For example, infected larvae may ascend to the tips of plants before death, thereby facilitating distribution of virus over the foliage below. Virus is released from decaying larval corpses and can be spread by physical forces, such as rain splash. Viruses may also be distributed passively by animals that eat the remains of the insect and subsequently defaecate the virus in other sites. Parasitic Hymenoptera may also have a role in the spread of baculovirus disease (Entwistle, 1982; Kaya, 1982). Virus ingestion by birds, rodents and beneficial insects such as carabids does not lead to infection of those species, instead the baculovirus passes directly through their intestines and is deposited in their faeces.

The impact of baculoviruses on insect populations is greatly affected by their capacity to retain infectivity over long periods of time, and to occur in situations where they can infect a new host. Persistence of the virus in the environment is necessary in order for the virus to infect subsequent generations of insect, generations that may not occur until the next, or subsequent, years. The prolonged presence of infectious baculoviruses on plant surfaces between host generations is probably a result of the release of virus from slowly decaying larval corpses and the persistence of viral occlusion bodies on plants and in soil (Carruthers *et al.*, 1988). Once released on to a leaf or another surface,

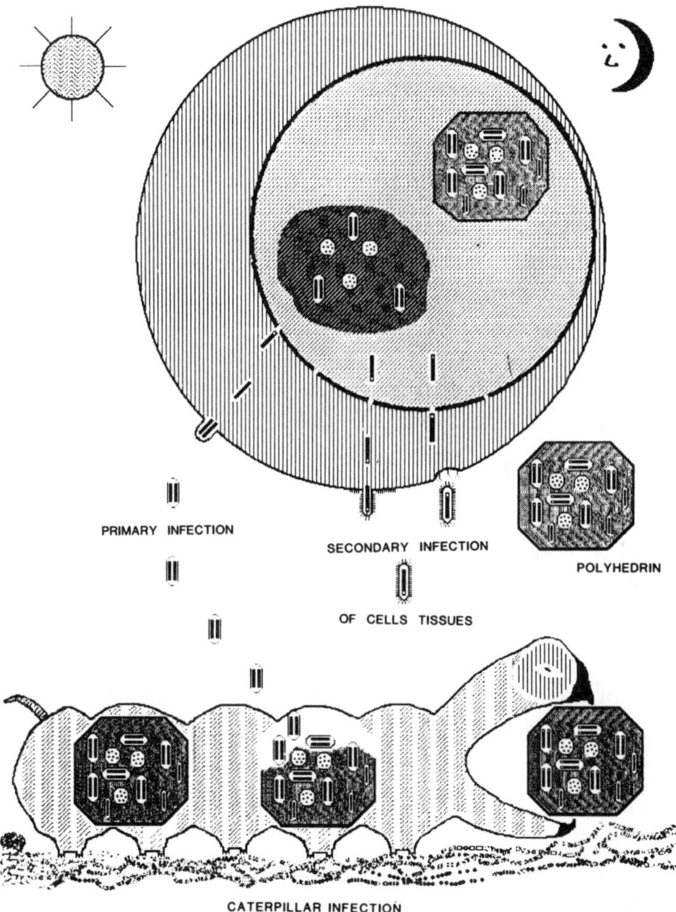

Fig. 1. Schematic infection of a lepidopteran larva by a naturally-occurring, occluded, baculovirus such as AcNPV. Polyhedra (PIBS) are ingested with foliage (or other diet) and dissolve in the alkaline midgut releasing infectious virus particles. The viruses infect columnar cells in the midgut epithelium producing progeny that distribute the infection via the haemocoel to other cells and tissues. At the end of the infection large numbers of polyhedra are produced that are distributed from the larval corpse in the environment.

baculoviruses are subject to slow physical loss and degradation as well as to ultraviolet light-induced inactivation (Killick (1987), particularly by light of wavelength 290–315 nm). The polyhedrin protein of NPVs provides a degree of protection against inactivation by ultraviolet light and probably also facilitates the retention of virus on particular plant and other surfaces. Baculoviruses may persist for several years in the soil. Such viruses can be acquired by seedlings

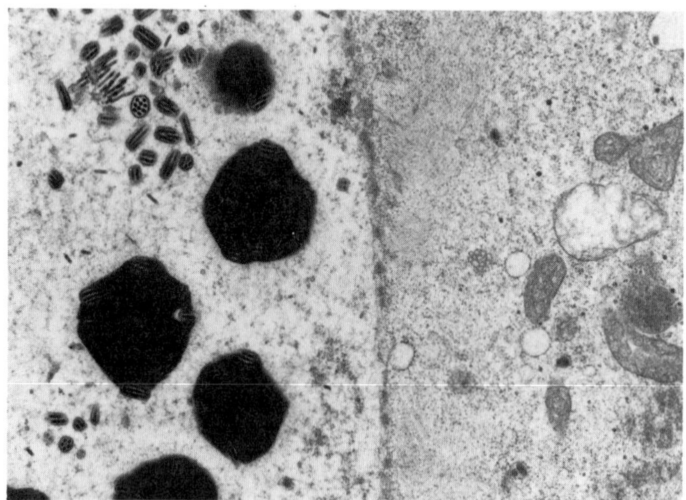

Fig. 2. Electron micrograph of an AcNPV infection in nervous tissue of *T. ni*. Naked bacilliform nucleocapsids, enveloped viruses and polyhedra containing occluded viruses are visible within the nucleus of a cell.

growing out of the soil, thereby bringing virus into contact with the target host. The subject of insect virus persistence in the environment has been reviewed (Evans and Harrap, 1982).

RISK ASSESSMENT AND ENVIRONMENTAL IMPACT ANALYSIS

As mentioned previously, we consider that it would be irresponsible to construct and release into the environment a genetically-engineered baculovirus insecticide that expresses a new, foreign, gene without a thorough understanding of the possible consequences. In this regard, attention has to be paid to the host range of the modified virus and whether previously non-susceptible insects become liable to infection. This is of particular concern where there is a rich insect fauna at a site chosen for the release. Another consideration is the possible spread of the virus from the site of application. As noted above, spread may be mediated by environmental conditions such as rain, or by the movement of the host species, or by other animals including birds, rodents, predatory insects, etc. Such spread has been documented by the study of natural epizootics of viral infection (Entwistle *et al.*, 1983) as well as epizootics induced when viruses are applied to caterpillar infestations in the wild. However, in such investigations it was never entirely possible rigorously to determine whether a virus recovered from the environment originated from the applied virus, or whether it represented a naturally-occurring isolate of the same virus. Latent viruses, present in insect populations, complicate this issue, since they may be 'stressed out' of an insect host after it is challenged with a different virus (Longworth and Cunningham, 1968). It would be valuable, therefore, if an introduced virus, particularly a genetically-engineered one, could be identified against such backgrounds. Another consideration of potential risk consequent to the release into the environment of a genetically-engineered virus, is the question of the possible exchange of genetic information between viruses, or to another organism. Although genetic exchange is frequent and well-documented among bacteria (involving plasmids, conjugation, etc.), such exchange is probably a low frequency event for baculoviruses, except for closely related baculoviruses in ecologically intimate circumstances. The exchange of a foreign gene from a virus into a dead cell would be an event without consequence. Gene exchange involving the acquisition by a baculovirus of genetic information from the host species may, however, occur, albeit probably infrequently. It could be sought by genome analyses of virus passaged in permissive cells, or in permissive host species. By such analyses it should be possible to determine whether an engineered virus exhibited any detectable change in its genome or any enhanced acquisition of foreign genetic information by comparison to that which occurred naturally in a non-engineered virus. As mentioned, a more likely gene exchange would be between genetically compatible baculoviruses, for example when two closely-related baculoviruses co-infect the same host. Marker rescue studies indicate that rates can be as high as 1–3% for sibling viruses (unpublished data). The likelihood of gene exchange occurring between closely-related viruses will depend on whether any such virus already exists in the target host, or whether another virus coinfects the same target host at the same time. Evidence has been obtained which indicates that gene transfer from a distantly related baculovirus only occurs at very low frequencies, if at all. Transfer of PfNPV polyhedrin gene sequences to a polyhedrin-negative AcNPV was not detected with plasmids containing PfNPV sequences (unpublished data, i.e. frequencies of less than 10^7–10^8). A fourth consideration is the question of induced mutation and the genetic stability of the engineered virus. In question is whether the introduced gene induces genetic instability. Again, this could be tested by genome analysis of passaged virus.

These, and other questions, are important in considering whether measurable risks are associated with the introduction into the environment of a genetically-engineered organism. Many of these questions can be analysed in the laboratory before undertaking a release. The task, however, is not trivial, and often involves several years of investigation. Eventually when the data are obtained and evaluated, including independent review, field trials are required to confirm the results and establish the validity of the predictions.

Each risk-assessment analysis we have performed with a genetically-engineered baculovirus has involved the following: (i) the construction of the candidate virus insecticide, (ii) verification at the nucleotide (sequence) level of the precise genetic change, (iii) laboratory analyses of the phenotype and genetic stability (mutability) of the virus, (iv) host range determinations and (v) analyses of the physical stability of the virus in simulated systems (plant surfaces and soil). The data have been submitted to all the relevant regulatory authorities for independent evaluation especially in relation to analysis of environmental impact, *viz*: the characteristics of the site proposed for the release, the insect fauna of the area, the possibility for spread beyond the immediate ecosystem, the proposed physical constraints imposed at the release site, etc.

Permission to undertake the releases described below were given in the form of licences to conduct the studies under certain prescribed conditions, including site arrangements, procedures to be followed for the introduction of the engineered virus, disinfection of the site after use, etc. Since baculoviruses are insecticides, the licences were issued by the (UK) Ministry of Agriculture, Fisheries and Food (MAFF) under the Food and Environment Protection Act 1985 and Regulation of the Control of Pesticides Regulations 1986. Primary evaluation of the proposals was made by the Health and Safety Executive (HSE) Advisory Committee on Genetic Manipulation (ACGM), in consultation with MAFF, the Department of the Environment (DOE) and the Nature Conservancy Council (NCC). In addition, other interested parties were informed for the 1986, the 1987, as well as for the prospective 1988 studies. These included the Natural Environment Research Council's senior management, the owners of the field site (Oxford University), senior University authorities, the University Safety Officer, the Oxford HSE factory inspector, the Vale of the White Horse Environmental Health Officer, the Environmental Services Committee of the Vale of the White Horse District Council and agencies such as 'Friends of the Earth'. Press coverage (national and local papers, including, for the 1988 releases, a 'Public Notice' placed in local papers and displayed in the local Post Office), as well as radio and television interviews also aided in informing the general public of the proposed studies.

PREPARATION AND ANALYSIS OF A GENETICALLY-MARKED AcNPV

The first release, undertaken in 1986, involved a genetically marked AcNPV that was otherwise in every way identical to the parent AcNPV (Bishop, 1986). The virus was designed to be distinguishable from all other baculoviruses as well as from genome sequences of the target, or another, host. The virus contained a synthetic oligonucleotide, 80 base-pairs in length, inserted into its genome. The oligonucleotide was placed *downstream* of the AcNPV polyhedrin coding region. Details of how the marked virus was prepared will be given elsewhere.

Hind III digest of viral DNA

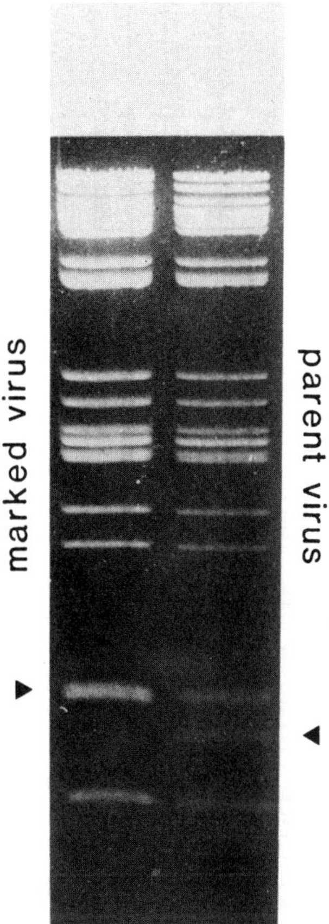

Fig. 3. Restriction endonuclease analysis of marked (left panel) and non-engineered (right panel) AcNPVs. DNA was extracted from purified virus preparations, digested with *Hind*III and analysed in a 0.6% agarose gel. The normal position of the AcNPV *Hind*III V fragment is indicated by the arrow (right panel), its position in the marked viral DNA [due to the inserted sequence (comigrating with the *Hind*III T fragment)] is also indicated (left panel).

The polyhedrin gene of AcNPV has been mapped on the virus genome in previous studies (Adang and Miller, 1982; Rohel *et al.*, 1983; Smith *et al.*, 1983a; Possee, 1986) and subsequently sequenced (Hooft van Iddekinge *et al.*, 1983). The identity of the 5′ and 3′ ends of the polyhedrin mRNA species have been determined (Smith *et al.*, 1983b; Howard *et al.*, 1986), together with analysis of the promoter sequences required for high level expression of the polyhedrin gene,

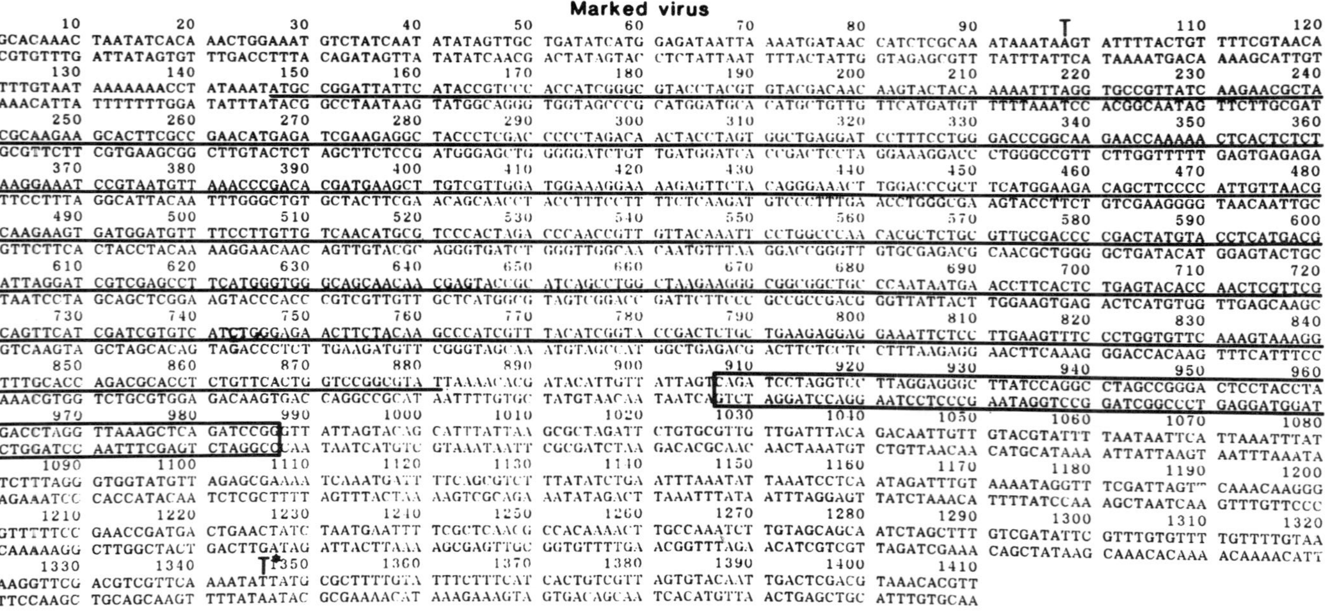

Fig. 4. Nucleotide sequence of a portion of the engineered AcNPV that includes the marker oligonucleotide sequence (boxed) downstream of the polyhedrin coding region (underlined). The positions of the polyhedrin gene mRNA transcription initiation (T) and termination (T*) sites are indicated.

or a foreign gene inserted in its place (see Matsuura *et al.*, 1987; Possee and Howard, 1987; Luckow and Summers, 1988). It has been established that the polyhedrin gene and some of its flanking sequences can be deleted totally from the virus (Smith *et al.*, 1983a; Matsuura *et al.*, 1987). Derivative viruses, although infectious for permissive hosts (see below), are non-occluded.

The synthetic oligonucleotide was introduced into the AcNPV genome so that it was located just after the end of the polyhedrin gene coding sequence, but within the immediate 3′ non-translated sequences. It served only as a marker, or flag. The position of the marker was verified by restriction enzyme (Fig. 3) and by sequence analyses (Fig. 4). The marker contained translation stop codons in all six reading frames but no ATG codon and no other genetic signal likely to affect the replication, or gene product expression, of the virus. The sequence of the marker, relative to that of the coding region of the polyhedrin gene, the transcription initiation and transcription termination sites of the polyhedrin gene are shown in Fig. 4. Because of the precise location of the insertion (*viz*: a dispensable, intergenic, non-regulatory position in the AcNPV genome), the oligonucleotide neither added to, nor detracted from the constitution, or synthesis, of any of the natural AcNPV gene products. This was verified by analysis of the viral phenotype in terms of infectivity and proteins synthesized by comparison with that of the unmodified virus.

Host-range analysis

The effect of the oligonucleotide addition on the host range of the virus was assessed by focussing particularly on UK Lepidoptera. The species chosen were selected in consultation with the Nature Conservancy Council. Shown in Table 1 are the species of insect that were analysed for sensitivity to the marked virus and to the plaque-cloned parent virus used to derive the marked virus. No difference in host range was found between the marked and the unmarked viruses. Not shown in the list are the results of attempts to infect other insects, such as ants, or honey bees (Hymenoptera), or lacewings (Neuroptera), or hoverflies (Diptera), or various beetles and ladybirds (Coleoptera). As expected, none of these insects was found to be susceptible to infection by the marked, or the unmarked AcNPV. The list of non-permissive species shown in Table 1 included all 17 butterfly species that were analysed, representing five families of butterflies, several microlepidoptera and Hymenoptera and some 58 moth species, representing 11 families. The tests involved collecting the adult insect species, allowing them to lay eggs or collecting eggs from native plants, surface sterilizing eggs with formalin (Hunter *et al.*, 1984), and permitting the larvae to hatch. Larvae were then infected by allowing them to ingest known numbers of polyhedral inclusion bodies (PIBS) together with their preferred diet. Control

Table 1. Host range of occluded AcNPV*.

Non-permissive species

	Butterflies	
Hesperiidae	*Thymelicus flavus*	Small skipper
Lycaenidae	*Lycaena phlaeas*	Small copper
	Polyommatus icarus	Common blue
Nymphalidae	*Aglais urticae*	Small tortoiseshell
	Argynnis paphia	Silver-washed fritillary
	Clossiana selene	Small pearl bordered fritillary
	Cynthia cardui	Painted lady
	Inachis io	Peacock
	Polygonia c-album	Comma
	Vanessa atalanta	Red admiral
Pieridae	*Pieris brassicae*	Large white
	Pieris napi	Green-veined white
	Pieris rapae	Small white
Satyridae	*Maniola jurtina*	Meadow brown
	Melanargia galathea	Marbled white
	Pararge aegeria	Speckled wood
	Pyronia tithonus	Gatekeeper
	Microlepidoptera	
	Cydia pomonella	
	Ephemerea chaerophyllella	
	Yponomeuta cagnegella	
	Hymenoptera	
	Nematus ribesii	
	Neodiprion sertifer	
	Gilpinia hercyniae	
	Moths	
Arctiidae	*Arctia caja*	Garden tiger
	Spilosoma lutea	Buff ermine
	Tyria jacobaeae	Cinnabar moth
Drepanidae	*Cilix glaucata*	Chinese character
Geometridae	*Biston betularia*	Peppered moth
	Campaea margaritata	Light emerald
	Chloroclysta truncata	Common marbled carpet
	Cyclophora albipunctata	Birch mocha
	Cyclophora punctaria	Maiden's blush
	Ecliptopera silaceata	Small phoenix
	Hemistola chrysoprasaria	Small emerald
	Idaea aversata	Riband wave
	Odontopera bidentata	Scalloped hazel

	Operophtera brumata	Winter moth
	Opisthograptis luteolata	Brimstone moth
	Ourapteryx sambucaria	Swallowtail moth
	Selenia dentaria	Early thorn
	Timandra griseata	Blood-vein
	Xanthorhoe montanata	Silver-ground carpet
	Xanthorhoe fluctuata	Garden carpet
	Xanthorhoe spadicearia	Red twin-spot carpet
Lasiocampidae	*Malocosoma neustria*	The lackey
	Philudoria potatoria	The drinker
Lymantriidae	*Dasychira pudibunda*	Pale tussock
	Dasychira sp. (Tenerife)	
	Euproctis chrysorrhoea	Brown tail
Noctuidae	*Acronicta aceris*	The sycamore
	Agrotis segetum	Turnip moth
	Anaplectoides prasina	Green arches
	Autographa pulchrina	Beautiful golden Y
	Autographa gamma	Silver Y
	Craniophora ligustri	The coronet
	Diachrysia chrysitis	Burnished brass
	Diarsia mendica	Ingrailed clay
	Diataraxia oleracea	Bright line brown eye
	Heliothis zea	Corn ear worm
	Mamestra brassicae	Cabbage moth
	Melanchra persicariae	Dot moth
	Mythimna separata	Rice armyworm
	Noctua fimbriata	Broad-bordered yellow underwing
	Panolis flammea	Pine beauty
	Peridroma saucia	Pearly underwing
	Phlogophora meticulosa	Angle shades
	Polia nebulosa	Grey arches
	Spodoptera littoralis	Mediterranean brocade
	Xestia baja	Dotted clay
	Xestia c-nigrum	Setaceous hebrew character
Notodontidae	*Cerura vinula*	Puss moth
	Chaonia ruficornis	Lunar marbled brown
	Phalera bucephala	Buff tip
	Harpyia furcula	Sallow kitten
Saturniidae	*Pavonia pavonia*	Emperor moth
Sphingidae	*Hyloicus pinastri*	Pine hawkmoth
	Manduca sexta	Tobacco hornworm
	Laothoe populi	Poplar hawkmoth
	Smerinthus ocellata	Eyed hawkmoth
Yponomeutidae	*Plutella xylostella*	Diamond backed moth

Table 1. *Continued.*

Zygaenidae	*Zygaena filipendulae*	Sixspot burnet
Permissive UK species		
Sphingidae	*Mimas tiliae*	Lime hawkmoth
	Sphinx ligustri	Privet hawkmoth
Noctuidae	*Spodoptera exigua*	Small mottled willow
Permissive non-UK species		
Noctuidae	*Heliothis armigera*	Cotton bollworm
	Spodoptera frugiperda	Fall army worm
	Trichoplusia ni	Cabbage looper

*Species tested *per os* (doses 10^2–10^6 PIBS/individual) and infection monitored by appropriate DNA hybridization. Total tested: 86 species.

caterpillars, not infected with virus, were treated similarly. For convenience of handling, usually second, or early third instar larvae were employed, depending on availability, in batches of 10–30 per dose of virus (10^2, 10^3, 10^4, 10^5, 10^6 PIBS per individual). When limited numbers of larvae were available, only the higher doses of virus (10^4–10^6) were used. The caterpillars were fed and kept until pupation or death and analysed for virus infection. If sufficient numbers of larvae perished as the result of virus infection, the data were used to calculate the lethal dose that caused death in 50% of the insects (LD_{50}) using probit analysis (Finney, 1971). This was possible only in the tests performed with *T. ni* and *Spodoptera exigua*. On occasion, pupae were allowed to develop into adults of the species, although other than observing that adults had a normal gross morphology and, after mating, were capable of laying fertile eggs, no systematic analysis of the effects of viruses on subsequent insect generations was undertaken.

Caterpillars of three moth species that are not native to the UK were found to be permissive for virus infection. Similar results were obtained with these species and the marked and unmarked viruses. All three moths are members of the Noctuidae (Table 1). The most sensitive species was *T. ni*, cabbage looper moth (LD_{50} : 10^2 PIBS/individual). The other two species were only sensitive at doses of 10^5–10^6 PIBS/individual, although insufficient larvae died to allow a probit analysis. One UK member of that family, *Spodoptera exigua*, small mottled willow moth, was quite sensitive to the viruses (LD_{50} : 10^5 PIBS/individual). Two UK members of the Sphingidae, *Mimas tiliae*, lime hawkmoth, and *Sphinx ligustri*, privet hawkmoth, were also permissive, although only when infected with high doses of the viruses (10^5–10^6 PIBS/individual). Again, insufficient larvae died to allow a probit analysis. In each case, permissive and non-permissive species, individual caterpillars or pupae were extracted and analysed for PIBS (Wigley, 1980) and the presence of virus verified by dot-blot analysis with a polyhedrin gene probe *and* an oligonucleotide marker probe (Possee and Kelly, 1988).

In summary, no difference in host range was detected for the marked virus by comparison with the plaque cloned parent virus used to prepare it.

Genetic and physical stability of the marked virus

In order to investigate the stability of the marked virus by comparison to the unmarked parent, virus was passaged for some 50 cycles of replication in *S. frugiperda* cells by standard plaque assay techniques (Brown and Faulkner, 1977). Virus was also passaged in *T. ni* larvae. With neither procedure was a change in genotype, or phenotype, detected by analysis with a variety of restriction enzymes, or by protein analysis. The marker oligonucleotide also remained stable, as determined by sequence analysis. No difference in genetic stability was detected in these studies between the marked and unmarked viruses. No new gene sequences, such as host DNA, were identified in the passaged virus.

In order to assess virus persistence, soil was collected from the proposed field site and steam sterilized. Polyhedra from the marked, or unmarked, virus stocks were mixed with soil and left in the laboratory at room temperature (*ca.* 18°C) for 2 weeks. Samples were taken throughout this period, virus recovered by differential centrifugation and tested for residual infectivity using *T. ni* insect larvae as biological indicators. There was no significant decline in virus infectivity for either the marked, or unmarked, virus over the entire period of the experiment.

Field studies

It was decided to use caterpillars of the small mottled willow moth (*S. exigua*) as the target pest. This is a noctuid moth with practically a world-wide distribution; it is a seasonal immigrant to the UK. It eats a range of herbaceous dicotyledons, including sugar beet and has a wide distribution in communities of low growing plants, both wild and cultivated. Larvae of this moth pass through five instars.

The site chosen and used for the release of the marked virus and, in 1987, the self-destructive virus, was situated in a field of light loam bounded by agricultural land at the Oxford University field station at Wytham, Oxfordshire. The area has been extensively studied with regard to fauna and flora for many decades. Moth species native to the region were collected at night with two ultraviolet light traps. The moths were trapped throughout the late spring,

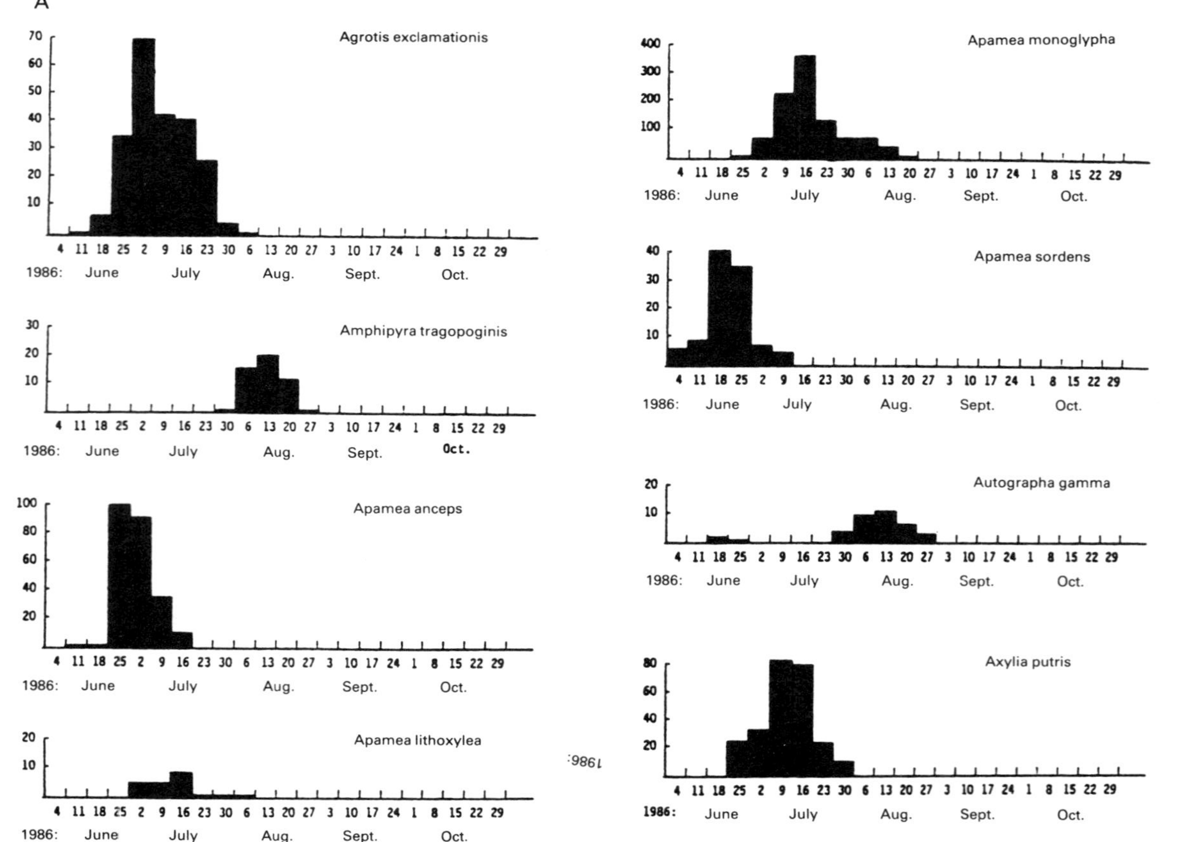
A
Agrotis exclamationis
Amphipyra tragopoginis
Apamea anceps
Apamea lithoxylea
Apamea monoglypha
Apamea sordens
Autographa gamma
Axylia putris
4 11 18 25 2 9 16 23 30 6 13 20 27 3 10 17 24 1 8 15 22 29
1986: June July Aug. Sept. Oct.

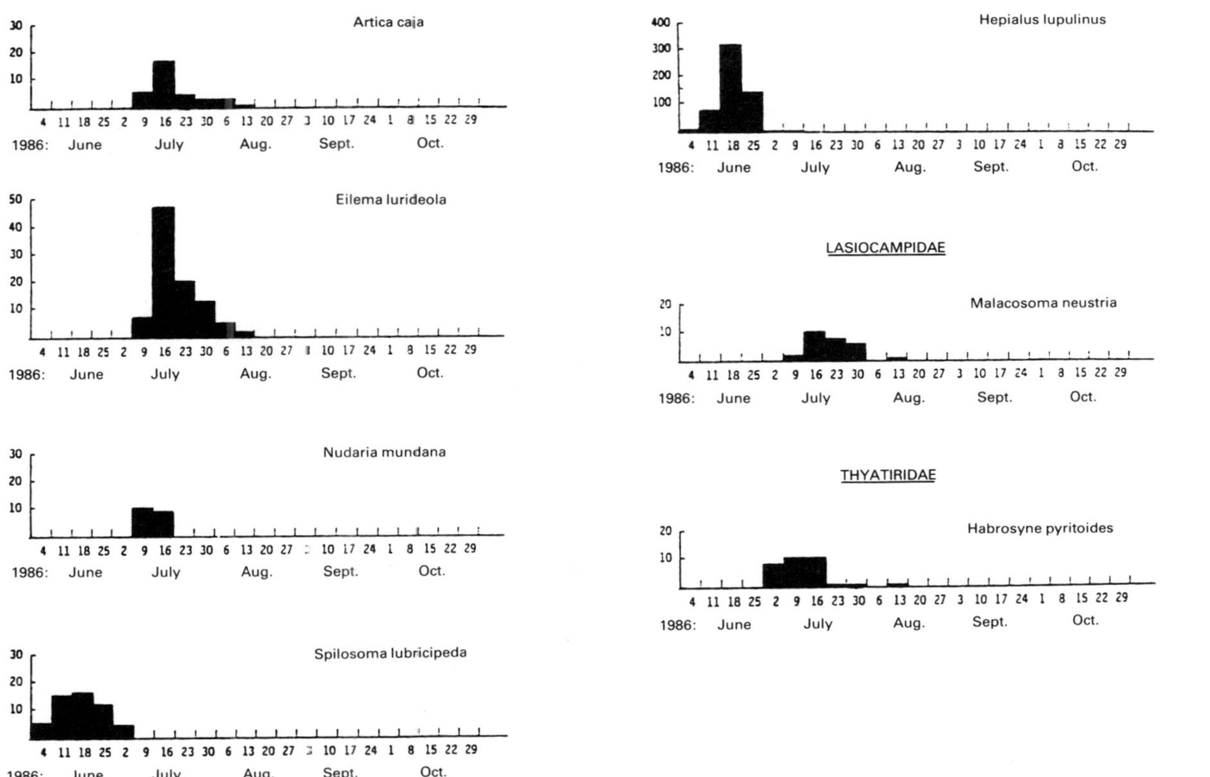

B
ARTCTIIDAE
Artica caja
Eilema lurideola
Nudaria mundana
Spilosoma lubricipeda
HEPIALIDAE
Hepialus lupulinus
LASIOCAMPIDAE
Malacosoma neustria
THYATIRIDAE
Habrosyne pyritoides
4 11 18 25 2 9 16 23 30 6 13 20 27 3 10 17 24 1 8 15 22 29
1986: June July Aug. Sept. Oct.

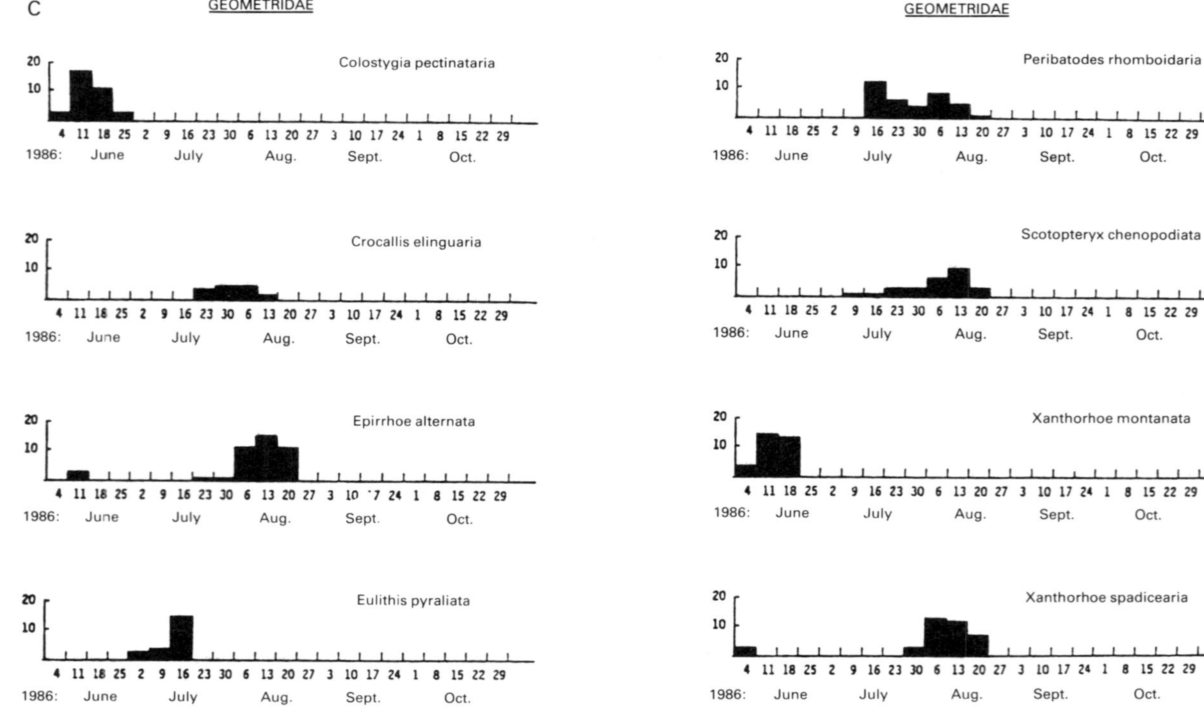
C
GEOMETRIDAE
Colostygia pectinataria
Crocallis elinguaria
Epirrhoe alternata
Eulithis pyraliata
GEOMETRIDAE
Peribatodes rhomboidaria
Scotopteryx chenopodiata
Xanthorhoe montanata
Xanthorhoe spadicearia
20
10
4 11 18 25 2 9 16 23 30 6 13 20 27 3 10 17 24 1 8 15 22 29
1986: June July Aug. Sept. Oct.

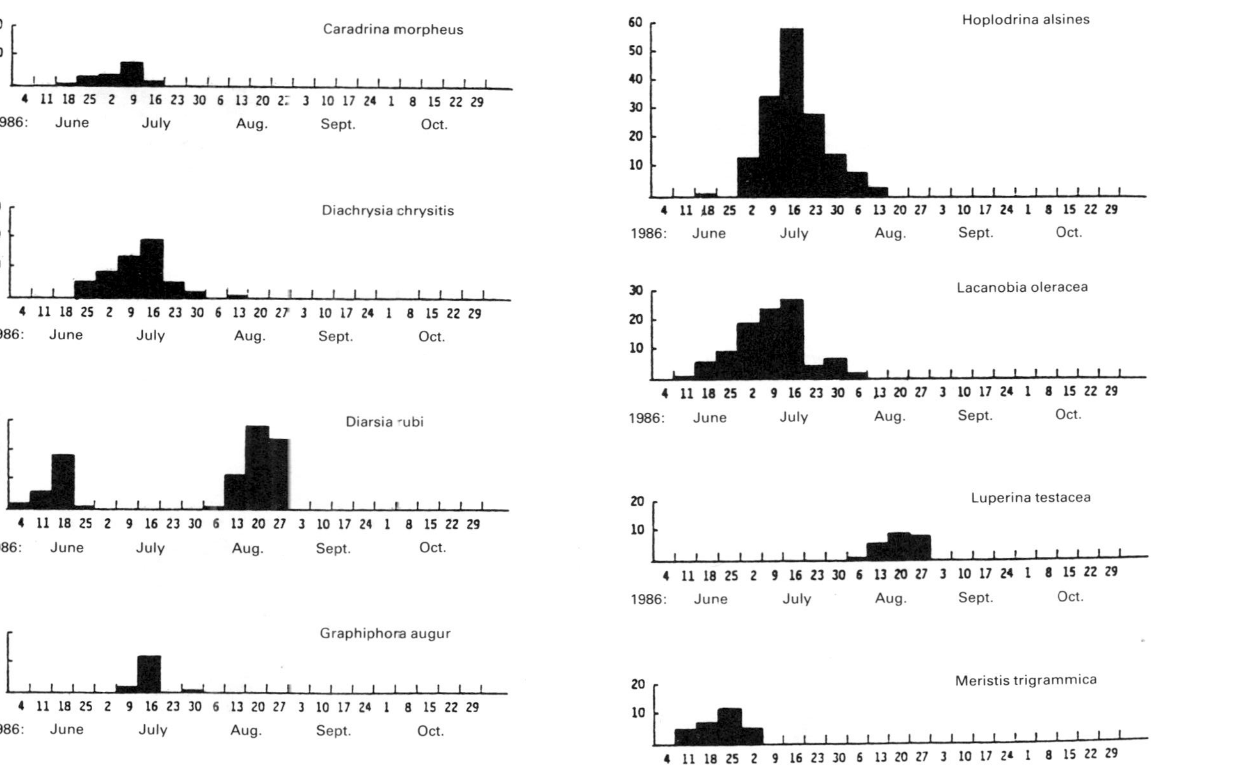
D
NOCTUIDAE
Caradrina morpheus
Diachrysia chrysitis
Diarsia rubi
Graphiphora augur
NOCTUIDAE
Hoplodrina alsines
Lacanobia oleracea
Luperina testacea
Meristis trigrammica
4 11 18 25 2 9 16 23 30 6 13 20 27 3 10 17 24 1 8 15 22 29
1986: June July Aug. Sept. Oct.

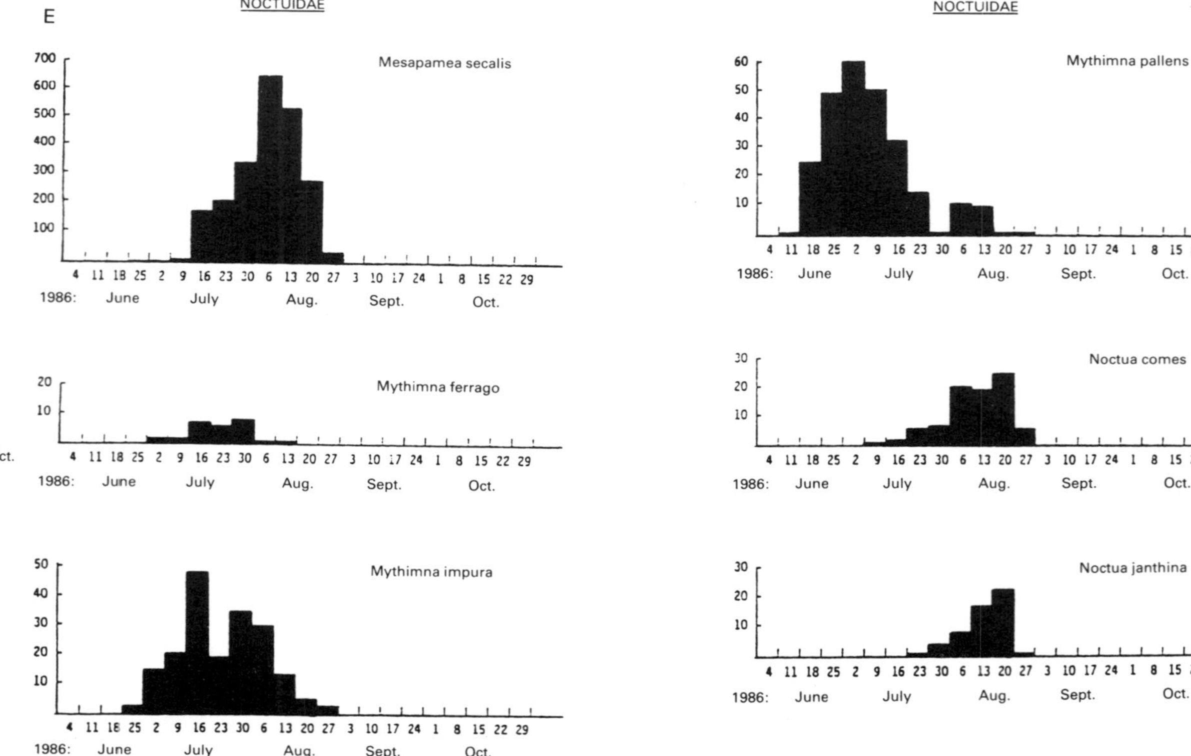
E
NOCTUIDAE
Mesapamea secalis
Mythimna ferrago
Mythimna impura
NOCTUIDAE
Mythimna pallens
Noctua comes
Noctua janthina
4 11 18 25 2 9 16 23 30 6 13 20 27 3 10 17 24 1 8 15 22 29
1986: June July Aug. Sept. Oct.

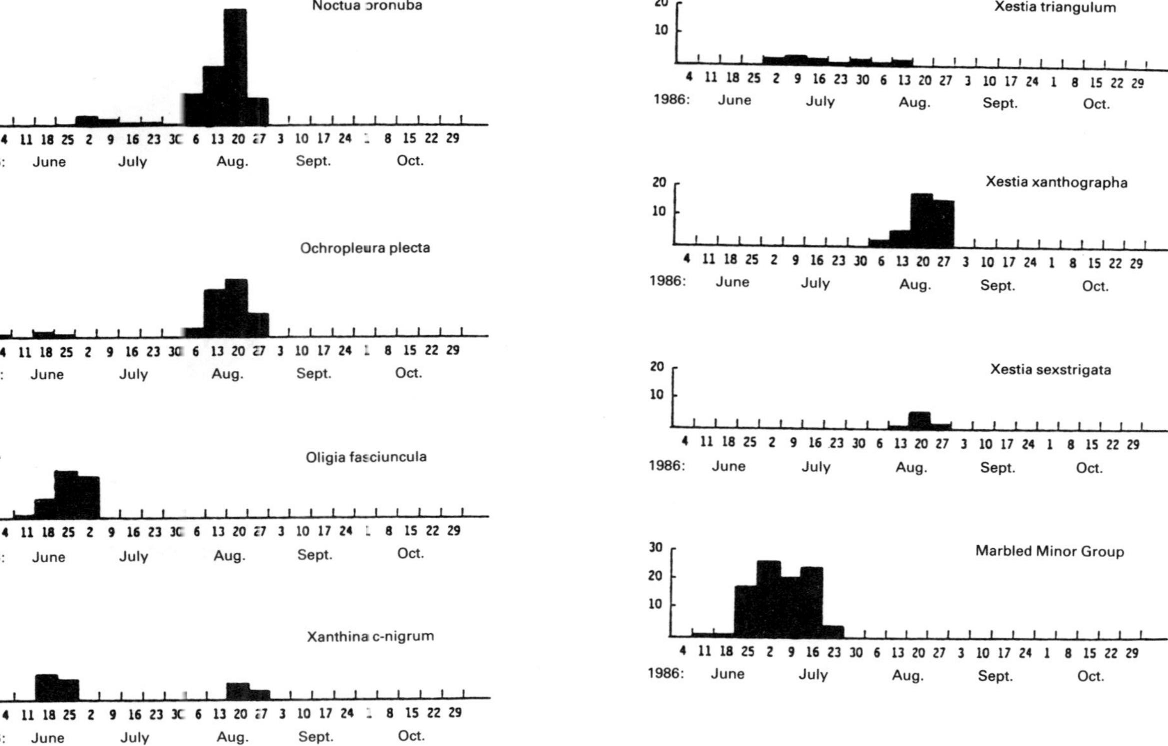
F
NOCTUIDAE
Noctua pronuba
160
120
80
40
4 11 18 25 2 9 16 23 30 6 13 20 27 3 10 17 24 1 8 15 22 29
1986: June July Aug. Sept. Oct.
Ochropleura plecta
30
20
10
4 11 18 25 2 9 16 23 30 6 13 20 27 3 10 17 24 1 8 15 22 29
1986: June July Aug. Sept. Oct.
Oligia fasciuncula
20
10
4 11 18 25 2 9 16 23 30 6 13 20 27 3 10 17 24 1 8 15 22 29
1986: June July Aug. Sept. Oct.
Xanthina c-nigrum
20
10
4 11 18 25 2 9 16 23 30 6 13 20 27 3 10 17 24 1 8 15 22 29
1986: June July Aug. Sept. Oct.
NOCTUIDAE
Xestia triangulum
20
10
4 11 18 25 2 9 16 23 30 6 13 20 27 3 10 17 24 1 8 15 22 29
1986: June July Aug. Sept. Oct.
Xestia xanthographa
20
10
4 11 18 25 2 9 16 23 30 6 13 20 27 3 10 17 24 1 8 15 22 29
1986: June July Aug. Sept. Oct.
Xestia sexstrigata
20
10
4 11 18 25 2 9 16 23 30 6 13 20 27 3 10 17 24 1 8 15 22 29
1986: June July Aug. Sept. Oct.
Marbled Minor Group
30
20
10
4 11 18 25 2 9 16 23 30 6 13 20 27 3 10 17 24 1 8 15 22 29
1986: June July Aug. Sept. Oct.

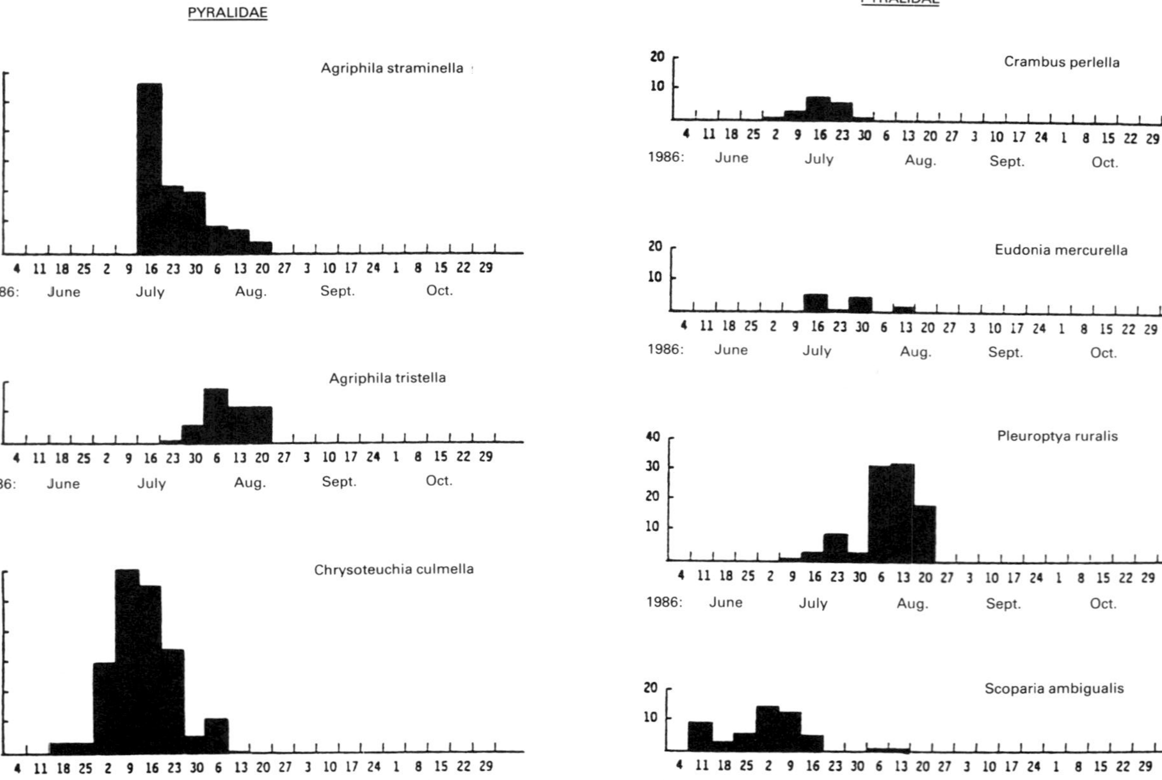
G
PYRALIDAE
Agriphila straminella
60
50
40
30
20
10
4 11 18 25 2 9 16 23 30 6 13 20 27 3 10 17 24 1 8 15 22 29
1986: June July Aug. Sept. Oct.
Agriphila tristella
20
10
4 11 18 25 2 9 16 23 30 6 13 20 27 3 10 17 24 1 8 15 22 29
1986: June July Aug. Sept. Oct.
Chrysoteuchia culmella
120
100
80
60
40
20
4 11 18 25 2 9 16 23 30 6 13 20 27 3 10 17 24 1 8 15 22 29
1986: June July Aug. Sept. Oct.
PYRALIDAE
Crambus perlella
20
10
4 11 18 25 2 9 16 23 30 6 13 20 27 3 10 17 24 1 8 15 22 29
1986: June July Aug. Sept. Oct.
Eudonia mercurella
20
10
4 11 18 25 2 9 16 23 30 6 13 20 27 3 10 17 24 1 8 15 22 29
1986: June July Aug. Sept. Oct.
Pleuroptya ruralis
40
30
20
10
4 11 18 25 2 9 16 23 30 6 13 20 27 3 10 17 24 1 8 15 22 29
1986: June July Aug. Sept. Oct.
Scoparia ambigualis
20
10
4 11 18 25 2 9 16 23 30 6 13 20 27 3 10 17 24 1 8 15 22 29
1986: June July Aug. Sept. Oct.

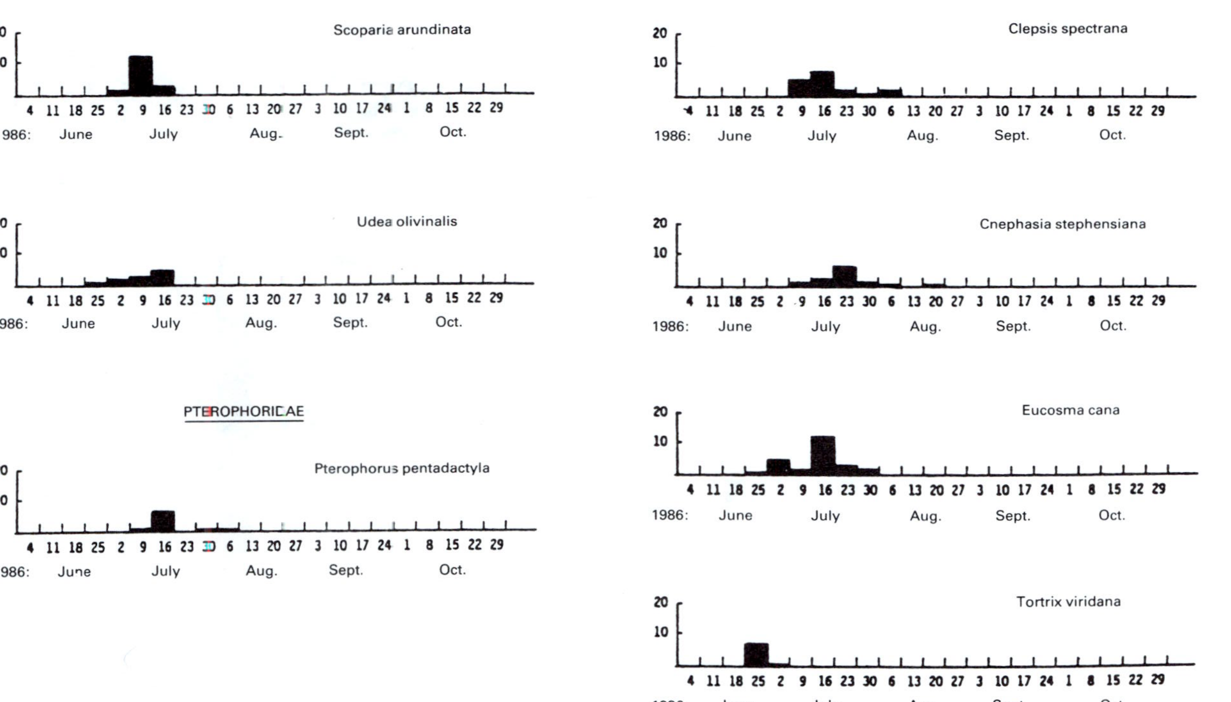

Fig. 5. Moth species trapped near the Wytham field site between June and October, 1986. The moth populations that were obtained from two ultraviolet light night traps were enumerated and, after identification, expressed in the bar charts (A–H) as the weekly collection rates.

summer and autumn of 1986 and 1987. Representative results obtained in 1986 are shown in Fig. 5.

The nearest human population to the field site is some 0.5 km distant. A plan and elevation of the facility are shown in Fig. 6. A 10-m square netted, insect-proof enclosure was constructed and surrounded by a 2-m-high wire fence to prevent access by large animals. The netted facility restricted arthropods from entering or leaving the enclosure. No moth or butterfly species were detected in the facility during the study. Some carabids (Coleoptera) and spiders appeared within the netted enclosure; presumably they developed from eggs or other life

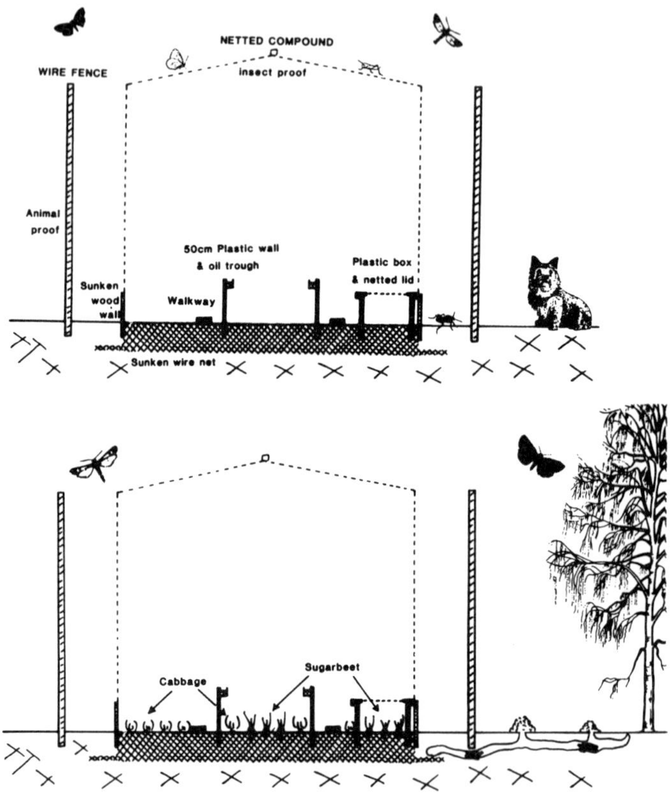

Fig. 6. Field facility used for the release of the genetically-engineered AcNPV. Schematic arrangement of the facility showing the outer, large animal-proof wire fence and the inner, insect- and mole/rodent-proof, netted compound (dashed line) (upper panel). The central area of the compound contained the 2 × 2 m plot used for the release. Within the netted compound, including the central area, sugarbeet and cabbages were planted (lower panel). The central area was bounded by a plastic wall with an upper oil trough to prevent escape of larvae.

stages dormant in the soil. Carabids were more numerous outside the facility. Formalin traps were used to collect carabids, spiders were caught by hand and removed. Wire netting was inserted into the ground all round the enclosure to prevent moles and rodents gaining entry (Fig. 6). Many fresh mole hills were formed in the field throughout the summer and autumn of 1986 and again 1987; none occurred within the facility during, before or after the study. Within the netted facility a plot of 4 m^2 (2 × 2 m) was surrounded by a 50 cm high Plexiglass wall with an upper, encompassing, oil trough (Fig. 6), but open to the air. The Plexiglass was inserted 25 cm into the soil to restrict egress of insect larvae, or entry of other arthropods. As indicated (Fig. 6), the plot was planted with sugar beet and cabbages for the virus release experiments.

After obtaining the requisite permissions the first field trials with the marked AcNPV were undertaken. In September 1986, early third instar *S. exigua* larvae were fed overnight in the laboratory with diet containing ten LD_{50}PIB equivalents of the marked AcNPV incorporated into an artificial diet plug. The infected larvae (*ca.* 200) were refed with diet containing similar quantities of virus and, after they had eaten all the diet, taken in sealed containers to the site and placed on the sugar beet plants in the central, Plexiglass enclosed, area of the containment facility (Fig. 7). Uninfected larvae were released onto nearby

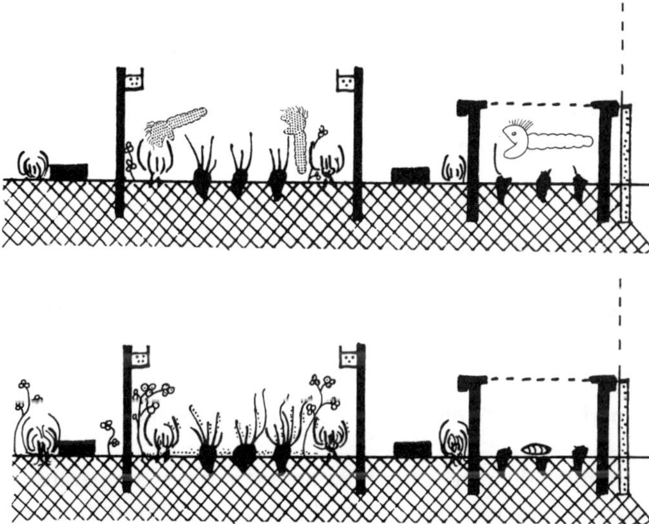

Fig. 7. Release of laboratory infected, early third instar *S. exigua* caterpillars (spotted, see text) in the central area of the netted compound of the field facility described in Fig. 6. Uninfected caterpillars (unspotted) were placed on plants in the right hand, totally netted enclosure. The diagram in the top panel indicates the initiation of the release, the bottom panel illustrates the end result with virus in soil and on foliage (including chickweed that grew during the course of the experiment) and uninfected pupae in the completely netted enclosure.

plants in an adjacent totally netted enclosure (Fig. 7). After 1 week none of the infected caterpillars remained alive. Dead caterpillars were evident on many of the plant surfaces (estimated to be *ca.* 55). Other dead caterpillars were observed in or on the surface of the soil. Samples of soil and foliage from cabbages, sugar beet, and chickweeds that grew within the facility, were returned at 1–2 week intervals for laboratory analysis for virus. These analyses were continued over a 6-month period until February 1987.

Virus was assayed on the recovered plants by allowing groups of permissive *T. ni* second instar larvae to eat the foliage completely. The larvae were reared on artificial diet until pupation or death and then assayed for virus polyhedra by DNA probing. Although, as expected for plants that originally had only a limited number of infected larvae, not all the insects died; virus was identified at every sampling period throughout the entire 6-month period, both on the cabbages and sugar beet plants that were present in the site at the initiation of the experiment as well as those, such as chickweed, that grew up in the facility after the caterpillars had died. The amounts of virus on the infected plants or in the soil were not quantitated. Virus identified after replication in *T. ni* larvae retained the marker element in its original form. Chickweed recovered from outside the facility, or from plants collected outside the enclosed area, did not yield virus.

The presence of virus in soil was assayed by bringing soil samples back to the laboratory and planting cabbage seeds in these samples. The derived seedlings were then fed to *T. ni* larvae and the insects reared until pupation, or death, and similarly analysed for the marked virus. Again, although not all these caterpillars died of virus infections, marked virus was recovered in caterpillars at each sampling period.

In the completely netted enclosure where uninfected caterpillars were placed, the larvae consumed substantial quantities of foliage for 3 weeks after the release. By the end of that period caterpillars were removed, all the plants destroyed, by autoclaving, and the area flooded with water to retrieve pupae. Virus was never isolated from these insects.

Survival of marked AcNPV on leaf surfaces at the field site

In order to estimate the survival of PIBS on leaf surfaces a separate experiment was conducted. In an area of the netted enclosure separate from the walled, central area used for the release of infected larvae, known concentrations of the marked AcNPV, expressed as *T. ni* LD_{50} equivalents, were placed on the upper surfaces of cabbage leaves (Fig. 8). Leaves from these plants were harvested at intervals and fed to second instar *T. ni* larvae in the laboratory to determine the rate of virus loss. Virus loss and/or inactivation were estimated from this

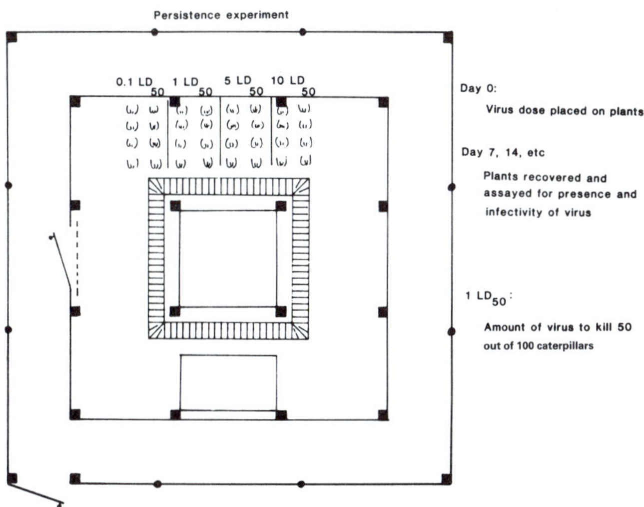

Fig. 8. Persistence of marked virus on cabbage leaves. At one side of the netted compound (Fig. 6) cabbage plants were grown. Multiple portions of each concentration of the marked virus were placed on the upper surfaces of the cabbage leaves (expressed as 0.1–10 LD_{50} doses for second instar *T. ni*). The positions of the viruses were marked. Similar size portions of leaves were recovered from each set of cabbages at weekly intervals and fed to *T. ni* larvae in the laboratory to estimate the loss of infectious virus. Note the double netted door of the compound.

experiment to be of the order of 1 log value of *T. ni* LD_{50} equivalents of infectious virus per week. Since many times during this period of the experiment plant surfaces came into contact with rain, some loss may have been due to virus elution from the leaf surfaces. Such loss, if it occurred, was not quantitated.

Disinfection of the field site and subsequent monitoring

At the completion of the experiment, in February 1987, the field site was cleared and foliage removed and sterilized to destroy any remaining virus. The site was treated with 5% formalin on three separate occasions, twice in February and again in March, 1987. This concentration of formalin is known to inactivate polyhedra very effectively (Hunter *et al.*, 1984). Soil samples were taken before and after each treatment and used in the laboratory to propagate 200–400 cabbage seedlings. The seedlings were fed to *T. ni* larvae to act as a biological indicator for the presence of virus. The larvae were reared until pupation, or death. All insects, infected or healthy, were processed to detect the presence of the marked virus as described above. The results showed that before the first formalin treatment infectious virus was present in the soil, in amounts sufficient to infect most of the *T. ni* larvae. No attempt was made to quantitate the numbers of infectious virus. After the third formalin treatment virus was not detected in the soil samples; all the insects that fed on the cabbage remained healthy, pupated and developed normally. None contained any detectable quantity of the marked virus.

These tests demonstrated that virus had survived in the soil throughout the 6-month period and that the site had been successfully disinfected by formalin treatment.

PREPARATION AND ANALYSIS OF A MARKED AND CRIPPLED AcNPV (SELF-DESTRUCTIVE VIRUS)

A second release of a genetically-engineered baculovirus insecticide was undertaken in 1987. This time the AcNPV polyhedrin gene and its transcription promoter were removed in their entirety from the virus and replaced by another, different, synthetic oligonucleotide. The construction of the recombinant virus will be reported elsewhere. The oligonucleotide sequence (Fig. 9) was positioned so that it would neither add to, nor detract from, the expression of any AcNPV gene product. No attempt was made to exclude ATG codons from the sequence. The size and constitution of the oligonucleotide were such that it would serve only as a unique marker for the recombinant virus. The sequence and position of the oligonucleotide in the viral genome was verified by sequence analyses (Fig. 9). No alteration to the phenotype of the virus was detected by protein analysis, other than that predicted by the loss of the polyhedrin protein and the inability of the virus to form occlusion bodies, as shown in Fig. 10.

The host range of the recombinant virus was determined as before (see above). This analysis involved testing a large selection of Lepidoptera, several Hymenoptera, Coleoptera, and individual species of Neuroptera and Diptera. Of the UK species tested, the only Lepidoptera that were identified as permissive for the non-occluded recombinant virus were *S. exigua* ($LD_{50}: 10^6$ plaque forming units (PFU)/individual) and *M. tiliae* and *Sp. liguistri* species. The latter species were only infected at doses of 10^6 PFU/individual. No other insect species among the UK fauna that were tested were found to be susceptible to infection by the recombinant virus. For *T. ni*, the LD_{50} was calculated to be 10^3 PFU/individual; higher, as in the case of *S. exigua*, than with the occluded AcNPVs.

Successive passage in *T. ni* larvae and consecutive plaque-to-plaque passage involving some 50 cycles of replication, followed by genotype (restriction enzyme) and phenotype (protein and infectivity) analysis, demonstrated that, as before, the recombinant virus was genetically stable and did not acquire novel sequences, or exhibit genome rearrangements, or any other altered property.

Marked, polyhedrin-negative virus

```
        10         20         30         40         50         60         70         80         90        100        110        120
ACGCACAAAC TAATATCACA AACTGGAAAT GTCTATCAAT ATATAGTTGC TGATATCATG GAGATAATTA AAATGATAAC CATCTCGCCC GGATCCCCAT CCCGGGTTAG TGAGTTAGTT
TGCGTGTTTG ATTATAGTGT TTGACCTTTA CAGATAGTTA TATATCAACG ACTATAGTAC CTCTATTAAT TTTACTATTG GTAGAGCGGG CCTAGGGGTA GGGCCCAATC ACTCAATCAA
       130        140        150        160        170        180        190        200        210        220        230        240
ATCTAGATTA GCTAAGTTAA CTTTAAACTA GTCATGACCT TAGGTGAGAT CTAGTTAGGA TCCCGGTGTT ATTAGTACAG CATTTATTAA GCGCTAGATT CTGTGCGTTG TTGATTTACA
TAGATCTAAT CGATTCAATT GAAATTTGAT CAGTACTGGA ATCCACTCTA GATCAATCCT AGGGCCACAA TAATCATGTC GTAAATAATT CGCGATCTAA GACACGCAAC AACTAAATGT
       250        260        270        280        290        300        310        320        330        340        350        360
GACAATTGTT GTACGTATTT TAATAATTCA TTAAATTTAT AATCTTTAGG GTGGTATGTT AGAGCGAAAA TCAAATGATT TTCAGCGTCT TTATATCTGA ATTTAAATAT TAAATCCTCA
CTGTTAACAA CATGCATAAA ATTATTAAGT AATTTAAATA TTAGAAATCC CACCATACAA TCTCGCTTTT AGTTTACTAA AAGTCGCAGA AATATAGACT TAAATTTATA ATTTAGGAGT
       370        380        390        400        410        420        430        440        450        460        470        480
ATAGATTTGT AAAATAGGTT TCGATTAGTT CAAACAAGGG TTGTTTTTCC GAACCGATGA CTGAACTATC TAATGAATTT TCGCTCAACG CCACAAAACT TGCCAAATCT TGTAGCAGCA
TATCTAAACA TTTTATCCAA AGCTAATCAA GTTTGTTCCC AACAAAAAGG CTTGGCTACT GACTTGATAG ATTACTTAAA AGCGAGTTGC GGTGTTTTGA ACGGTTTAGA ACATCGTCGT
       490        500        510        520        530        540     T*550        560        570        580        590        600
ATCTAGCTTT GTCGATATTC GTTTGTGTTT TGTTTTGTAA TAAAGGTTCG ACGTCGTTCA AAATATTATG CGCTTTTGTA TTTCTTTCAT CACTGTCGTT AGTGTACAAT TGACTCGACG
TAGATCGAAA CAGCTATAAG CAAACACAAA ACAAAACATT ATTTCCAAGC TGCAGCAAGT TTTATAATAC GCGAAAACAT AAAGAAAGTA GTGACAGCAA TCACATGTTA ACTGAGCTGC
       610
TAAACACGTT
ATTTGTGCAA
```

Fig. 9. Nucleotide sequence (boxed) of part of the marked, polyhedrin-negative AcNPV (genetically-crippled virus) in the region of the genome from which the polyhedrin gene had been removed. The area of the AcNPV genome is the same as that shown in Fig. 4. The polyhedrin protein coding region and mRNA promoter have been removed and replaced with a synthetic oligonucleotide (text).

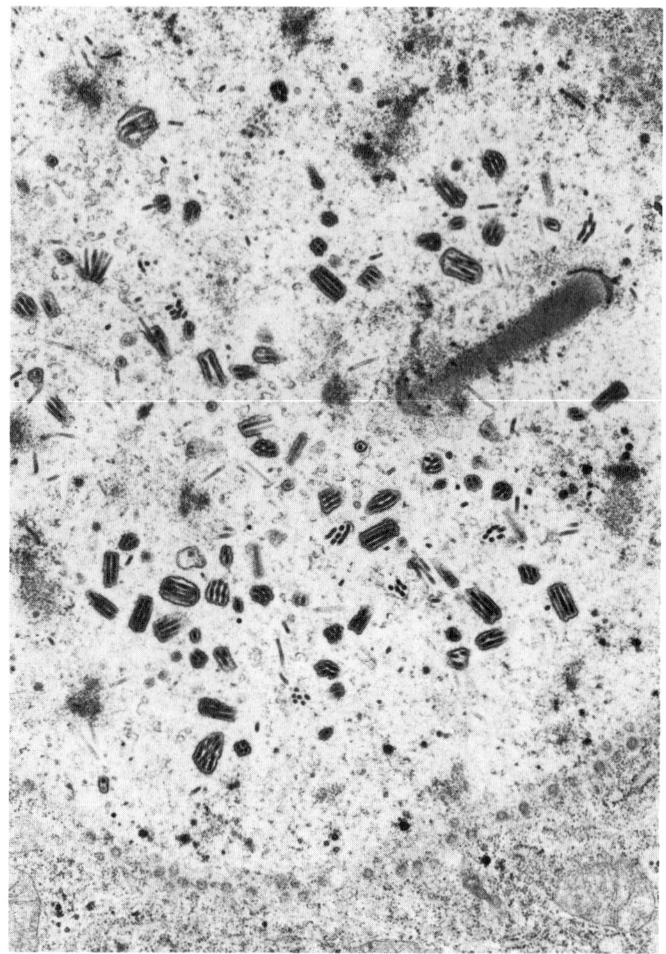

Fig. 10. Electron micrograph of *T. ni* fat body cells infected with the marked, polyhedrin-negative, virus. Viral nucleocapsids and enveloped virus particles are visible, but no occlusion bodies.

Physical stability tests were undertaken as before. Soil was collected from the proposed field site and steam sterilized. The marked, polyhedrin-negative, recombinant virus was mixed with soil and left in the laboratory at room temperature (*ca.* 18°C) for two weeks. Samples were taken throughout this period, and virus quantitated by plaque assays. There was a highly significant decline in virus infectivity with over 2 log values of virus infectivity lost in 3 days of incubation (Fig. 11). Since prior experiments indicated that *occluded* AcNPV remained infective in soil (see above), these results showed that, by comparison,

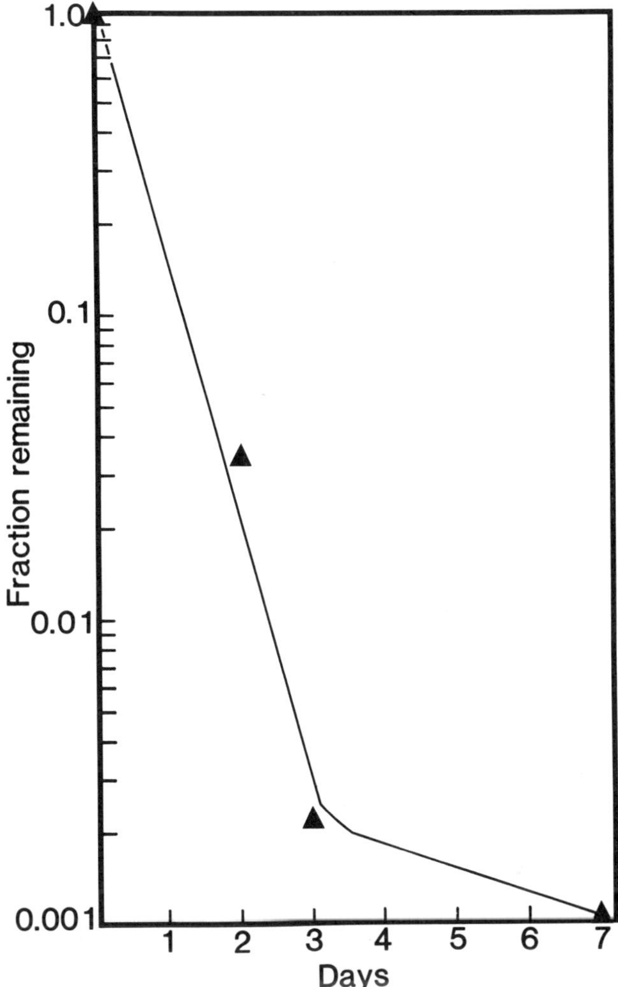

Fig. 11. Physical stability of the genetically-crippled virus in soil. The loss of virus PFU as a function of time is expressed as the fraction remaining (see text for experimental conditions).

the *non-occluded* virus was phenotypically crippled as a consequence of the removal of the polyhedrin gene and was inactivated or degraded in soil. Similarly, a greatly increased sensitivity to ultraviolet light inactivation was demonstrated.

After review of the laboratory data and independent evaluation, a licence was issued by MAFF to undertake a release under the same conditions and using the same protocols and facility that were employed for the occluded, marked recombinant AcNPV. The objective was to determine whether virus released from dead caterpillars persisted in the environment, in soil and on plant surfaces. As before, caterpillars were infected in the laboratory and taken to the field site and released in the central enclosed area on to sugar beet. By 1 week all the caterpillars died. One week later, no virus capable of infecting second instar *T. ni* larvae was identified either on the plant surfaces, or in soil samples. Even the corpses of dead caterpillars did not yield infectious virus.

In summary, the data obtained with the marked, polyhedrin-negative virus, the so-called 'self-destructive' virus, demonstrated that the recombinant virus did not persist in the environment.

DISCUSSION

In considering the results of the studies reported here, it is important to emphasize that the objectives were to assess and minimize any risk that might be involved in the release to the environment of a genetically-engineered virus insecticide. The study is part of a programme of scientific research, rather than the commercial development of a product. As such, the results and information derived from the studies *apply only to baculovirus insecticides*, and only to the particular virus that was used. There is, however, good reason to believe that other baculoviruses, similarly marked or manipulated will behave in a similar manner.

Various factors were considered in order to minimize the risk to the environment of the proposed releases. These included the choice of virus, the choice of the target pest, the design of the engineered viruses, the design of the field site and the procedures employed to conduct the experiment, including the final disinfection of the site. After preparing the engineered viruses, they were subjected to extensive laboratory analysis. From the data obtained no *measurable* risk associated with the proposed experiments could be identified. This does not exclude the possibility of an unforeseen risk but none was identified in the laboratory studies undertaken before the field trials. With the results obtained in the laboratory analysis, we therefore applied to the appropriate authorities for permission to proceed with the field trials. The data were independently evaluated and, after obtaining the requisite licences, the field trials were initiated and the predictions of the laboratory studies were verified.

In the first phase of our analysis, involving the development and deliberate release into the environment of a genetically altered virus insecticide, a marked baculovirus was constructed. This contained a unique oligonucleotide which allowed us to distinguish it from any other baculovirus, including the unmodified parental AcNPV, as well as insect host nucleic acids. The marked virus could be assayed by a simple 'dot-blot' hybridization procedure with crude extracts of infected caterpillars. The marker, inserted in the immediate 3′ non-coding region of the polyhedrin gene, did not affect the phenotype or replication of the virus in cell culture or in insects. Furthermore, its design and site of insertion was such that it could not and, from the studies conducted, did not alter the synthesis of any other virus gene product. After passage in cell culture or in insects, the marker remained stable. It is predicted that it should remain so indefinitely. No alteration in the viral genome was detected after repeated passage of the virus.

The host range of the marked virus was compared with that of a plaque-cloned isolate of the parental virus. Most of the UK insects tested (*ca.* 80 species) were not susceptible to either the marked or the parent virus. The lime and privet hawkmoths succumbed, but only *at very high doses*. Such doses are probably seldom significant in epizootic initiation and maintenance. Epizootics depend on a balance between virus infectivity, pathogen productivity and persistence, host-range population densities and behaviour, etc. NPVs which cause epizootic disease usually have low LD_{50} values, e.g. in *Lymantria dispar* (Lewis *et al.*, 1981) and *Orgia pseudotsugata* (Hughes, 1978). The small mottled willow moth (*S. exigua*) was susceptible to AcNPV, although also only at high doses. In view of this and their susceptibility, caterpillars of *S. exigua* were considered to be suitable as the host insect species for the field studies. As discussed, the field arrangements (time of year, etc.) minimized the opportunity for contact between the virus released from the infected caterpillars and other species of Lepidoptera. In addition the site was disinfected when the experiment was terminated. All of these attributes were important considerations in relation to the design and conduct of the study.

Before the field release the physical stability of the marked virus was determined. Studies in the laboratory showed that the marked virus was not affected by a period of 2 weeks in contact with soil, retaining full infectivity throughout this time. In view of this it was considered inevitable that soil at the field site would become contaminated with virus. This was found to be the case, consequently the site was disinfected at the termination of the study.

Approval for the release of the marked virus was sought from a number of regulatory authorities and interested parties. Consultations on the design of the field site and the host range tests were undertaken before the issue of a licence. Eventually, the marked virus was released into the field site in September 1986 by infecting *S. exigua* larvae in the laboratory and then placing them on sugar beet that had been planted in a netted facility. The insects died from the virus infections. Virus was recovered from plant surfaces and the soil, over a 6-month period after the release, demonstrating the resistance of baculoviruses to environmental degradation. The site was disinfected with formalin to inactivate the remaining virus. In a separate experiment the rate of loss of infectious virus from plant surfaces was shown to be of the order of 1 log value of virus per week. Whether this was due only to the effects of ultraviolet light, or other causes of inactivation, or to elution from the plants, is not known.

A second release was undertaken in 1987 to determine whether an environmentally safe, self-destructive, baculovirus could be engineered that would be a suitable substrate for further engineering. The objective was to make a virus that could not persist; *viz*: a genetically crippled virus that is quickly degraded in the environment. Such a virus should be even *safer* than the natural virus. A marked, polyhedrin-negative virus was prepared and tested with regard to its genetic and physical stability as well as host range among UK insects. The virus, as predicted, was found to be labile in soil and on leaf surfaces. After obtaining the requisite licence, a field release experiment confirmed these laboratory results and showed that the genetically crippled virus was unstable in the environment.

In summary, the results from this study demonstrate that an innocuous

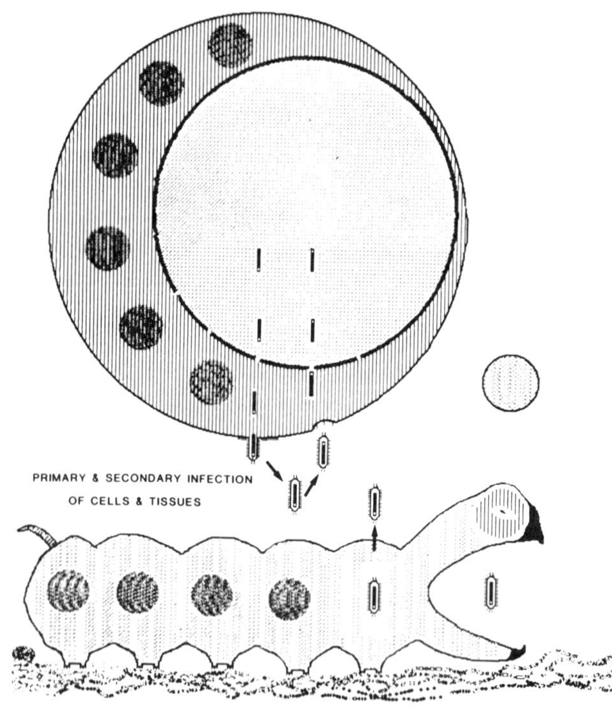

Fig. 12. Schematic infection cycle by a genetically-crippled AcNPV that codes for a new (foreign) gene product. Naked, polyhedrin-negative, viruses are ingested with food and establish an infection expressing the foreign gene at the same time.

genetic marker is a suitable tool for the identification of an engineered virus released into the environment. They also showed that a genetically crippled virus can be prepared that degrades in the environment after its effect on a caterpillar host. Such a virus is a suitable substrate for further engineering and the preparation of viruses which act more quickly than the natural virus through the incorporation of genes of insect specific toxins, or hormones, or other genes (Fig. 12).

ACKNOWLEDGEMENTS

We thank Mr C. F. Rivers, Mr P. H. Sterling, Mrs M. E. K. Tinson, Mr T. Carty and Mr M. L. Hirst for their assistance in collecting and/or rearing the insects for the host-range studies. In 1986 moth collections and data analyses were undertaken by Mr S. M. Eley. In 1987 similar collections were undertaken by Ms C. Doyle and Mr A. C. Forkner of the Institute. Carabid surveys and analyses were undertaken by Dr J. S. Cory. Ultraviolet light measurements in the facility were conducted by Mr H. J. Killick. The support of these and other members of the Institute's staff, particularly Mr J. S. Robertson, are gratefully acknowledged. Dr J. I. Cooper and his staff kindly provided the plants. We are indebted to Mr R. Broadbent and Mr R. M. MacKenzie who constructed the field facility. The photographic services of Mr C. D. Hatton and electron microscopy performed by Ms M. K. Arnold are also gratefully acknowledged.

REFERENCES

Adang, M. J. and Miller, L. K., 1982. Molecular cloning of DNA complementary to mRNA of the baculovirus *Autographa californica* nuclear polyhedrosis virus: Location and gene products of RNA transcripts found late in infection. *Journal of Virology* **44,** 782–793.

Bishop, D. H. L., 1986. UK release of genetically marked virus. *Nature* **323,** 496.

Brown, M. and Faulkner, P., 1977. A plaque assay for nuclear polyhedrosis viruses using a solid overlay. *Journal of General Virology* **36,** 361–364.

Carruthers, W. R., Cory, J. S. and Entwistle, P. F., 1988. Recovery of pine beauty moth (*Panolis flammea*) nuclear polyhedrosis virus from pine foliage. *Journal of Invertebrate Pathology* **2,** in press.

Doyle, C. J. and Entwistle, P. F., 1988. Aerial application of mixed virus formulations to control joint infestations of *Panolis flammea*, and *Neodiprion sertifer* on lodgepole pine. *Bulletin of Entomological Research*, in press.

Entwistle, P. F., 1982. Passive carriage of baculoviruses in forests. *Proceedings 3rd International Colloquium on Invertebrate Pathology and Microbial Control; Brighton, UK*, pp. 344–351.

Entwistle, P. F. and Evans, H. F., 1985. Viral Control. In *Comprehensive Insect Physiology, Biochemistry and Pharmacology*, Gilbert, L. I. and Kerkut, G. A. (eds), Vol. 12, pp. 347–412. Oxford: Pergamon Press.

Entwistle, P. F. and Evans, H. F., 1987. Trials on the control of *Panolis flammea* with a nuclear polyhedrosis virus. In *Population Biology and Control of the Pine Beauty Moth*, Leather, S. R., Stoakley, J. T. and Evans, H. F. (eds), Forestry Commission Bulletin 67, pp. 61–68.

Entwistle, P. F., Adams, P. H. W., Evans, H. F. and Rivers, C. F., 1983. Epizootiology of a nuclear polyhedrosis virus (Baculoviridae) in European spruce sawfly (*Gilpinia hercyniae*): spread of disease from small epicentres in comparison with spread of baculovirus diseases on other hosts. *Journal of Applied Ecology*, **20,** 473–487.

Entwistle, P. F., Evans, H. F., Harrap, K. A. and Robertson, J. S., 1985. Control of European pine sawfly *Neodiprion sertifer* (Geoffr.) with its nuclear polyhedrosis virus in Scotland. In *Site Characteristics and Population Dynamics of Lepidopteran and Hymenopteran Forest Pests*, Bevan, D. and Stoakley, J. T. (eds), Forestry Commission Research and Development Paper 135, pp. 36–46.

Evans, H. F. and Harrap, K. A., 1982. Persistence of Insect Viruses. In *Virus Persistence*, Minson, A. C. and Darby, G. K. (eds), SGM Symposium 33, pp. 57–96. Cambridge: Cambridge University Press.

Finney, D. J., 1971. *Probit Analysis*. Cambridge: Cambridge University Press.

Hooft van Iddekinge, B. J. L., Smith, G. E. and Summers, M. D., 1983. Nucleotide sequence of the polyhedrin gene of *Autographa californica* nuclear polyhedrosis virus. *Virology* **131,** 561–565.

Howard, S. C., Ayres, M. D. and Possee, R. D., 1986. Mapping the 5′ and 3′ ends of *Autographa californica* nuclear polyhedrosis virus polyhedrin mRNA. *Virus Research* **5,** 109–111.

Hughes, K. M., 1978. Virus in biological control. In *The Douglas-fir Tussock Moth: A Synthesis*, Brookes, M. A., Stark, R. W. and Campbell, R. W. (eds), pp. 133–136. Washington DC: US Department of Agriculture.

Hunter, F. R., Crook, N. E. and Entwistle, P. F., 1984. Viruses as pathogens for the control of insects. In *Microbial Methods for Environmental Biotechnology*, Grainger, J. M. and Lynch, J. M. (eds), pp. 323–347. New York: Academic Press.

Kaya, H. K., 1982. Parasites and predators as vectors of insect diseases. *Proceedings 3rd International Colloquium Invertebrate Pathology and Microbial Control;* Brighton, UK pp. 39–44.

Kelly, P. M., Sterling, P. H., Speight, M. R. and Entwistle, P. F., 1988. Preliminary spray trials of a nuclear polyhedrosis virus as a control agent for the Brown-tail moth, *Euproctis chrysorrhoea*, (L.) (Lepidoptera: Lymantriidae). *Bulletin of Entomological Research*, in press.

Killick, H. J., 1987. Ultraviolet light and *Panolis* nuclear polyhedrosis virus: a non-problem? In *Population Biology and Control of the Pine Beauty Moth*, Leather, S. R., Stoakley, J. T. and Evans, H. F. (eds), Forestry Commission Bulletin 67, pp. 69.

Lewis, F. B., Rollinson, W. D. and Yendol, W. G., 1981. Laboratory evaluations. In *The Gypsy Moth: Research Toward Integrated Pest Management*, Doane, C. C. and McManns, M. L. (eds), pp. 455–461. Washington DC: Department of Agriculture.

Lomer, C. J., 1986. Release of *Baculovirus oryctes* into *Oryctes monoceros* populations in the Seychelles. *Journal of Invertebrate Pathology* **47,** 237–246.

Longworth, J. F. and Cunningham, J. C., 1968. The activation of occult nuclear polyhedrosis viruses by foreign nuclear polyhedra. *Journal of Invertebrate Pathology* **10,** 361–367.

Luckow, V. A. and Summers, M. D., 1988. Trends in the development of baculovirus expression vectors. *Biotechnology* **6,** 47–55.

Matthews, R. E. F., 1982. *Classification and Nomenclature of Viruses*. Basel: Karger.

Matsuura, Y., Possee, R. D., Overton, H. A. and Bishop, D. H. L., 1987. Baculovirus expression vectors: the requirements for high level expression of proteins, including glycoproteins. *Journal of General Virology* **68,** 1233–1250.

Podgewaite, J. D., 1985. Strategies for field use of baculoviruses. In *Viral Insecticides for Biological Control*, Maramorosch, K. and Sherman, K. E. (eds), pp. 775–797. New York: Academic Press.

Possee, R. D., 1986. Cell surface expression of influenza virus haemagglutinin in insect cells using a baculovirus vector. *Virus Research* **5,** 43–59.

Possee, R. D. and Howard, 1987. Analysis of the polyhedrin gene promoter of the *Autographa californica* nuclear polyhedrosis virus. *Nucleic Acids Research* **15,** 10 233–10 248.

Possee, R. D. and Kelly, D. C., 1988. Physical maps and comparative DNA hybridization of *Mamestra brassicae* and *Panolis flammea* nuclear polyhedrosis virus genomes. *Journal of General Virology* **69,** in press.

Rohel, D. Z., Cochran, M. A. and Faulkner, P., 1983. Characterization of two abundant mRNAs of *Autographa californica* nuclear polyhedrosis virus present late in infection. *Virology* **124,** 357–365.

Small, D. A., 1984. The use of a baculovirus for the control of cabbage pests. *NERC, Institute of Virology Report*, pp. 143.

Smith, G. E., Fraser, M. J. and Summers, M. D., 1983a. Molecular engineering of the *Autographa californica* nuclear polyhedrosis virus genome: deletion mutations within the polyhedrin gene. *Journal of Virology* **46,** 584–593.

Smith, G. E., Vlak, J. M. and Summers, M. D., 1983b. Physical analysis of *Autographa californica* nuclear polyhedrosis virus transcripts for polyhedrin and 10 000-molecular weight protein. *Journal of Virology* **45,** 215–225.

WHO/FAO, 1973. Expert Group (GTRES): *The Use of Viruses for the Control of Insect Pests. WHO Technical Report Series 531.*

Wigley, P. J., 1980. Diagnosis of virus infections-staining of insect inclusion body viruses. In *Microbial Control of Insect Pests*, Kalmakoff, J. and Longworth, J. F. (eds), pp. 35–39. New Zealand Department of Scientific and Industrial Research Bulletin 228.

13 Pre-release Testing Procedures: US Field Test of a *lac*ZY-engineered Soil Bacterium

D. J. DRAHOS[1], G. F. BARRY[1], B. C. HEMMING[1], E. J. BRANDT[1], H. D. SKIPPER[2], E. L. KLINE[2], D. A. KLUEPFEL[2], T. A. HUGHES[2] and D. T. GOODEN[2]

[1]*Monsanto Co., St Louis, MO 03198, USA,* [2]*Clemson University, Clemson, SC 29634, USA*

INTRODUCTION

The practical application of genetic engineering in agriculture not only offers tremendous promise but also demands a careful and responsible approach as initial field testing begins. There is a perception that genetically-engineered micro-organisms (GEMs) may behave in some manner differently from the non-recombinant parental strain, either in the environment or under laboratory conditions. In particular, basic questions have been raised about the survival and possible dissemination of the field-introduced GEMs, the stability of the recombined genes and the potential for gene exchange with similar indigenous micro-organisms. This account will focus on the development of a specific means for addressing these questions, as well as the predictive value of data obtained from certain pre-release testing procedures.

The recently initiated field test of the recombinant soil bacterium, *Pseudomonas aureofaciens* 3732RNL11, marked with the *E. coli lac*ZY tracking system, serves as a very appropriate 'real world' example. This field study was initiated, after regulatory approval, on 2 November 1987, in collaboration with the Monsanto Co., St Louis, Missouri, and Clemson University, Clemson, South Carolina. The 18-month study is being conducted on a 0.4 hectare (1 acre) test plot by Clemson scientists at the Edisto Experimental Research Station, Blackville, South Carolina. This is a 930 hectare (2300 acre) university research facility in the rural south-central part of the state.

The field test proposal was extensively reviewed by federal regulators at the

RELEASE OF GENETICALLY-ENGINEERED MICRO-ORGANISMS ISBN 0–12–677521–4

US Environmental Protection Agency (USEPA) and US Department of Agriculture (USDA), as well as by the Institutional Biosafety Committee (IBC) at Monsanto and at Clemson. As part of the regulatory compliance effort, a Premanufacture Notification (PMN) was submitted on 18 June 1987 to the EPA Office of Toxic Substances. The EPA granted approval for the test on 19 October 1987. This, therefore, represents the first test approved by the EPA under the Toxic Substances Control Act (TSCA) for the environmental release of a live genetically-engineered micro-organism and the first field test in the US of a live recombinant microbe containing genes from two different strains.

BACKGROUND OF *lac*ZY MARKER SYSTEM

The introduction of genetically-modified bacteria into natural ecosystems to control insects and diseases, enhance development of agronomically important plant varieties, for waste treatment, pollution control and mineral leaching is of considerable potential. The ability of certain fluorescent psuedomonad soil bacteria effectively to colonize the root surfaces of a variety of crop species has made them prime candidates for use in plant growth promotion and pest control. Engineered strains of these bacteria may soon be available to act as efficient delivery vehicles for plant beneficial agents, such as natural pesticides, fungicides, or plant hormones. Such use would not be practical, or even possible, without the precise, targeted, continuous input afforded by these bacteria. Critical evaluation of the performance capabilities of such bacteria is essential, even to determine which are the most efficient host organisms to use for given environmental conditions or crop species. For this analysis to be realistic in predicting, for example, pesticidal effectiveness, it is crucial to determine the competitive ability (location and population densities on crop plants), of the engineered bacteria under actual field conditions. However, few genetic tools exist rapidly and accurately to monitor environmental distribution of either released recombinant organisms or the engineered genes they contain.

A sensitive tracking method has been developed that largely bypasses problems associated with conventional antibiotic-marking methods and which is particularly effective for the fluorescent pseudomonads (Hemming *et al.*, 1984; Drahos *et al.*, 1986). This marking system relies on the introduction and expression of the *E. coli* K12 *lac* operon genes *lac*Z and *lac*Y. The unexpected inability of most, if not all, fluorescent pseudomonads efficiently to utilize lactose as a sole carbon source permits efficient metabolic selection of marked strains on defined laboratory media after inoculation into natural, non-sterile field soil.

Furthermore, rapid and efficient methods were developed to insert the *lac*ZY genes directly into the host bacterial chromosome (Barry, 1986, 1988). This chromosomal insertion provides great stability for these genes within the engineered strain under natural, non-selective field conditions. In contrast to some other systems, these insertion methods effect an essentially permanent, or one-way, gene transfer into the chromosome, providing a high degree of containment for the *lac*ZY element within the engineered strain. Finally, chromosomal introduction of these genes, which have been sequenced and their products well characterized, may eventually aid the implementation of direct bacterial detection methods, such as nucleic acid and antibody hybridization probes, once these techniques become more practically applicable.

PRE-RELEASE TESTING

Before any field study utilizing a recombinant micro-organism may be initiated in the US, it is necessary to obtain approval from appropriate government regulatory bodies, as well as the Institutional Biosafety Committee (IBC). Which government regulatory bodies are involved depends on the nature and intended use of the recombinant organism being tested. In-depth reviews of the field test proposal were carried out by both the EPA and USDA. The focal point of the EPA review was a Pre-manufacture Notification (PMN) which the Monsanto Company submitted to the EPA Office of Toxic Substances in voluntary compliance with Section 5 of the Toxic Substances Control Act (TSCA) and the recent policy guidance of the Agency regarding biotechnology products subject to TSCA as set out in the 'Coordinated Framework for Regulation of Biotechnology' (Federal Register, 26 June 1986).

An integral part of the PMN review process is an assessment of the potential hazard presented to human health and the environment as a consequence of field testing of micro-organisms. The degree of hazard is a function of the toxicity/pathogenicity of the test organisms and the level of human and environmental exposure to them. The pre-release testing was designed properly to address these issues, and focused on the following areas:

(i) Identity of bacterial host strain to species level;
(ii) Non-toxicity/non-pathogenicity of microbe to animals;
(iii) Non-pathogenicity of microbe to plants;
(iv) Potential for microbial survival;
(v) Stability of the introduced genes and containment within host organism;
(vi) Efficacy of the monitoring system to track the GEMs in the environment in the presence of native field microbes.

PRE-RELEASE TESTING RESULTS

Bacterial strain identification was accomplished by three principal methods: (a) standardized micro-methods for phenotypic characterization, (b) gas chromatographic analysis of free fatty acids and (c) independent laboratory verification (American Type Culture Collection, ATCC). Phenotypic characterization was achieved by combining 21 biochemical and assimilation tests for the identification of Gram-negative, non-fermentative bacteria, with the Rapid NFT *In Vitro* Diagnostic Test Kit (DMS Industries) and the API20E microtube system (Analytab Products, Inc.). Results of this testing provided a genus/species identification of Ps. 3732RN as *Pseudomonas fluorescens*.

Identification by gas chromatographic analysis of free fatty acids was carried out by the Hewlett-Packard 5898A Microbial Identification System (MIS). This system analyses and identifies micro-organisms isolated in pure culture on artificial media through use of a standardized sample preparation procedure and gas chromatography to yield quantitatively reproducible fatty acid composition profiles. These profiles are then compared with a library of reference organisms stored in the system of the computer to determine the identity of the strain. By this method, strain Ps. 3732RN was identified as either of two very closely related sub-species: *Pseudomonas fluorescens* biovar D (also referred to as *P. chlororaphis*) or *Pseudomonas fluorescens* biovar E (also referred to as *P. aureofaciens*). Independent laboratory analysis by the ATCC identified Ps. 3732RN as *Pseudomonas aureofaciens*.

The importance of this identity verification is to aid in an accurate assessment of the non-pathogen/pathogen status of a given test strain. Extensive search of the literature confirms that *Pseudomonas fluorescens* strains of either biovar D or E (*P. chlororaphis* or *P. aureofaciens*) are not known to be disease-causing agents in humans or animals.

To provide even further confirmation that the host organism Ps. 3732RN was non-pathogenic to animals a series of toxicology studies was carried out to evaluate the potential of this organism to induce toxicity and be pathogenic in various species, or to colonize animal tissues and induce pathological effects. Results of these studies showed that Ps. 3732RN is non-toxic, non-pathogenic and non-infective to mice, birds, fish, daphnids and earthworms. There was no treatment-related mortality in any of these studies. Further, there is no reason to expect that the chromosomal insertion of the *lac*ZY genes from *E. coli* K12, a certified non-pathogenic organism, into *P. aureofaciens*, which is also a non-pathogenic organism, confers any pathogenic properties. Thus, there is no significant concern about the pathogenicity of *lac*ZY-marked *P. aureofaciens* organisms.

In order to verify that a given test organism is non-pathogenic to plants, a standard series of tests, the LOPAT series (Lelliott *et al.*, 1966) is typically performed. This series includes tests for levan formation, oxidase production, potato rotting ability, arginine dihydrolase production and tobacco hypersensitivity. Ps. 3732RN was found to be positive for levan production, positive in the oxidase test, negative for potato rotting ability, positive for arginine dihydrolase production, and negative for tobacco hypersensitivity. Identical properties were also found for the *lac*ZY-marked strain Ps. 3732RNL11, showing that both organisms are non-pathogenic.

This conclusion is supported by tests performed independently by Dr Arthur Kelman in the Department of Plant Pathology, University of Wisconsin, Madison. These tests led to the conclusion that the strains Ps. 3732RN and Ps. 3732RNL11 do not possess the pectinolytic ability to cause potato rotting or maceration, or to form pits on solid media containing pectin. This confirms that these strains do not possess pectin esterase activity, a property shared by most plant pathogens. In addition, it was also demonstrated by that laboratory that neither strain induces a hypersensitivity response on tobacco.

Root-colonizing *Pseudomonas* strains typically are highly dependent upon plant root exudates for survival in natural field conditions. When such strains are introduced onto crop plants as a seed inoculum, these bacteria usually decline from a relatively high initial population density to much lower levels in the course of the typical growing season. This is probably in response to intense competition by very similar native bacterial strains which are highly adapted to the specific environmental conditions of that field. The field test being conducted provides a more definite, and practical, means to analyse such population levels, both on the initial inoculated crop and on subsequent crops planted in the same field plot.

Before field introduction, extensive survival tests were performed under laboratory conditions for both the parental strain Ps. 3732RN and the *lac*ZY-engineered strain Ps. 3732RNL11. These were carried out on several crop varieties and weed species, which were analysed for the presence of bacteria both on the root surfaces and upper plant parts. Further, field trials of the non-engineered parent strain Ps. 3732RN were carried out over a 2-year period to provide data about survival both during the growing season and after the overwintering period. Results confirmed that this strain was an efficient root colonizer during early parts of the growing period, declined gradually and was not detectable after the winter period. Colonization did not take place on above-ground plant or weed parts, or on the roots of adjacent plants which did not receive the initial bacterial inoculum. Finally, the inability of Ps. 3732RN to survive in natural aquatic environments, such as lake and river water and sewage influent, was demonstrated. These experiments were designed to determine the resistance of this microbe to starvation, as well as its ability to compete in natural biotic communities. The results of these tests confirmed a low competitive ability of Ps. 3732RN in typical aquatic environments, probably reflecting the biotic effects of predation, parasitism and competition.

Importantly, similar growth-chamber tests were also carried out by workers at Clemson University with soil from the field test site in order to (a) verify that effective colonization by the introduced bacteria would be achieved and (b) to determine the predictive value of such a chamber test to indicate actual field performance of a released organism.

The inserted sequences in Ps. 3732RNL11 can be viewed as being permanent additions to the genotype of the isolate. As such, their genetic transfer can be regarded as no more likely to occur than that of the Ps. 3732RN genes located adjacent to them. A mating system is not known in Ps. 3732RN, though this may reflect on a general lack of knowledge of mating systems for all but a small number of bacteria. Conjugative plasmids are thought frequently to be responsible for genetic exchange in nature. Plasmids of Inc groups P-1, W and N cannot be introduced into and/or replicate in Ps. 3732RN and no evidence for endogenous plasmids could be found. The Tn*7*-*lac* element is fixed in the bacterial chromosome because of the absence of the required transposition genes. These are unlikely easily to be supplied given the rarity of the Tn*7* element, except in clinical settings after heavy use of trimethoprim (Datta and Richards, 1981; Pulkkinen *et al.*, 1984). This is supported by an experiment in which the Tn*7* element was used as a labelled hybridization probe to detect the presence of Tn*7*-like DNA in a large collection of native soil bacteria. Such elements were not detected after extensive hybridization analysis of more than 500 individually isolated, fluorescent pseudomonad strains.

A series of laboratory tests was performed which strongly support the argument that the chromosomally inserted *lac*ZY gene element is stable and will remain within the original introduced strain Ps. 3732RNL11. Unselected growth of Ps. 3732RNL11 for over 110 generations failed to yield a *lac* derivative in greater than 10 000 isolates which were analysed. The potential for conjugal transfer of the inserted *lac*ZY element was then analysed first by testing for transfer between identical strains of Ps. 3732RN, differentially marked with separate antibiotic resistances. No genetic exchange was observed. The tests were now repeated with four independent pseudomonad strains isolated from the actual field test site. Once again, genetic exchange was not detected.

As the field test progresses, a large number of native bacterial strains are being isolated from the roots of the inoculated plants which also carry the *lac*ZY-marked organism Ps. 3732RNL11. These native isolates will be extensively probed by colony hybridization analysis to determine whether any have acquired the *lac*ZY genes after their close association with the engineered microbe in the field.

Tests were performed at the field site before GEM release to confirm that there would be no interference by indigenous microbes with the ability to detect *lac*ZY-marked introduced bacteria. Wheat was planted in the field site a month before the actual test, harvested after 2 weeks and analysed on various selective media. Similar analyses were carried out for surface water in a pond near the test site. No interfering native strains were found which were able to grow on either of the two selective media to be used in the field study.

In addition, for this particular field test, it was considered whether the *lac*ZY-marked strain Ps. 3732RNL11 might interfere with certain presumptive water analysis tests for coliform bacteria. A series of studies was carried out both at Clemson University, and independently by the South Carolina Department of Health and Environmental Control (DHEC), to determine whether such interference might occur. All results demonstrated conclusively that neither the parent strain Ps. 3732RN nor the *lac*ZY-containing strain Ps. 3732RNL11 caused any detectable interference in these tests. Another study is underway by Monsanto to verify that the *lac*ZY element does not confer any growth advantage to Ps. 3732RNL11 in stored milk products.

INITIAL FIELD TEST RESULTS

The initial results of the Monsanto-Clemson field test in South Carolina of the recombinant bacterium Ps. 3732RNL11 have demonstrated the effectiveness of the *lac*ZY marker in tracking the engineered bacterium in the field and provided important information about the longevity and migration of such an introduced organism. The data obtained during ten weeks of field testing indicate that:

(i) The recombinant micro-organism is as effective as the non-engineered strain in establishing high population levels on the inoculated wheat plant roots. All inoculated wheat rows of all replications were generally well-colonized even during 10 weeks post-planting by the *lac*ZY-marked *Pseudomonas* strain (Fig. 1).

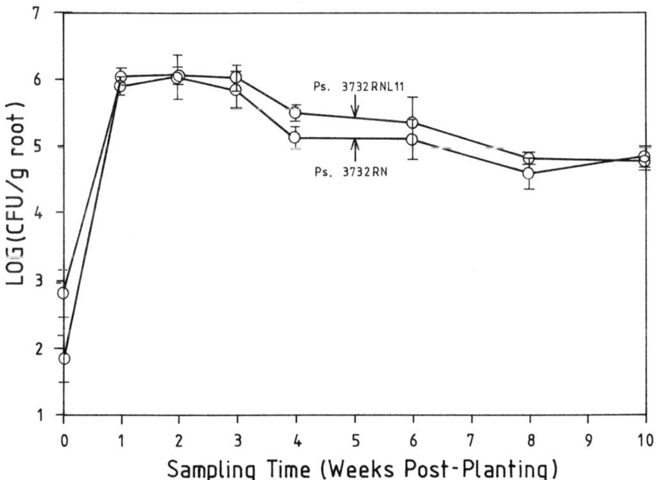

Fig. 1. Comparison of the level of colonization of wheat roots in the field by the *lac*ZY-marked recombinant (Ps. 3732RNL11) and its parental strain (Ps. 3732RN). Analysis was by growth on PAF/Rif/Nal medium. Bars indicate a difference of 1 standard deviation.

(ii) The *lacZY* tracking system is very effective for monitoring the presence and relative population level of the engineered fluorescent *Pseudomonas* bacterium (Fig. 2). Some differences in recovery were observed for the *lacZY*-marked strain on the two media, particularly at weeks 4 and 10. This was found to be largely reversible by a slight modification in the minimal M9 medium and incubation at 30°C rather than at 23°C.

(iii) Very limited movement of the introduced engineered microbe has been observed. The furthest migration of the bacteria from the inoculated rows of wheat was approximately 18 cm. This was observed in only one sample of 96 taken at this distance and at only one sampling time (Week 2). Neither the engineered strain nor the non-engineered parent have been found in any of the samples taken from non-planted border areas, drainage areas or the pond water. These data indicate that the introduced engineered or non-engineered bacteria do not out-compete the native microbe population and are largely confined to the actual place of introduction.

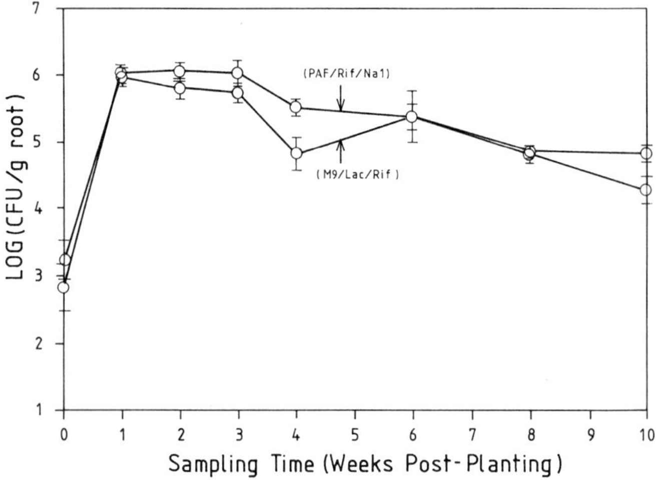

Fig. 2. Comparison of bacterial recovery (Ps. 3732RNL11) on PAF/Rif/Nal and M9/Lac/Rif media after growth on wheat roots in the field. Bars indicate a difference of 1 standard deviation.

(iv) The pattern of survival for the recombinant bacterium in the field was predicted very well by the contained growth chamber tests (Fig. 3).

These results are in very good agreement with the significant findings obtained earlier by L. Watrud and co-workers, who isolated the parental strain Ps. 3732RN and performed extensive preliminary testing (Obukowicz *et al.*, 1986, 1987).

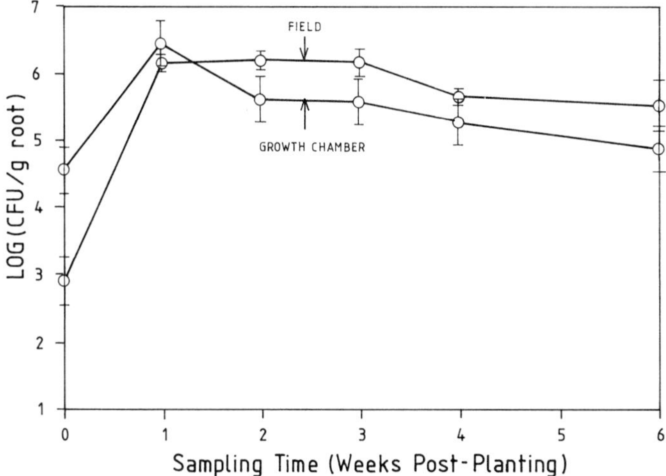

Fig. 3. Comparison of bacterial colonization (Ps. 3732RNL11) in the growth chamber with that in field tests. Analysis was by growth on PAF/Rif/Nal medium. Bars indicate a difference of 1 standard deviation.

Data will continue to be accumulated throughout the remainder of the 18-month study. After harvest of the first wheat crop, uninoculated, no-till soybeans will be planted in the same field plot and then also will be extensively assessed for the presence of the original inoculated bacteria. In the autumn of 1988 a second winter wheat crop will be planted on the site without any further bacterial inoculation.

SUMMARY

As environmental introduction and testing of genetically-engineered micro-organisms begins, it is essential to provide adequate assurance that such introductions will be environmentally responsible and carry no unacceptable risk. Inherent in this goal is a two-fold mandate. First, initial field tests must be designed to aid in the evolution of responsible and practical pre-release analyses. Second, appropriate methods must be developed which would allow a more accurate assessment of basic microbial behaviour under actual field conditions. For these purposes, an approved US field test has been initiated of a recombinant soil bacterium (*Pseudomonas fluorescens* 3732RNL11), which has been engineered to carry on its chromosome the *lac*ZY marker genes from *E. coli* K12.

Before field use, the determination of certain basic characteristics of the test organism was considered essential. These include: host strain identification to the species level, verification that this species was considered non-pathogenic, and verification that the microbe is not a plant pest. In addition, analysis was performed to estimate the potential for survival of the microbe under simulated field conditions, to verify the stability and permanence of the inserted genetic material, and to determine that the intended tracking method will function efficiently without interference from native microbial strains. Initial field data indicate that the engineered microbe is equally effective in populating the roots of inoculated plants as the non-engineered parent strain, the lateral and vertical dissemination of the GEM is very limited, the *lac*ZY tracking system functions effectively as an environmental monitoring tool under field conditions and that many of the basic tests performed before the field test have a significant predictive value for strain performance.

ACKNOWLEDGEMENTS

We would like to express our sincere thanks to: Arthur Kelman for generously offering to perform plant pathogenicity tests; Lidia Watrud, Mark Obukowicz, Fred Perlak, Kuniko Kusano-Kretzmer, Ernest Mayer, Ellen Lawrence, Margan Miller-Wideman, Minhtien Tran and Suzanne Bolten for providing the host bacterial strain Ps. 3732RN, as well as important initial data on strain toxicity and growth characteristics; and Juliet Johnson for her expert assistance in manuscript preparation.

REFERENCES

Barry, G. F., 1986. Permanent insertion of foreign genes into the chromosomes of soil bacteria. *Bio/Technology* **4**, 446–449.

Barry, G. F., 1988. A wide host-range shuttle system for chromosomal gene insertion in bacteria. *Gene*, submitted.

Datta, N. and Richards, H., 1981. Trimethoprim-resistant bacteria in hospital and in the community: spread of plasmids and transposons. In *Molecular Biology, Pathogenicity, and Ecology of Bacterial Plasmids*, Levy, S. B., Clowes, R. C. and Koenig, E. L. (eds), pp. 21–30. New York: Plenum Press.

Drahos, D. J., Hemming, B. C. and McPherson, S., 1986. Tracking recombinant organisms in the environment: β-galactosidase as a selectable non-antibiotic marker for fluorescent pseudomonads. *Bio/Technology* **4**, 439–444.

Hemming, B. C. and Drahos, D. J., 1984. β-Galactosidase, a selectable non-antibiotic marker for fluorescent pseudomonads. *Journal of Cellular Biochemistry Supplement* **8B**, 252.

Lelliott, R. A., Billing, E. and Hayward, A. C., 1966. A determinative scheme for the fluorescent plant pathogenic pseudomonads. *Journal of Applied Bacteriology* **29**, 470–489.

Obukowicz, M. G., Perlak, F. J., Kusano-Kretzmer, K., Mayer, E. J., Bolton, S. L. and Watrud, L. S., 1986. Tn5-mediated integration of the delta-endotoxin gene from *Bacillus thuringiensis* into the chromosome of root-colonizing pseudomonads. *Journal of Bacteriology* **168**, 982–989.

Obukowicz, M. G., Perlak, F. J., Bolten, S. L., Kusano-Kretzmer, K., Mayer, E. J. and Watrud, L. S., 1987. IS50L as a non-self transposable vector used to integrate the *Bacillus thuringiensis* delta-endotoxin gene into the chromosome of root-colonizing pseudomonads. *Gene* **51,** 91–96.

Pulkkinen, L., Huovinen, P., Vuorio, E. and Toivanen, P., 1984. Characterization of trimethoprim resistance by use of probes specific for transposon Tn7. *Antimicrobial Agents and Chemotherapy* **26,** 82–86.

14 Round Table 1: Applications

Chairman
K. M. TIMMIS *(Geneva, Switzerland)*

Rapporteur
GARETH J. WARREN *(Oakland, USA)*

Contributors
GARETH J. WARREN *(Oakland, USA)*
H. R. WHITELY *(Seattle, USA)*
NEAL GUTTERSON *(Oakland, USA)*
KENNETH M. TIMMIS *(Geneva, Switzerland)*
M. J. GASSON *(Norwich, UK)*

Editor
C. H. COLLINS

Gareth J. Warren (*Oakland, USA*) discussed the applications of ice nucleation that involved the deliberate release of genetically-engineered micro-organisms (GEMs). These are diverse and encompass agriculture, food processing, medicine and research. Deliberate release of recombinant organisms is necessary or desirable if each of these applications is to achieve its potential.

Frost protection of sensitive crop plants

Ice-nucleation activity (INA) by epiphytic pseudomonads is responsible for the susceptibility to frost damage of certain crop plants within temperature ranges close to 0°C. Arny and Lindow developed the concept of using competitor bacteria to displace or exclude the naturally-occurring INA^+ bacteria from the plant. It is necessary that such competitors should themselves be INA^- and that their own ecological niche should correspond closely to that of the naturally-occurring INA^+ strains. Unfortunately, there is only very imprecise knowledge of what constitutes the ecological niche of the INA^+ organisms in terms of

RELEASE OF GENETICALLY-ENGINEERED
MICRO-ORGANISMS ISBN 0–12–677521–4

nutrients, pH, temperature optimum, temperature range, and a host of other variables. Therefore, there are two options for obtaining competitors to fill this niche. The first would be experimentation to find naturally-occurring competitors which displace INA^+ bacteria in a series of empirical trials. There would be no theoretical basis for expecting that a competitor which 'worked' in one environment could compete effectively in another. It is therefore more satisfactory to adopt the second option of deriving INA^- from INA^+ bacteria by minimal genetic engineering. Here one can expect the engineered INA^- to occupy *precisely* the same niche as the naturally-occurring INA^+ and to compete effectively in all environments. Since the INA^- must be used in agricultural fields to create the desired effect, this application necessitates the release of GEMs into the environment.

The first such release was carried out by J. Lindemann and T. Suslow in 1987. The second release, only days later, was carried out by S. Lindow.

Improvement of spray-freezing efficiency

In Arctic-ice engineering and in snow-making for recreational purposes, water is sprayed into the air to be converted to ice. By preventing the sprayed water droplets from supercooling, ice nuclei can greatly improve the efficiency of spray freezing at temperatures that are otherwise marginal (e.g. −5°C). At temperatures above 0°C the nuclei are of no help, while at temperatures below −10°C mechanical agitation effectively nucleates most of the sprayed water droplets.

The efficacy of an ice-nucleating additive is related to the number of ice nuclei that are active *at the temperature of use*. Over-production of the ice-nucleating protein can be achieved by genetic manipulation and this increases the frequency of ice nuclei active at any temperature, but particularly of those active at the warmer temperatures that are marginal for spray freezing. Unfortunately, the protein alone has almost no activity, and even killing the producing cells reduces both the activity and stability of the ice nuclei which are active in the most useful temperature range. The ability to use the over-producing GEMs as a viable additive in spray freezing would reduce the amount of material required and increase its efficacy. There is no reason why the GEM could not, in this and the following cases, be a disabled strain that would not survive in the environment for even short periods.

Improvement of frozen food manufacture and storage

In the manufacture of certain frozen foods, such as ice cream, many small ice crystals are formed and their stability during storage is necessary for the maintenance of acceptable quality. What tends to happen during storage is that, because of differences in water vapour pressure for crystals of differing size, water diffuses from small crystals and accretes on larger ones, causing a deleterious increase in mean crystal size. This is accelerated by fluctuations in storage temperature.

Ice nuclei cause a greater number of ice crystals to form independently and thus reduce mean crystal size at the time of manufacture. For this reason it takes longer for storage to cause a given degree of quality deterioration, even if no other effects act subsequently. In fact, we can expect that ice nuclei will re-nucleate repeatedly during storage, tending to stabilize the number of ice crystals present and thereby keep their mean size low.

As before, the efficacy of an ice-nucleating additive is determined by the number of ice nuclei present which are active at the relatively warmer temperatures. At present, it is not possible to maintain and stabilize such ice nuclei in cell-free extracts. The elegant and economical solution would be to use viable food-grade micro-organisms to provide the ice nuclei.

Molecular detection in clinical diagnostics

Single ice-nucleation events can be detected easily and reliably. This means that if a single cell responds to a stimulus by synthesizing an ice nucleus, one would be able to detect that stimulus with great sensitivity. Theoretically, the binding of a single analyte molecule by a cell receptor could be coupled to the synthesis of an ice nucleus and its subsequent detection. The mechanism by which binding could be coupled to induce the appropriate transcription is not trivial. It would necessarily be different for each analyte to be detected and is beyond the scope of this discussion. What is important is that a living cell, engineered for the appropriate response, is required. The major application of such an approach would be in the detection of medically-significant molecules in clinical samples. This would require transport of the system to reference laboratories for use on a large scale. The probability of accidental release in such use is high enough to consider the approach as a type of deliberate release, although as before, there would be no reason for the GEMs not to be environmentally disabled.

Research uses

Because the nucleation activity can be easily monitored and can detect very small numbers of bacteria, T. Suslow has proposed that INA be used as a marker for released GEMs to aid in their subsequent detection. Additionally, transcriptional fusion of *ina* genes to regions of interest allows the study of genetic regulation in these regions. This has been facilitated by the construction of a suitable transposon by B. Staskawicz. The main use of such a system is for the study of regulation under field conditions, where the harvestable bacterial population may be extremely small.

Discussion

In conclusion, in each application of ice nucleation, the use of GEMs appears scientifically or practically preferable to its alternative. It may be difficult, however, to compare the potential benefits of the different types of application, since the benefits would accrue in very different arenas. Nevertheless, consistent analysis of benefits will be essential for the conduct of the risk/benefit analysis that forms part of the regulatory process of considering the release of GEMs.

H. R. Whiteley (*Seattle, USA*) then considered the properties and genetic engineering of the insecticidal crystal proteins of *Bacillus thuringiensis* (BT).

Bacillus thuringiensis is notable because it synthesizes large insecticidal proteinaceous crystals during sporulation (Whiteley and Schnepf, 1986). Larvae of three orders of insects are susceptible to the crystals produced by different subspecies of BT. Some crystals are lethal to *Lepidoptera*, others to *Coleoptera* and still others to some *Diptera*. Many of the insects that can be killed by BT crystals cause serious damage to crops and forests. Some insects, mosquitoes and black flies are also carriers of human and animal disease agents. The economic losses due to these insect pests is enormous. For example, the world-wide cost of losses of crops to lepidopteran pests alone has been estimated to be $1–2 billion each year.

Preparations of crystals and spores have been used for about 35 years to control lepidopteran pests. There is also a more limited use of the dipteran-specific crystals to control mosquito and black fly larvae. BT has a number of important advantages over conventional chemical pesticides. Firstly, the crystals are effective at fairly low doses and are highly specific. They are toxic to lepidopteran larvae but are non-toxic to all animals, plants and all insects other than lepidopterans. Similarly, the dipteran and coleopteran toxins are toxic only to these insects. Secondly, the toxins can be more easily combined with other biological control measures (integrated pest management) than chemical pesticides. Lastly, only one instance of insect resistance to crystal proteins has, so far, been observed. Despite these advantages, BT has not been widely used as a pesticide because of cost, need for repeated spraying because of inactivation and because insects are not killed immediately on contact.

A gene designation is proposed based on the pathogenicity of crystals to different orders of insects: *cry*A genes are those that code for proteins toxic only to *Lepidoptera*, *cry*B genes code for proteins toxic to both *Lepidoptera* and

Diptera, specifically mosquitoes, *cry*C codes for a peptide that is toxic to *Coleoptera* and cryD proteins are toxic to certain *Diptera* (mosquitoes and black flies).

Most crystal protein genes are located on plasmids and several genes have been cloned and sequenced. The best studied are genes coding for cryA proteins. Depending on the BT subspecies and strain, each crystal is composed of 1–3 polypeptides having a molecular mass of *ca.* 130–160 kDa. Each peptide is a protoxin that is cleaved in the insect gut to a toxic peptide of 55–68 kDa. Lectin-binding studies (Knowles and Eller, 1986) suggest that the toxic peptide interacts with a glycoprotein receptor on the epithelial gut cell surface. This interaction results in release of K^+ and other components and is followed by cell lysis. The cryB proteins are smaller (71 kDa) and are present in small cuboidal crystals; the cryC protein is of about the same size (72 kDa) and is located in square, flat crystals. The cryD proteins are found in a multi-component crystal consisting of at least four major proteins (135, 128, 65 and 27 kDa plus minor proteins such as one of 58 kDa). The mechanisms of action of the latter three types of proteins have not been investigated except for the 27 kDa dipteran toxin. The latter protein binds to phospholipids in the cell membrane and causes lysis by a detergent-like action (Thomas and Ellar, 1983).

Examples of each of the four crystal genes have been cloned and sequenced. Four different lepidoptera-specific protoxins can be identified from the deduced amino acid sequences (Adang *et al.*, 1985; Schnepf *et al.*, 1985; Geiser *et al.*, 1986; Brizzand and Whitely, 1988) and combinations of these genes are found in different subspecies of BT (Kronstad and Whiteley, 1981). When more genes are sequenced, additional types of lepidoptera-specific protoxins will undoubtedly be found. Comparisons of the amino acid sequences of six genes representative of the four proposed classes, two lepidoptera-specific, two diptera-specific, one coleoptera- and one lepidoptera–diptera-specific, show that thcy contain regions of similarity suggesting an evolutionary relationship.

BT genes have been cloned into a variety of bacteria in addition to *Escherichia coli* and *Bacillus subtilis*: (1) into a strain of *Pseudomonas* which has then been encapsulated and heat-killed to increase stability of the toxin; (2) into a species of root-colonizing *Pseudomonas* with the idea that seeds treated with the recombinant strain would be resistant to cutworm; (3) into an endophytic bacterium that colonizes the xylem of plants, thus making the plant resistant to *Lepidoptera* and (4) into blue-green algae which might serve as food for mosquito larvae in ponds and streams. Genetic engineering has also been used to introduce BT toxin genes into tobacco and tomato plants to produce pest-resistant plants. These experiments were reported in 1987 by Fischhoff *et al.* (1987), Vaeck *et al.* (1987) and Barton *et al.* (1987). Each used the same cloning strategies and reported essentially the same results, i.e. the plants were resistant to several different *Lepidoptera* larvae and the genes were transmitted to the next generation of plants. These experiments open the possibility of producing a variety of pest-resistant plants, thus by-passing the use of toxic chemical pesticides.

Since the BT toxins are non-toxic to all forms of life, except the target insects, and unexpected toxicities have not been revealed in about 35 years of use of crystals as pesticides, it seems unlikely that the introduction of these genes into other bacteria or plants would endow them with new and unexpected pathogenicities. However, pest-resistant plants could be retarded in growth. A more serious question is whether the existence of large numbers of pest-resistant plants would lead to the selection of BT-resistant insects. Selection pressure, the number of factors involved in toxicity to the target insect and whether all toxins of a given class have the same mechanism of toxicity, are factors that might affect the rate of development of insects resistant to BT.

Neal Gutterson (*Oakland, USA*) followed with an account of genetically-engineered micro-organisms in the development of microbial fungicides.

Disease control in agriculture by GEMs is still in a developmental phase and his group are working on control by microbial fungicides (MF). A principal motive for this is the problem of chemical toxicity, both in production and in the environment. In the USA there have been pressures from the public and regulating agencies and, for example, both dibromochloropropane and ethylene dibromide have been withdrawn from the market. Biological disease-control agents do not have these toxicity problems. Moreover, some diseases are not amenable to chemical control.

The requirements and limitations in the development of an MF include:

(1) It must proliferate at the desired location, such as seed and root. In the latter site it must compete with the indigenous microbial population.
(2) It must produce substances that inhibit the growth of the pathogen; antibiotics are elaborated by many soil micro-organisms.

The genetics of proliferation in the seed and root are poorly understood as are those of competition between introduced and indigenous bacteria. In addition, it is likely that the deficient performance of MFs may be due to variability in production of the antibiotics. The application of genetic engineering is critical, both for a better understanding of these processes and for improving the performance of MFs by altering the genes responsible for proliferation and antibiotic synthesis.

The role of GEMs and their release into the environment may be illustrated by the analysis of antibiotic synthesis *in vitro* and *in situ* in a strain of *Pseudomonas fluorescens* (Hv37a) being developed in their laboratories. This protects cotton seedlings from seed root rot mediated by *Pythium ultimum* by the production of an antibiotic oomycin A. This reduces significantly the root infections caused by the *Pythium* so that more, and healthier, plants emerge from planted seeds.

The production of oomycin A cannot be studied directly on seeds or roots as antibiotics are efficiently absorbed by soil particles. Genetic engineering offers an indirect method for monitoring antibiotic biosynthesis in this environment by using the expression of the genes at a level correlated with the biosynthesis. The expression is monitored with a reporter phenotype, that of bioluminescence encoded by the *lux* genes of *Vibrio fischeri*, programmed by appropriate promoters for antibiotic biosynthetic genes.

The genes involved in the oomycin biosynthesis were isolated from a cosmid by complementation of mutants deficient in oomycin A biosynthesis. Four operons were identified, including a minimum of five genes. Four of these genetic loci (*afu* A, B, P and R) appear to be required for regulation and only the *afu*E operon is expressed at levels correlated with antibiotic biosynthesis. For example, oomycin A biosynthesis is enhanced 200-fold by glucose, as is expression of the *afu* operon.

Transcriptional fusions in which the *afu*E promoter programmes the *lux* expression were constructed by inserting the Lux Km cassette of Shaw and Kade into a plasmid containing the *afu*E region in a wide host-range plasmid (Fig. 1).

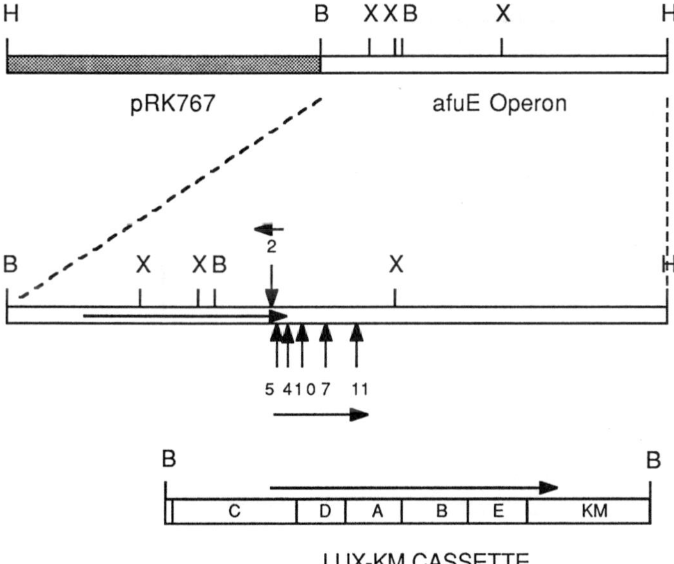

Fig. 1. Isolation of *afu*E-*lux* fusions.

These fusions were then mobilized to the *P. fluorescens* Hv37a strain and bioluminescence measured under various conditions *in vitro* and *in vivo*.

To provide background information for understanding *in vivo* antibiotic biosynthesis *afu*E expression was monitored in broth culture. *afu*E expression is at a low basal order under optimal conditions until late log phase, at which time expression is induced 50–100-fold (Fig. 2). The effect of nutrients common to seed and root exudates on *afu*E expression was also measured. Glucose and galactose induced expression, but the TCA cycle intermediate citrate and malate repressed it even at only 4% of the concentration of glucose (Fig. 3). Thus, both nutritional and temporal factors could affect *afu*E expression on seeds and roots.

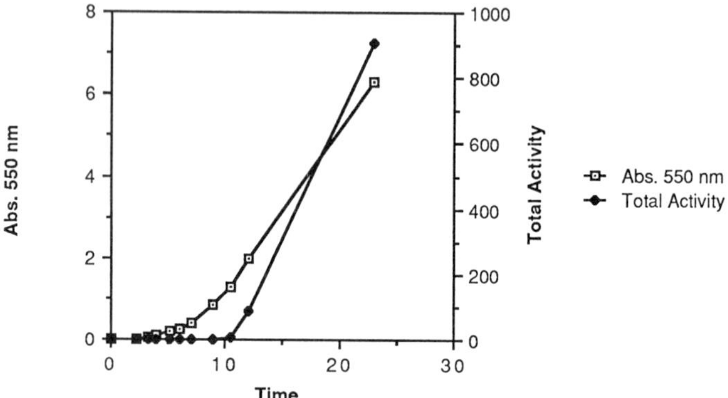

Fig. 2. Time course for *afu*E expression.

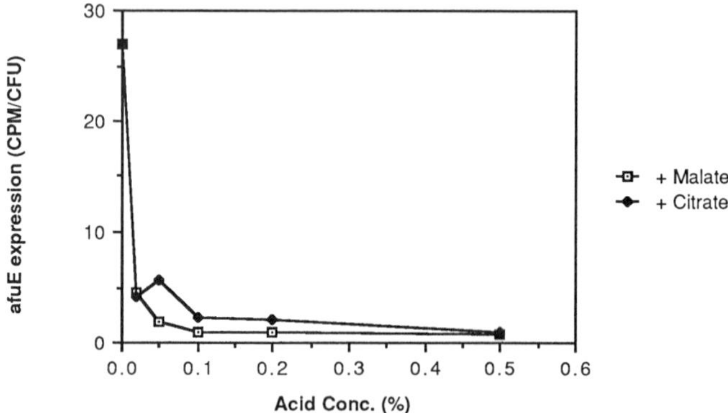

Fig. 3. Inhibition of *afu*E expression by acids.

One example of the use of the *afu*E-*lux* system in soil experiments is an examination of the timing of oomycin A biosynthesis on the seed in comparison with the timing of infection by *Pythium ultimum*. By 6 h after planting 60% of cotton seeds were infected, by 10 h about 100%, by 20 h after planting cotton seeds coated with Hv37a harbouring the *afu*E-*lux* fusion *afu*E expression was induced to nearly maximal levels. Thus, it appears that induction of oomycin A biosynthesis occurs early after germination but not as early as *Pythium* infection. The maximal level of expression on seeds was less than 10% of that observed under optimal *in vitro* conditions.

There are several additional questions about the release of the *afu*E-*lux* system, both in greenhouse and field conditions, e.g. the effects of soil type, moisture level and soil pH on oomycin A biosynthesis. In general, the system may be used to monitor biosynthesis in the field to determine whether failure of disease control can be attributed to lack of oomycin A biosynthesis despite adequate bacterial growth.

GEMs are being developed not only for analysis of MF function but also for improving their efficiency. The approach requires the cloning of all the genes encoding enzymes of the antibiotic biosynthetic pathway. These genes can be assembled into a single DNA element which can either be returned to the original host to alter the pattern of regulation or to a heterologous host with desirable traits of proliferation on seeds or roots. The genes can be integrated into the host genome by either site-specific recombination or transposon-mediated events. In either case, the introduced genes would be expected to be at no greater risk of gene transfer than any other chromosomal locus.

In conclusion, the release of GEMs is essential to the development of MFs. Firstly for essential studies on the mechanisms and limitations of performance and, secondly, to develop strains with improved properties of antibiotic biosynthesis and environmental proliferation. The released organism will be designed so that they will not possess introduced extrachromosomal information that could be transmitted to other organisms.

Kenneth M. Timmis (*Geneva, Switzerland*) followed with a discussion of the prospects for producing GEMs that are able to degrade environmental pollutants.

Large quantities of industrial chemicals are released into the environment. Their persistence varies from weeks, in the case of some chemicals that are readily degraded by soil and water micro-organisms, to years. The more persistent compounds tend to be those having structural elements rarely found in nature (xenobiotics), which are catabolized poorly by naturally-occurring micro-organisms. Certain of the persistent xenobiotics are also toxic; these present a major problem of environmental pollution.

Soil and water micro-organisms collectively exhibit a wide range of degradative activities that are versatile and in some instances are able to evolve new pathways. There is a need, however, to accelerate the evolution of degradative capabilities in many cases and this can be done by genetic engineering.

The evolution of a new metabolic potential involves both the acquisition of new or modified enzymatic activities and natural selection of appropriate regulation to ensure appropriately-timed expression of the new activities. The recruitment of enzymes that exhibit broad substrate-specificities and the ability of enzymes and regulators to undergo a change in specificity without loss of function are critical to both natural and accelerated evolution of new biodegradative pathways.

Two strategies are available for the development of a particular novel degradative capacity. Firstly, if the chemical has a substantial structural homology to degradable compounds it is possible to identify which steps of the known degradation pathway are unable to handle it and specifically and sequentially to modify these so that they become permissive (pathway restructure). Secondly, where an existing pathway for related compounds is not known, new pathways can be conceived by piecing together various types of enzymatic activity and the appropriate component enzymes can then be sought in soil bacteria (pathway assembly). When the feasibility of a new route is established, the corresponding enzymes can be assembled in a single host, or a consortium, and placed under appropriate metabolic regulation.

The following example of pathway modification was discussed.

Pseudomonas putida containing TOL plasmid pWWO is able to grow on benzoate and alkyl benzoates such as 3- and 4-methyl benzoate, 3, 4-dimethyl benzoate and 3-ethyl benzoate but not 4-ethyl benzoate (4EB). The TOL plasmid-specified pathway involves dioxygenation of the ring, decarboxylation to yield alkyl-catechols and further oxidation involving *meta* cleavage. Other soil bacteria can be isolated that degrade methyl benzoates but generally they too cannot degrade 4EB. 4EB thus seem inherently more resistant to microbial attack. Analysis was conducted to see which step(s) of the TOL pathway were non-permissive for 4EB. Firstly, it was discovered that an important regulator, the XylS protein, was not activated by 4EB, so that 4EB did not induce synthesis of the catabolic enzymes. Mutant *xylS* genes, whose products could be activated by 4EB, were selected in a system specifically designed for this purpose. The system involved only the *xylS* gene and a XylS-inducible promoter fused to a tetracycline-resistance gene. Tetracycline selection, and screening for appropriate inducibility, yielded the desired type of 4EB-inducible system. The mutant *xylS* gene was transferred to *P. putida* (TOL) and the next non-permissive step in the pathway was identified. This turned out to be *meta*-cleavage of the catechol derivative of 4EB, because this derivative constitutes a suicide substrate for the *meta* cleavage enzyme. Selection for utilization of 4EB as a carbon source was then used to obtain mutants less sensitive to substrate inhibition and thus able to perform this step and complete the degradation.

The example discussed was the patchwork assembly of an *ortho* cleavage pathway for the degradation of mixtures of substituted aromatics such as methyl and chlorophenols. These occur in industrial wastes and can threaten the microbial communities in soil and waste-water treatment plants. The reason for this is that such compounds are typically degraded first to catechols for which both *ortho* and *meta* cleavage routes occur. When both are induced simultaneously, compounds which can be properly degraded by only one pathway are also processed by the other, poisoning the enzymes of that pathway.

The solution envisaged to this problem was the design of a pathway for mixtures of compounds that employed only one mode of catechol ring cleave. An organism possessing only *ortho* cleavage activity (*Pseudomonas* sp. B13) was found to perform ring cleavage on all relevant catechols, although the high substrate specificity of its initial catabolic enzyme restricted its catabolic activity to 3-chlorobenzoate. This activity was first expanded by recruitment of a relaxed-specificity benzoate dioxygenase from *P. putida*. This recruitment permitted the hybrid bacterium to form and cleave catechols from a number of methylated and chlorinated aromatics but to generate dead-end products from some. An *Alcaligenes* sp. capable of mineralizing these dead-end products was found, and the appropriate gene identified from a library of its DNA. This gene, transferred to *Pseudomonas* sp. B13, permitted the degradation of mixtures of methyl- and chlorobenzoates and phenols through the newly constructed pathway.

Kenneth Timmis concluded that laboratory procedures, including genetic engineering, can vastly accelerate the evolution of micro-organisms capable of degrading environmental pollutants. Understanding the degradative and regulatory pathways permits certain changes to be engineered or selected in isolation from other components of the pathway, then returned to a complete system. Other targets, whose manipulation may be needed to improve microbial degradative capacities, may be envisaged. For example, it may be desirable to engineer appropriate chemotactic responses, to improve cellular uptake of chemicals, and to increase an organism's resistance to a chemical's toxicity.

Finally, **M. J. Gasson** (*Norwich, UK*) considered the genetic manipulation of food fermenting micro-organisms.

The application of genetic engineering technologies in the food industries raises questions about the acceptability of food commodities intended for human consumption as well as the current issue of environmental release. The major opportunities for the development and use of recombinant microbes in food processing are within the traditional food fermentation industries which rely upon a variety of carefully selected strains of yeast and lactic acid bacteria.

Release in this context is a matter of use in the various food manufacturing fermentations and this is of relevance to the debate since neither brewing nor dairy fermentations are closed processes; both are hygienic rather than aseptic. As well as the opportunity for environmental release at manufacture, some commodities contain live microbes which may be an essential component of the product, as in the case of yoghurt. Attention was drawn to some foreseen opportunities for genetic manipulation of yeasts predominantly for brewing and of lactic acid bacteria for use in dairying.

The use of classical genetics to improve brewers' yeast is limited by the fact that those strains of *Saccharomyces cerevisiae* are usually polyploid and do not sporulate.

Genetic engineering is thus a valuable alternative strategy for strain improvement. Yeast is used to ferment sugars released from barley starch by the malting process. Yeast is not capable of starch degradation, the necessary amylase being provided by germinating barley grain. A major target for genetic engineering of yeast is the construction of amylolytic yeast strains. Interest is predominantly in enzymes that would further degrade the dextrins that are generated by the barley amylase. The glucoamylase gene from *Saccharomyces diastaticus* (DEX) has been cloned and expressed in brewers' yeast. Similar enzymes from food grade *Aspergillus* strains and enzyme genes from the cellulase complex of *Trichoderma reesei* are also being investigated. A distinct interest in β-glucanase genes exists. β-Glucan is another polysaccharide from the barley grain that can cause downstream problems by blocking filters and generating haze formation in the final product. β-Glucanase genes from *Bacillus subtilis* (Guinness) and from barley (Carlsberg) have been cloned in order to engineer yeast strains capable of degrading β-glucan.

Some aspects of yeast metabolism are important for product flavour. One problem concerns the generation of vicinal ketone compounds such as diacetyl and pentanedione non-enzymically from intermediates of the isoleucine–valine pathway. Auxotrophic mutations, increasing flux through the pathway, and the introduction of an enzyme to eliminate the precursors of the off-flavour are solutions under investigation.

Another area of interest is the spent yeast, most of which represents a waste product. The possibility of producing a secondary product by switching on a silent cloned gene after completion of the brewery fermentation has been demonstrated with the gene for human serum albumin with a galactose-inducible promoter (Delta Biotechnology). Food grade products such as enzymes might equally be produced in this way.

Genetic engineering has been possible in the lactic acid bacteria only since 1985 and the use of genetic manipulation for strain improvement is relatively new. Targets include aspects of metabolism that are important for dairy fermentations including lactose catabolism, proteolysis, and diacetyl production from citrate. All of these properties can be unstable and plasmid encoded genes are often present. One objective is to clone and stabilize the key genes by chromosomal integration. Proteolytic enzymes are being manipulated to effect desirable product flavour profiles and to accelerate mature flavour generation in cheese. A major problem of dairy fermentations is the susceptibility of starter lactic acid bacteria to bacteriophage attack. The use of bacteriophage-resistance genes to protect starter cultures has been demonstrated by conjugal transfer of plasmids and gene cloning offers a more sophisticated approach.

It is likely that the immediate targets for genetic manipulation in the food fermentation industries will involve homologous genes or genes from other food grade microbes. To this end in both yeast and lactic acid bacteria homologous food grade vector systems and the use of chromosomal integration of cloned genes are being developed.

SUMMARY

The applications discussed varied widely. The fields of potential use for released GEMs included agriculture, food technology, and the detoxification of organic pollutants.

There are two clear reasons for the use of GEMs in agriculture. The more obvious is that genetic engineering can produce combinations of traits that are not readily identified in nature, even if they exist somewhere. This is exemplified by the desire to create novel combinations of insect specificity by combining the syntheses of existing *Bacillus thuringiensis* toxins in one organism. It may even be possible to create completely novel specificities by the construction of hybrid genes, or to change the delivery site of toxins by expressing them in other bacterial, or even yeast and cyanobacterial species. It was also exemplified by the desire to synthesize fungicides in aggressively root-colonizing bacteria, to protect plants from several important fungal diseases.

A second reason for using GEMs concerns their scientific preferability even where non-recombinant microbes are or may become available as alternatives. This is that genetic engineering produces alterations that are often simpler and more easily understood than changes produced by other techniques: as a result their properties may be more predictable. Moreover, the rapidity with which genetically-engineered changes can be produced reduces the time in culture during which a microbial strain may lose 'fitness' or 'virulence'. This is an additional reason for engineering root-colonizers for fungicide production, and for engineering rather than mutagenizing to obtain INA^- bacteria.

In food technology and in degradation of organic chemicals, genetic engineering again makes possible combinations of traits that are not readily found in naturally-occurring microbes. The combination of biodegradative capacities from different organisms is an impressive creative feat that offers hope for practical solutions to some serious problems of pollution. In this case the objective is to create bacteria whose novel characteristics will give them an inherent selective advantage in a particular environment, and while it is true that such microbes might evolve without human intervention, it is likely that we should have to wait a long time for them.

Many characteristics of micro-organisms used in food manufacture might be improved to affect flavour, storage stability, appearance, and usefulness of process by-products. It is interesting that some organisms, selected through food production over long periods, are without naturally-occurring counterparts—they are already highly selected by human intervention. And it seems appropriate that genetic engineering could solve some of the problems that may have been created by long growth in culture: instability due to the presence of many plasmids, and susceptibility to bacteriophages.

The applications of GEMs that would involve their deliberate release are impressive in their breadth and ingenuity. We can look forward to a still further broadening of the field, and to eventually obtaining some experimental data on practical feasibility.

REFERENCES

Adang, M. J., Staver, M. J., Rocheleau, J., Leighton, R. F. *et al.*, 1985. *Gene* **36,** 289–300.

Barton, K. A., Whiteley, H. R. and Yang., N. S., 1987. *Plant Physiology* **85,** 1103–1109.

Brizzard, B. L. and Whiteley, H. R., 1988. *Nucleic Acids Research.* In press.

Fischhoff, D. A., Bowdish, K. S., Perlak, F. J., Marrone, P. G., McCormick, S. M. *et al.*, 1987. *Bio/Technology* **5,** 807–813.

Geiser, M., Schweitzer, S. and Grimm, C., 1986. *Gene* **48,** 109–118.

Knowles, B. H. and Ellar, D. J., 1986. *Journal of Cell Science* **83,** 89–97.

Schnepf, E. H., Wong, H. C. and Whiteley, H. R., 1985. *Journal of Biological Chemistry* **260,** 6272.

Thomas, W. E. and Ellar, D. J., 1983. *FEBS Letters* **154,** 362–368.

Vaeck, M., Reynaerts, Hofte, H., Jansens, S., DeBeuckeleer, M. *et al.*, 1987. *Nature* **328,** 33–37.

Whiteley, H. R. and Schnepf, H. E., 1986. *Annual Review of Microbiology* **40,** 549–576.

15 Round Table 2: Detection Methods including Sequencing and Probes

Chairmen
MANFRED G. HOFLE *(Plön, FRG)*
DAVID A. STAHL *(Urbana, USA)*

Rapporteur
GARY SAYLER *(Knoxville, USA)*

Contributors
RONALD M. ATLAS *(Louisville, USA)*
GERARD F. BARRY *(St Louis, USA)*
GERARD MUYZER *(Leiden, Netherlands)*
GARY SAYLER *(Knoxville, USA)*
DAVID A. STAHL *(Urbana, USA)*
ROBERT J. STEFFAN *(Louisville, USA)*

Editor
D. E. STEWART-TULL

The release of genetically-engineered micro-organisms (GEMs) into natural or man-made environments will add new dimensions, for instance, to food biotechnology, the microbial treatment of waste streams and biomining. Because these habitats are often so complex, with different populations of micro-organisms present, it may be necessary to identify individual bacterial species and determine their relative abundance in these natural samples, so as to assess their possible role in biotechnology processes. In addition, the introduction of GEMs into the field requires monitoring of their survival, their interaction with the indigenous microflora and their possible spread to surrounding areas. Accurate methods for identifying and quantifying individual bacterial taxa in natural samples are essential for these studies. Conventional microscopical counting techniques or the use of special culture conditions have been applied but these lack the selectivity and precision necessary to follow individual

RELEASE OF GENETICALLY-ENGINEERED
MICRO-ORGANISMS ISBN 0–12–677521–4

bacteria in complex microbial communities. Today, the application of pyrolysis–gas chromatography combined with mass spectrometry are better alternatives for fulfilling these needs. These methods have provided valuable information about the understanding of the role of individual bacterial species in complex microbial consortia. The experience with these methods may make it possible to follow the fate of GEMs in natural samples and to design and apply these techniques to this specific purpose.

OVERVIEW

The theme of the session was established by **Gary Sayler** (*Knoxville, USA*) in an overview of the available approaches for the determination of GEMs in the environment. In a concise talk, akin to a good after-dinner speech, he brought together the available technologies, and those still being developed, for the detection of engineered or specific recombinant organisms in the environment. It is necessary to discuss the issues relating to the dislocation of genes from, what may be considered, the host itself. The basic criteria for monitoring methods for detection of these organisms must be laid down, e.g. some of the methods may be specific for risk assessment or biotechnical needs in terms of the application of a microbial process in the environment. For these reasons the methods employed may vary depending upon the particular application of the detection or monitoring method established.

There are a number of requirements for environmental monitoring methods of recombinant DNA (rDNA):

	Purpose and criteria
Sensitivity	Quantitative risk analysis, should approach a zero threshold
Specificity	Detection of specific rDNA sequence or novel genotype, organismally independent and unambiguous
In situ	
Analysis	Avoid laboratory bias of cultivation and gene maintenance, direct or indirect environmental quantitation
Speed/efficiency	Rapid analysis of impact or risk identification and dispersal, process many or large samples with short turn around
Cost	Determined value of product or process, or level of identified hazard may determine acceptable cost, risk dependent

Table 1. Comparative target and sensitivity of rDNA monitoring methods.

Methods	Principal target*		Sensitivity	Comments
Cultivation-dependent methods				
Conventional selection and enrichment	Host		Potentially† >1 to 100 ml^{-1}	Presumptive, assumes 100% growth, alternative host?
Antibiotic and heavy metal resistance	Host		Potentially† >1 to 100 ml^{-1}	Presumptive, assumes 100% growth, alternative host?
Genetically linked traits	Host	(rDNA)	Potentially† >1 to 100 ml^{-1}	Nonconventional techniques only as good as the cultivation method
Plasmid analysis	Host		Potentially† >1 to 100 ml^{-1}	Nonconventional techniques only as good as the cultivation method
Protein analysis	Host		Potentially† >1 to 100 ml^{-1}	Nonconventional techniques only as good as the cultivation method
NA-sequence analysis	rDNA	(host)	Potentially† >1 to 100 ml^{-1}	Nonconventional techniques only as good as the cultivation method
NA-hybridization	rDNA	(host)	Potentially† >1 to 100 ml^{-1}	Nonconventional techniques only as good as the cultivation method
Cultivation-independent methods				
Gene cassette product	Host	(rDNA)	1 ml^{-1}	Technology not developed, expense?
Immunofluorescence	Host	(rDNA)	10^5 ml^{-1}	Can be rDNA specific if expressed as antigen
NA-sequence analysis	rDNA	(host)	Unknown	Technology not developed, expensive
NA-hybridization	rDNA	(host)	>1 ml^{-1}	Requires standardization

*Assuming the rDNA gene product is not expressed.

†With concentration or enrichment, sensitivity may be pushed to less than one organism per ml, this may not be achievable on a routine basis.

The comparative target and sensitivity of the recombinant DNA monitoring methods are shown in Table 1.

A number of methods have already been introduced for the monitoring of GEMs with potential application for quantifying degradative populations and many of these can be classified under the broad headings:

(a) Immunological techniques
 (1) Immunofluorescence *in situ* analysis
 (2) Enzyme-linked immunosorbent assay (ELISA)
(b) Radioactive markers
(c) Fluorescent markers
(d) Plasmid epidemiology and restriction profile
 (1) Use of plasmid epidemiology
 (2) Use of restriction profile (RFLP)
(e) Selectable genotypic markers
(f) Nucleic acid sequence analysis
 (1) DNA sequence analysis
 (2) Ribosomal RNA sequence analysis
(g) Nucleic acid hybridization techniques
 (1) DNA:DNA colony hybridization
 (2) Southern blot hybridization
 (3) Nucleic acid hybridization with DNA extracts
 (4) DNA:RNA hybridization
 (5) Biotinylated probes
(h) New selective enrichments
(i) Protein and enzyme analysis
 (1) Isozymes (MLEE)
 (2) Protein gels (SDS–PAGE)
(j) Genetically-engineered markers
 (1) Metabolic changes
 (2) Surface marker changes
 (3) Susceptibility changes
 (4) Other changes
(k) Mass spectrometry

Nucleic acid hybridization techniques are being used extensively in classical colony hybridization approaches. It is also possible to do blot or slot-blot hybridizations or with DNA extracted directly from the environmental samples or from cells released from sample populations (Fig. 1). This particular example, coupling that to the case of a specific catabolic genotype of a catabolic plasmid pSS50 detected in extracts of biomass recovered from aquatic environmental samples. This demonstrates the ability to detect nucleic acids in a complex mixture of DNA. It might be possible to use DNA extracts in a reassociation

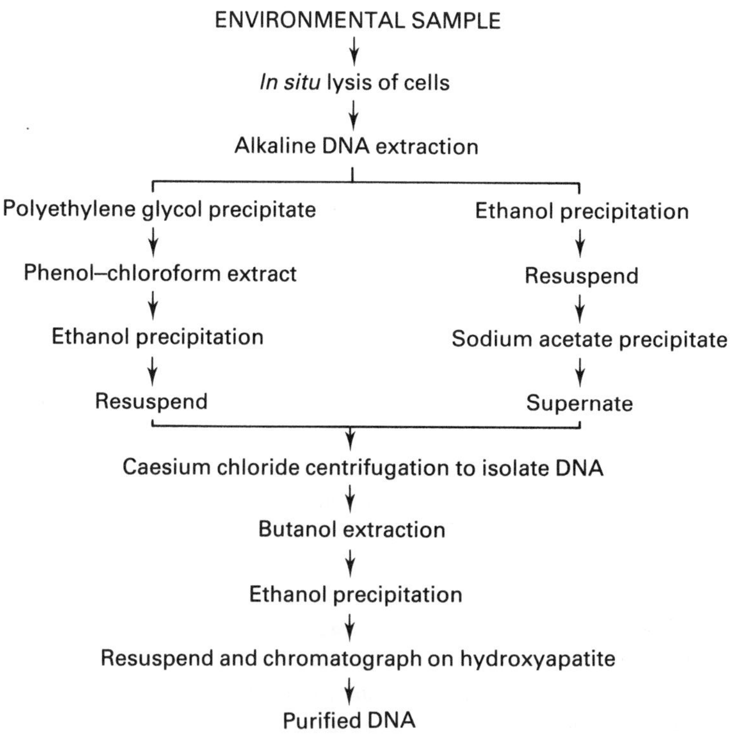

Fig. 1. Flow diagram of the extraction of DNA from environmental samples.

kinetic format as a measure of microbial population complexity in nature. Another possibility is the introduction of a 'reporter gene', e.g. a bioluminescence gene into a catabolic plasmid to develop a reporter strain as it would give off visible light. The regulated emission of light from a catabolic pathway might be used as a potential tool for measuring gene expression.

CONSTRUCTION OF REPORTER STRAINS

Gerard Barry (*St Louis, USA*) discussed the tracking and monitoring of released GEMs. The use of tracker genes and probe sequences in soil micro-organisms was described.

Fluorescent pseudomonads were used because they are high-level soil and plant root colonizers and important and exploited for agronomic purposes. They are potential hosts for the very precise and localized delivery of plant beneficial or plant pest antagonistic products. These bacteria are *Lac*$^-$ both by ONPG and

X-Gal tests and this was originally exploited in a series of plasmids which conferred on these bacteria the ability to grow on lactose and provide a differential means to recover specific introduced populations in non-sterile environments:

	Fluorescent pseudomonads	
(a)	Strain 701 El/pMON 5003 containing IncQ-*lac*ZY	Able to grow on lactose minimal plate and cleave X-gal to give the characteristic blue colour
(b)	Strain 701 El/pMON 5012 containing IncQ-*lac*Z	Growth is very poor showing a requirement for both Z and Y
(c)	Control strain 701 El/pKT 230 containing IncQ vector alone	Does not grow

While the *E. coli lac*ZY genes expressed from different promotors on broad host-range plasmids are highly effective selectable marker genes for the fluorescent pseudomonads (Drahos *et al.*, 1986), these genes are equally effective when delivered by a transposon Tn*7*-*lac* element and present in a single copy on the chromosome (Barry, 1986: other appropriate references are to be found in Barry, 1988; Craig, 1988). The Tn*7* is a large and complicated transposon but it has a number of features that make it useful (Fig. 2). The advantage of Tn*7* is that the transposition gene products function in *trans* and as little as 160 base pairs at each terminus are required to allow this transposition to occur. In addition, Tn*7* inserts with very high specificity and at a high frequency into the chromosomes of many Gram-negative bacteria. Tn*7* typically has only one insertion site per bacterial chromosome and is relatively rare in the environment. The Tn*7*-*lac* elements were developed as vectors for the insertion of additional genes onto the chromosomes of agronomically important bacteria.

The first approach consisted of placing the ends of the transposon on a large, unstable, broad host-range plasmid and placing the *lac*ZY between the Tn*7* termini. On a compatible, but also broad host-range, plasmid the transposition gene functions were cloned. The strains to be marked were mated with these two plasmids and initial selection was for the plasmids followed by selection for lactose utilization alone. The result, after a number of generations, was a plasmid-free cell in which the Tn*7*-*lac* element had transposed to the chromosome.

The Tn*7*-*lac* constructs were recoverable on lactose minimal medium from non-sterile environments. This was done by pipetting a culture on to seed at the time of planting, e.g. soybean, tomato, and plating out root-washings after 2–12 weeks. In no case, so far, has a *lac*-marked pseudomonad been found unable to colonize or to be slower-growing than the parent. There was no affect on the growth rates in laboratory media or on the colonization ability.

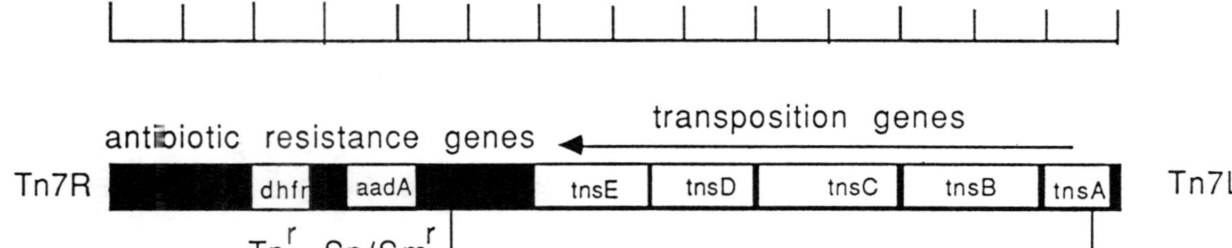

Fig. 2. Transposon Tn7.

The features that made this unsuitable for very common use was that any manipulation on the Tn*7* element would have to be done on a low copy plasmid not amenable to fast manipulations. Another problem with the earlier method was that the Tn*7*-*lac* element itself had a number of unknown regions within it as a result of the cloning process and it was necessary, before release into the environment, to know the source of every base pair. In particular, there was an unknown region of DNA that might have come from *E. coli* or from φ80 or λ at the end of the *lac*A gene. The manipulation of this element on the plasmid would be quite difficult. The *E. coli* Tn*7* insertion site is located downstream of the *glm*S gene, there is a five-base duplication on insertion but this is not always the same. This region of the chromosome can be sub-cloned and as few as 68 base pairs can retain the insertion function. This cloned fragment will act as a Tn*7* sink when placed on almost any replicon. The Tn*7* elements were transferred onto more easily manipulated plasmids.

The early scheme used to deliver the Tn*7*-*lac* element was composed of two unstable plasmids of different incompatibilities. Because of the size of the delivery plasmid, the cloning and co-insertion of additional genes and tracking sequences was extremely difficult. The first improvement involved the use of the smaller IncQ plasmids in the cloning and delivery system. Because of the decrease in size and the apparent broader host-range of the IncQ plasmids, the double IncQ system was used very effectively in the cloning of genes and in the introduction of Tn*7*-*lac* elements into a large number of fluorescent pseudomonad isolates. To further facilitate the cloning steps, and in particular the construction of more versatile and more widely applicable Tn*7*-*lac* elements, the elements were transferred to a very small, high-copy replicon. To do this a very small replicon was made by making a deletion of pUC8. A 500 base pair fragment was cloned into this from the *E. coli* chromosome containing the Tn*7* insertion site. This plasmid was introduced into an *E. coli* strain, into the chromosome of which the Tn*7*-*lac* element had been transposed and containing a helper plasmid. From the amp^R Lac$^+$ progeny was isolated a replicon of 2 kb with ~11 kb transposon. The Tn*7*-*lac* element was easier to manipulate in this form. Smaller cloning Tn*7*-*lac* elements were then constructed and the effectiveness of different promoters in the expression of the *lac* genes was determined for a number of soil bacteria.

Monocomponent Tn*7*-*lac* delivery systems were developed to expand the range of bacteria that may be marked with the lac genes. These 'suicide' delivery vectors eliminate the need for antibiotic sensitivities in the target bacteria and for the replication of the delivery replicon. The vectors are based on unstable IncQ replicons or on pBRS22.

The IncQ-based monocomponent delivery system (Fig. 3) was used to mark a fluorescent pseudomonad Ps. 3732RN. The Tn*7*-*lac* element in Ps. 3732RNL11 is composed of around 1700 bp of the termini of Tn*7* and the *lac*ZY genes (and a

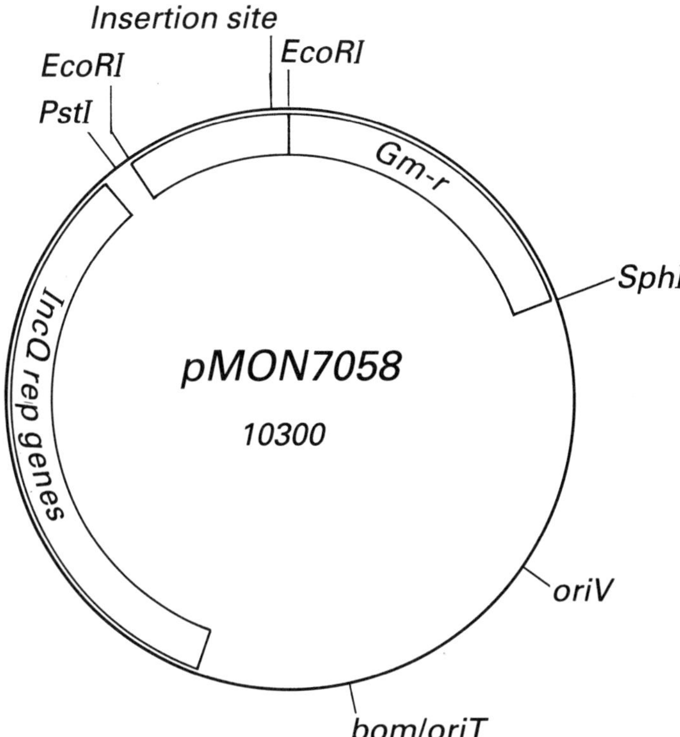

Fig. 3. To introduce the Tn*7*-*lac* elements into the IncQ plasmids, the *E. coli* Tn*7* insertion site (Tn*7*-*att*) was first cloned into a gentamicin-resistant IncQ derivative. This plasmid (pMON7058) was then introduced into a *P. fluorescens* isolate containing the Tn*7*-*lac* element (along with an IncQ Tn*7* helper plasmid) and the pMON7058::Tn*7*-*lac* derivative was isolated following retransformation into *E. coli*.

truncated *lac*A gene) promoted by the *iuc* operon promoter. The element used (Tn*7*-*lac*7117) contains a number of restriction enzymes sites to allow ease of cloning of additional genes, promoter replacements or substitution of *lac* with other selectable markers.

When the *lac*ZY genes were inserted into the chromosome of *P. fluorescens* 3732RN, this allowed the cleavage of the chromogenic dye X-Gal. This provided a marker that could effectively be used to track a micro-organism in the environment without relying on antibiotic resistance as the primary selecting agent. This Lac^+ engineered bacterium Ps.3732 RNL11 and its non-engineered parent Ps. 3732RN were released in an EPA-approved test in 1987. The degree of root colonization exhibited a significant increase over the first 3 weeks after inoculation. Both vertical and horizontal movement away from the site of inoculation were negligible (see also Chapter 13).

The procedure for *lac*-marking soil bacteria is shown below:

Triparental mating of:
(1) Soil bacterium
(2) *E. coli* HB101/pRK2013
(3) *E. coli* containing:
(a) pMON7190 or
(b pMON7197 (Fig. 4)

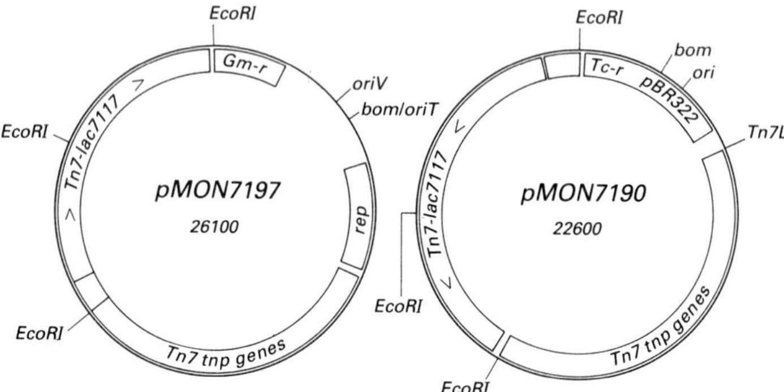

Fig. 4. Monocomponent delivery plasmids for the introduction of Tn*7*-*lac* elements into a wide range of soil bacteria. The unstable IncQ derivative pMON7197 and the pBR322-based pMON7190 both function in 'suicide' fashion in this delivery. Both plasmids encode the Tn*7* *tns* genes and carry the Tn*7*-*lac* element in the *E. coli* Tn*7*-*att* cloned into the plasmid sequences.

There is an initial selection for Lac^+ in the case of (a) and for Lac^+ or Gm^r+Lac^+ in the case of (b). The Lac^+ colonies appear in 2–5 days at 30°C. Gm^rLac^+ isolates quickly lose the Gm^r plasmid.

Tn*7* is rarely isolated except in human and animal infections after heavy use of the antibiotic trimethoprim. A DNA probe consisting of the *tns*BCD genes showed no homology in an examination of 597 bacterial isolates (Fig. 5) from soybean, wheat and the roots of other crop plants to determine the occurrence of Tn*7* in this population.

Individual colonies were grown in microtitre wells, replicated onto King's B agar plates with positive controls, transferred to nylon filters and probed with the 6100 bp Pst1-Cla1 fragment containing *tns*B, C and D. 'Stringent' hybridization conditions were employed. No hybridization of the probe with any isolate except the positive controls was found.

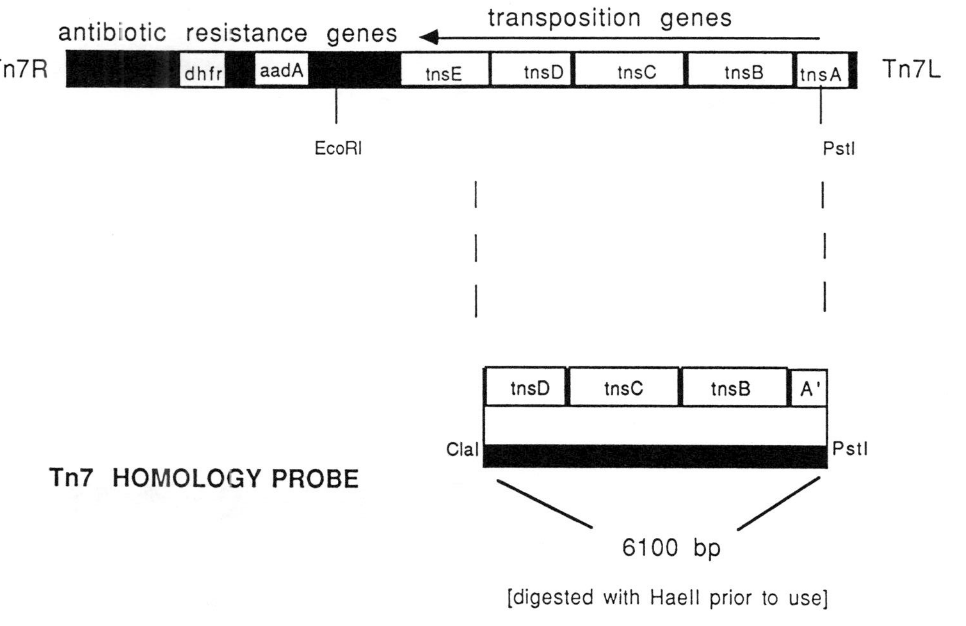

Fig. 5. The Tn7-*tns*BCD probe.

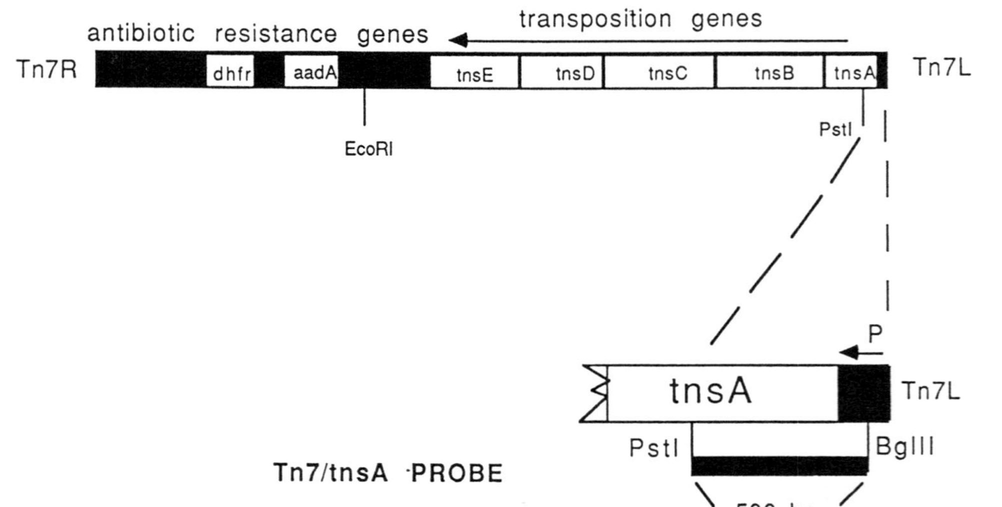

Fig. 6. The Tn*7*-*tns*A probe.

In the on-going field test, a portion of the Tn*7* *tns*A gene (Fig. 6) was employed both as a probe for Tn*7* and as an additional monitor for genetic exchange between the introduced bacterium and the native population.

A DNA fragment consisting of 360 bp of the *P. aeruginosa* 23S rRNA gene has been shown to hybridize specifically, in colony lifts, to bacteria belonging to the rRNA homology Group 1 pseudomonads (including the fluorescent pseudomonads) (Festl *et al.*, 1986). This same probe has now been examined in a colony lift screen of 597 characterized, soil isolates. These included over 56 isolates belonging to the rRNA homology Group 1. In every case the probe hybridized only with the DNA from these bacteria, Table 2. This probe may be used to determine the levels of certain populations in future studies.

APPLICATIONS OF IMMUNOFLUORESCENT TECHNIQUES

Gerard Muyzer (*Leiden, Netherlands*) outlined the use of immunological methods for the detection of micro-organisms in natural environments. The use of antibodies as biological probes allows the identification and enumeration of individual bacterial species in mixtures of different micro-organisms in complex systems, including natural and man-made habitats. Immunological methods have been used by many investigators for the identification of bacteria, fungi and other micro-organisms in soil, water and other natural samples. This approach was used to estimate the relative abundance of the acidophilic bacterium *Thiobacillus ferrooxidans* among mixed populations of micro-organisms in coal–water slurries.

Antibodies are produced against the different antigenic determinants on the bacteria. A single bacterium contains a wide range of antigenic determinants, and the resulting antiserum will be a 'polyclonal antiserum'.

Different tests are available to determine the specificity of the antiserum; however, those most commonly used at present are the 'solid phase immunoassays', such as the ELISA, the dot-blot immuno-binding assay and the indirect immunofluorescence assay. These assays are characterized by sensitivity, reproducibility and rapidity.

Although these immunoassays seem to be ideal to monitor individual bacteria in complex natural samples, problems can be foreseen with the specificity of the antiserum, the detection limit and the sample preparation. For instance, antibodies may be raised against antigens which are common to other micro-organisms too. This will result in false-positive reactions, when tested with natural samples. Absorption of the antiserum with pure strains of these undesirable species may remove this cross-reactivity; however, the production of antibodies against species-specific antigens of the desired micro-organism would be a better solution. Monoclonal antibodies offer this possibility.

Table 2. Bacteria tested and their reaction with the pHF360 probe.

Strain	Number of isolates tested	Hybridization with pHF360 probe
Pseudomonas putida biovar A	58	Positive
Pseudomonas putida biovar B	23	Positive
Pseudomonas chlororaphis	170	Positive
Pseudomonas tolaasii	41	Positive
Pseudomonas aureofaciens	28	Positive
Pseudomonas corrugata	13	Positive
Pseudomonas fragi	18	Positive
Pseudomonas marginalis	1	Positive
Pseudomonas syringae (multiple pathovars)	93	Positive
Pseudomonas fluorescens A	4	Positive
Pseudomonas fluorescens B	5	Positive
Pseudomonas fluorescens C (Inc. ATCC 10844)	12	Positive
Pseudomonas fluorescens G (Inc. ATCC 13525)	14	Positive
Pseudomonas coronafaciens	3	Positive
Pseudomonas aeruginosa (Inc. ATCC 15526)	9	Positive
fluorescent pseudomonads (incomplete i.d.)	61	Positive
Pseudomonas mendocina	1	Positive
Pseudomonas stutzeri	1	Positive
Pseudomonas alcaligenes	1	Positive
Total	**556**	
Pseudomonas testosteroni (Inc. ATCC 17409, 17510, 11996)	7	Negative
Pseudomonas cepacia ATCC 10856	1	Negative
Pseudomonas delafieldii ATCC 17505	1	Negative
Pseudomonas diminuta ATCC 11568	1	Negative
Pseudomonas acidovorans	3	Negative
Pseudomonas cruciuriae ATCC 13262	1	Negative
Pseudomonas methanolica ATCC 21704	1	Negative
Pseudomonas pickettii ATCC 27511	1	Negative
Pseudomonas vesicularis ATCC 11426	1	Negative
Xanthomonas maltophilia ATCC 13637	2	Negative
Agrobacterium tumefaciens	3	Negative
Erwinia herbicola	1	Negative
Enterobacter cloacae ATCC 13047	1	Negative
Enterobacter aerogenes ATCC 13048	1	Negative
Hafnia alvei	2	Negative
non-fluorescent (incomplete i.d.)	11	Negative
Bacillus thuringiensis	1	Negative
Bacillus licheniformis	1	Negative
Corynebacterium fascians	1	Negative
Total	**41**	

Monoclonal antibodies are homogeneous antibodies, specific for a single antigenic determinant. They are secreted by hybridoma cells, which are produced artificially by cell fusion of an antibody producing B-lymphocyte and a myeloma tumour cell. Apart from the desired specificity, they can be produced indefinitely and in unlimited quantities.

The advantages of the use of immunological methods for the detection of micro-organisms in natural samples are:

(1) Specificity*
(2) Sensitivity*
(3) Rapidity—'solid phase assays' are completed in hours and ELISA can be automated for rapid multisample analysis
(4) Accurate reproducibility
(5) Low costs
(6) Simplicity

A drawback of the use of antibodies is that only positive results can be interpreted. Negative results may be due either to absence of the micro-organism in the sample or to an alteration of the antigenic determinant.

Solid phase immunoassay has been used for the enumeration of *T. ferrooxidans* in a mixed population of acidophilic bacteria in the microbial desulphurization process.

Many studies have been done on biomining: the use of microbes for the extraction of valuable metals from ores. In addition, with the public concern about air pollution by combustion of fossil fuels the possibility was explored to remove sulphur from coal by microbes. It is generally thought that the chemolithotrophic bacterium *T. ferrooxidans* plays a prominent role in the leaching of iron pyrites from coal and metal ores.

This bacterium can oxidize pyrites, the main inorganic sulphur compound in coal according to the equation:

$$4FeS_2 + 15O_2 + 14H_2O \rightarrow 4Fe(OH)_3 + 8H_2SO_4$$

However, these habitats usually have a complex consortia of microbes. *T. ferrooxidans* may be accompanied with a variety of other acidophilic organisms, such as *T. thiooxidans*, *T. acidophilus*, *Acidophilium cryptum* and *Leptospirillum ferrooxidans*. Results of several studies have suggested that mixed cultures of acidophilic bacteria are more efficient in leaching of pyrites than are pure bacterial cultures. Piet Bos and co-workers (Delft University of Technology, Netherlands) and we are interested in the role of *T. ferrooxidans* in

---

*The detection limit of bacteria by these methods is 3.1×10^6 to 1.4×10^7 cells per gram dry sediment, by immunofluorescence (Strayer and Tiedje, 1978) or 10^5 to 10^6 cells per ml sample with an enzyme immunoassay (EIA) (Minnich *et al.*, 1982).

the microbial desulphurization process of coal. To evaluate the importance of *T. ferrooxidans* in this process the abundance of this bacterium in coal samples was estimated with a polyclonal antiserum directed against a pure strain of *T. ferrooxidans*. The antiserum was allowed to react with a variety of acidophilic and non-acidophilic bacteria in an ELISA and an immunofluorescence assay to determine the specificity of the serum. It was found that the antiserum was specific at the species level. The antiserum was also tested against cells of *T. ferrooxidans* which were grown on different substrates (*viz*, ferrous iron, tetrathionate, thiosulphate, iron pyrites and elemental sulphur). Positive reactions of equal intensity were obtained with all the cultures tested, which indicates that the reactivity of *T. ferrooxidans* with the antiserum does not depend on the culture conditions. A combined immunofluorescence–DNA-fluorescence staining technique was used to estimate the abundance of *T. ferrooxidans* in mixed populations of acidophilic bacteria in coal slurries. Sterile coal samples were inoculated with a pure strain of *T. ferrooxidans*, a mixture of micro-organisms, or both. Pyrites leaching was monitored by measuring the concentrations of solubilized iron; at the same time the numbers of bacteria in the leached samples were estimated by use of the combined immunofluorescence–DNA-fluorescence staining technique.

Pyrites leaching was observed in all samples inoculated with bacteria. It was found that leaching of pyrites was much higher in coal samples incubated with mixtures of acidophilic bacteria than in those incubated with a pure culture alone.

The dual fluorescence assay gave surprising results. High numbers of *T. ferrooxidans* were found only in the coal sample inoculated with a pure culture of *T. ferrooxidans*. In samples inoculated with mixtures of acidophilic bacteria or with both, *T. ferrooxidans* was not found, or only in reduced numbers (Fig. 7). From these results, it was concluded that other acidophilic bacteria may leach pyrites as well as, or even better than, *T. ferrooxidans* and that *T. ferrooxidans* may be out-competed by these bacteria.

Further investigation of the microbial composition of this ecosystem and the possible interactions between the different acidophilic species should be very rewarding. In this respect it would be convenient to consider carefully the genetic manipulation of *T. ferrooxidans* for the improvement of biomining and microbial desulphurization of coal, which has been suggested by some investigators.

David Stahl (*Urbana, USA*) considered the application of sequencing techniques to microbial ecology and release of GEMs. Of all databases that now exist of nucleic acid sequences the most comprehensive is that of rRNA sequences. Most intensively studied are the 5S and 16S; probably 600 5S RNA sequences

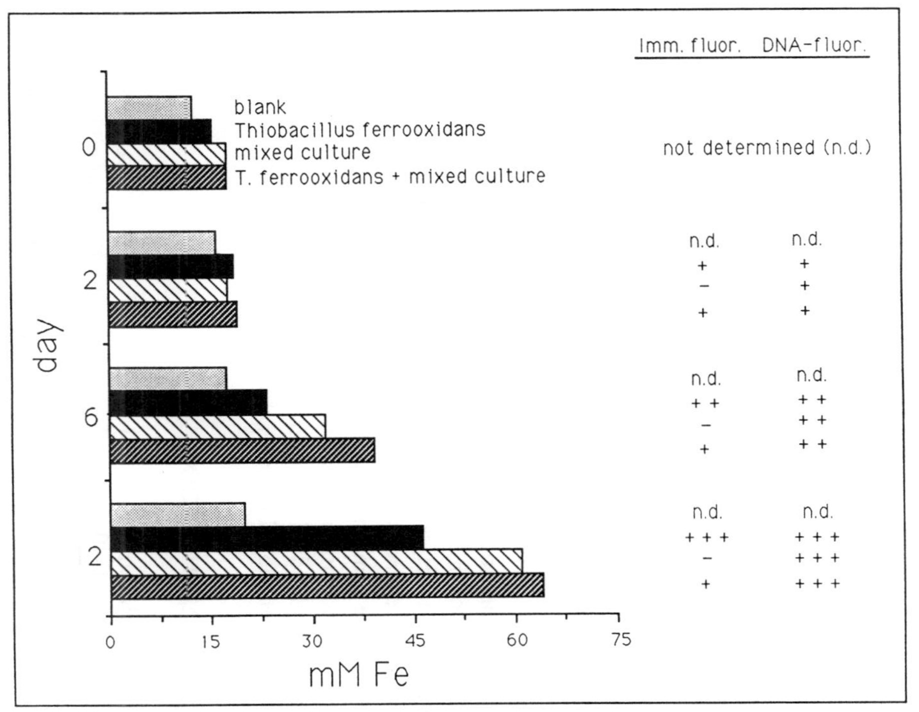

Fig. 7. Enumeration of bacterial cells during bioleaching of pyrites from coal.

and 1000 16S RNA sequences are either partially or completely known. This represents the most extensive database of homologous genes and has been used for studies in microbial phylogeny, taxonomy and evolution. It is a vast resource and is now being exploited for the study of natural microbial systems. The rRNA provides sequences for tracking individual organisms or for micro-organisms in the environment. The more diverse the background and ecosystem perturbation the more it affects the sensitivity and specificity of detection.

There are 1540 nucleotide bases in the 16S ribosomal RNA and these are highly conserved but also ubiquitous regions. Variable regions enable one to look for sub-species variations.

There is a unique sequence for the organism *Bacteroides succinogenes* in the bovine rumen, i.e. a 'signature probe'. Oligonucleotide probes can be made for these regions of the molecule to enumerate and probe similar organisms in the natural habitat. It is possible to separate a ruminant-type probe and a caecal-type probe. This is useful for genus-specific and species-specific probes. Genus-specific 16S rRNA-targeted oligonucleotide hybridization probes were used to enumerate cellulolytic bacteria (e.g. *B. succinogenes*). The combined use of general (genus) and specific (species) probes pointed out greater diversity within natural samples than is currently recognized in the pure culture collection.

Another application of this technique is to *in situ* studies in that there are enough copies of the ribosome per cell for fluorescently tagged oligonucleotide specifically to label cells fluorescently, e.g. *B. subtilis* and a methanococcus treated with a eubacterium-specific fluorescent probe specifically lights up the *B. subtilis*. The combination of this vast data collection and *in situ* techniques as well as hybridization to bulk nucleic acid offers a powerful approach not only for assessing individual organisms introduced into the environment but also for looking at more global changes that might arise from the release of GEMs.

TARGET DNA AMPLIFICATION

Ronald M. Atlas and **Robert J. Steffan** (*Louisville, USA*) have enhanced the detection of genetically-engineered bacteria by DNA amplification.

In the search for sensitive methods to detect GEMs, they used the polymerase chain reaction (PCR), which is an *in vitro* method for increasing the number of copies of a target nucleotide sequence (Fig. 8). The method involves melting the DNA, annealing short oligomer primers to regions flanking a target sequence and using *taq* polymerase to extend the DNA from the primers across the target region. The new duplexes are melted by heating and the process is repeated. The result is an exponential increase of the target sequence, such that the target sequence can be amplified by a factor of more than 10 million within a few hours.

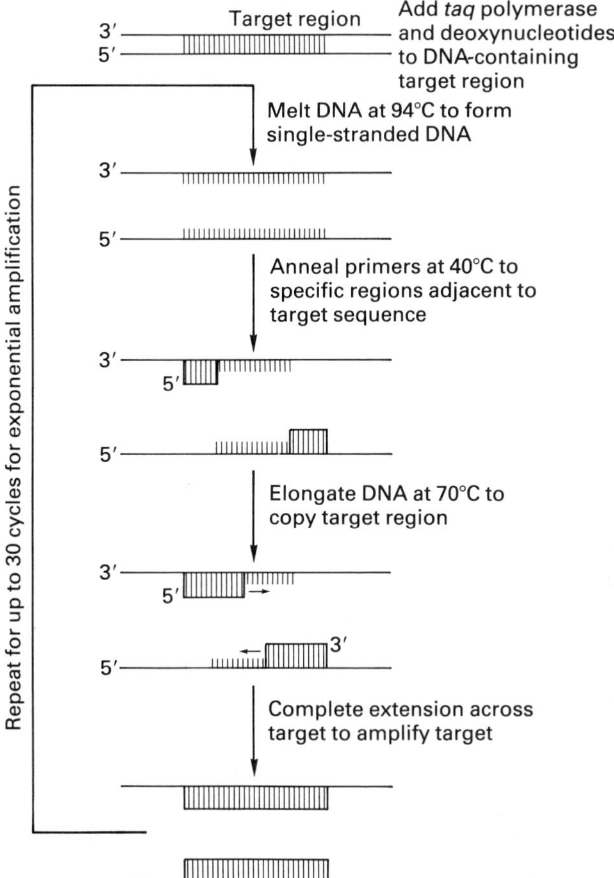

Fig. 8. The polymerase chain reaction (PCR) for increasing the number of copies of a target nucleotide sequence.

In these studies a 1.0 kb target was used; this is an integral portion of a larger 1.3 kb repeat sequence present as 15–20 copies in the genome of the 2,4,5-T GEM *Pseudomonas cepacia* AC1100. Twenty-four base oligonucleotide primers were synthesized and these were unique to the regions flanking the 1.0 kb region. The first six bases of the 3′-OH ends of the primers are sequences that do not occur elsewhere within the region to be amplified, which is essential for optimal amplification. After amplification the reaction mixture was transferred to nylon filters and hybridized against radiolabelled 1.0 kb fragment probe DNA. Less than 0.3 pg of target DNA was detected in the amplified sample, even against a background of more than 20 μg of non-specific DNA isolated from sediment.

By coupling PCR with direct extraction of DNA, *P. cepacia* was detected at concentrations as low as 1 cell per gram in sediments containing 10^9 diverse non-target bacteria per gram. This detection level is equivalent to finding one foreigner living in Japan! This represented at least a 1000-fold increase in detection sensitivity over non-amplified samples. Even greater sensitivity may be achieved by refining the DNA extraction procedure, by using an increased cycle of PCR, by using probes with more intense labels and by modifying the hybridization detection procedure.

The PCR method can readily be applied for the detection of GEMs, since *taq* polymerase and an automated thermal cycler are commercially available from Cetus–Perkin Elmer. An advantage of the PCR method is that cultivation of micro-organisms is not necessary, so that even viable but not culturable microbes and genes not contained within living cells can be detected with great sensitivity. The ability to detect a single target organism in a gram of complex environmental sample, such as sediment, represents a powerful tool for monitoring GEMs that are released into the environment.

DISCUSSION

Major points of audience discussion and concern about available and developing methods for detecting GEMs in the environment were voiced in relation to issues of quantitation, efficiency and specificity. Simple qualitative detection of GEMs is not a goal of current research, rather, the specific quantitative relationships of GEMs to the physical, chemical and ecological parameters affecting their maintenance, stability and effects is the primary target of research in the field.

Main interest was expressed for methods that are targeted to direct environmental analysis, in particular nucleic acid and fluorescent antibody techniques. This is no doubt due to the ability of these techniques to avoid the sampling bias that results from laboratory cultivation of environmental microbes and the potential of these methods to provide a more detailed analysis of microbial community complexity and diversity in natural field habitats. Current researches on DNA extraction and recovery from environmental samples focus on two approaches both of which require further analytical quantitation.

These methods focus on the release of cells from soil followed by cell concentration, lysis and DNA extraction (Torsvik, 1980; Holben and Tiedje, 1988) or the direct lysis of cell *in situ* and alkaline extraction and purification of DNA (Ogram *et al.*, 1988). Some investigators who compared these methods noted greater DNA recovery by *in situ* lysis but somewhat lower molecular weight sized DNA recovery. Neither method appears to recover DNA significantly greater than 100 kb in size. There is current concern as to whether all cell types are uniformly lysed in these procedures, in particular with Gram-positive cells. It was reported that apparent DNA recovery can exceed 90% and sometimes exceeded estimates of total cell densities in soils and sediments examined. These reported recoveries are based on direct and viable cell counts which are likely to be a poor basis for comparison and it is possible that DNA may be extracellular or from non-viable or starved cells that cannot be cultivated. While it was noted that extraction methods have been calibrated with reconstructed populations of Gram-negative and Gram-positive bacteria, some lysis methods, such as acetone treatment at elevated temperatures, will not lyse some *Bacillus* strains. However, physical lysis methods appear to be capable of lysing virtually all species. Chemical lysis modifications with lysozyme, lysostaphin, SDS, etc., are also available to ensure quantitative lysis of virtually any species.

There are remaining concerns about absolute quantitation of DNA recovery that require investigation. Furthermore, the quantitative detection of specific organisms or genes at low abundance or in highly complex mixtures of DNA recovered from natural communities also requires the development of a quantitative protocol. Thus far, MPN type determination, or dot or slot-blot hybridizations that compare the relative abundance of target sequences to total DNA have been the only attempts at direct quantitation. There is also concern that high concentrations of eucaryotic DNA may mask procaryotic DNA or specific sequences in soil and water column samples. At the very least, this problem can be assessed with an integrated approach with rRNA gene probes for quantitation of procaryotic and eucaryotic DNA or by the use of other eucaryotic or procaryotic specific gene probes.

Immunofluorescence antibody (IFA) techniques appear to be a high priority research tool and monitoring method with *in situ* applications. There is a need to evaluate the limits of sensitivity of these techniques with a currently reported range of sensitivity from 10^3 to 10^6 cell per ml per gram. This available sensitivity may be enhanced by increasing the sample size by filtration of large water samples but it is not likely to be greatly increased for soils and sediments and other solid matrices such as coal particles. It may be possible to separate cells from solid matrices and then concentrate cells for IAF analysis to increase sensitivity. The efficiency of this procedure is unknown especially for cells such as *T. ferrooxidans* tightly associated with particle surfaces and coal particles. Laser-based cell sorting technology may offer some hope for sensitivity improvement. It is unlikely that sensitivity limits of detection will approach a 1 cell per gram limit. Because of this detection threshold limit, absolutely negative results for detection cannot be assured.

The applications of IFA for GEM-detection remains a concern in two areas:

(1) the need for a rDNA antigenic product presented in an accessible mode to the IAF and (2) specificity of the antibodies involved. It was noted that the first problem may be solved in that recent evidence has suggested that occluded proteins, such as nitrogenase, are accessible to IAF in glutaraldehyde-fixed cells. Antibody specificity can be improved by the use of monoclonal antibodies and adsorption of immune serum to cross-reacting cells from natural populations. Specific mixtures of monoclonal or polyclonal antibodies may also be applied to enhance relative specificity. It was pointed out that environmentally-starved cells may present different antigenic determinants from cells grown under laboratory conditions and this may result in poor specificity of antibodies prepared against laboratory strains when applied to environmental populations. For studies on *T. ferrooxidans* involved in pyrites oxidation in coal slurries, this potential problem was solved by growing *T. ferrooxidans* in pyritic medium for more than a year. Comparison of their susceptibility to IFA to cells grown in different media showed no difference in specificity. The question remains as to the potential of directly extracting total protein and attempting to quantitate a target or recombinant protein directly from a population or environmental sample by ELISA. It appears that this approach has not yet been explored.

Analogies for the proposed direct protein detection exist for other cell constituents such as lipids, fatty acids and pigments. Gas chromatography/mass spectrometry (GC/MS), tandem and triple quadropole MS have been used by geochemists and microbiologists to detect specific biogenic materials in environmental samples. Detection limits of one organism, in indeterminate sized samples, have been reported for archaebacterial signature lipids/fatty acids with tandem and triple GC/MS systems. Mycobacterial populations have also been studied by GC/MS approaches. Pyrolysis GC techniques have also been investigated as possible sensitive fingerprinting techniques for microbial communities such as those associated with coal.

SUMMARY

The purpose of this round table session was to discuss the current availability and development of methods for detecting and monitoring GEMs in the environment, to define the current applications and limitations of these methods, and to identify requirements for improvements in these methods. A broad range of methods is available for specific and sensitive detection of GEMs, rDNA hosts, and rDNA in microbial communities. These include conventional enrichment and selection techniques, immunological methods, nucleic acid analysis and hybridization techniques, analytical methods for specific cell constituents and the development of reporter or marker strains for engineered traits.

Of critical concern in the application of these techniques are the criteria and objectives of sensitivity and specificity required for environmental measurement of GEMs in ecological studies, biotechnical applications and optimization and risk evaluation. All of these share central requirements for genotype-specific determination of the maintenance, stability, proliferation, die-off and effects on specific GEMs or rDNA in complex microbial communities.

These problems determine the need for highly specific and sensitive quantitation. Detailed analysis of IFA techniques and applications in monitoring autotrophic pyrites-oxidizing bacteria were examined, as were the development and application of *lacZ* marked rhizosphere-specific fluorescent pseudomonads and group-specific rRNA gene probes in differential isolation from soils. The examination of these approaches led to the discussion of direct environmental analysis and *in situ* techniques for monitoring GEMs.

Direct environmental analysis of nucleic acids, *in situ* applications of IAF, highly specific instrumental methods, such as triple quadropole GC/MS, are in use which approach nearly absolute levels of specificity and sensitivity. However, major requirements exist to increase the quantitative confidence that these methods offer. In particular, quantitation and efficiency of nucleic acid extraction and hybridization methods for environmental samples require innovative procedures for verification of current results. IAF methods require that improvements in specificity and instrumental methods must become more selective for broader ranges of organisms. These are the priority areas that must be addressed in order to interpret the factors affecting the fate of GEMs within the background of complex microbial communities into which they will be introduced. However, rapid advances in the application of these and other methods the past several years gives a clearly optimistic outlook for the continued development and improvement of approaches for quantitative analysis of GEMs in the environment.

REFERENCES

Atlas, R. M. and Steffan, R. J., 1988. DNA amplification to enhance the detection of genetically engineered bacteria in environmental samples. In preparation.

Barry, G. F., 1986. Permanent insertion of foreign genes into the chromosomes of soil bacteria. *Biotechnology* **4**, 446–449.

Barry, G. F., 1988. A wide host-range shuttle system for chromosomal gene insertion in bacteria. *Gene*, in press.

Craig, N. L., 1988. In *Mobile DNA*, Berg, D. and Howe, M. (eds). American Society for Microbiology. In press.

Drahos, D. J., Hemming B. C. and McPherson, S., 1986. Tracking recombinant organisms in the environment: beta-galactosidase as a selectable, non-antibiotic marker for fluorescent pseudomonads. *Biotechnology* **4**, 439–443.

Festl, H., Ludwig, W. and Schleifer, K., 1986. *Applied and Environmental Microbiology* **52,** 1190–1194.

Höfle, M. G., 1988. Taxonomic structure of bacterial communities in mixed cultures as measured by low molecular weight RNA profiles. *Archiv für Hydrobiologie*, supplement **31,** 71–77.

Holben, W. E. and Tiedje, J. M., 1988. Applications of nucleic acid hybridization in microbial ecology. *Ecology*, in press.

Jain R. K., Burlage, R. S. and Sayler, G. S., 1988. Methods for detecting recombinant DNA in the environment. *CRC Critical Review Biotechnology*, in press.

Minnich, S. A., Hartman, P. A. and Heimsch, R. C., 1982. Enzyme immunoassay for detection of Salmonellae in foods. *Applied and Environmental Microbiology* **13,** 877–883.

Muyzer, G., de Bruyn, A. C., Schmedding, D. J. M., Bos, P., Westbroek, P. and Kuenen, G. J., 1987. A combined immunofluorescence–DNA-fluorescence staining technique for enumeration of *Thiobacillus ferrooxidans* in a population of acidophilic bacteria. *Applied and Environmental Microbiology* **53,** 660–664.

Ogram, A., Sayler, G. S. and Barkay, T., 1988. DNA extraction and purification from sediments. *Journal of Microbiological Methods* **7,** 57–66.

Olsen, G. J., Lane, D. J., Giovannoni, S. J., Pace, N. R. and Stahl, D. A., 1986. Microbial ecology and evolution: a ribosomal RNA approach. *Annual Review of Microbiology* **40,** 337–366.

Stahl, D. A., Flesher, B., Mansfield, H. R. and Montgomery, L., 1988. The use of phylogenetically based hybridization probes for studies in ruminal microbial ecology. *Applied and Environmental Microbiology*, in press.

Strayer, R. F. and Tiedje, J. M., 1978. Application of the fluorescent-antibody technique to the study of a methanogenic bacterium in lake sediments. *Applied and Environmental Microbiology* **35,** 192–198.

Torsvik, V. L., 1980. Isolation of bacterial DNA from soil. *Soil Biology and Biochemistry* **12,** 15–21.

Woese, C. R., 1987. Bacterial Evolution. *Microbiology Reviews* **51,** 221–271.

16 Round Table 3: Survival, Persistence and Colonization

Chairman
S. KJELLBERG *(Goteborg, Sweden)*

Rapporteur
M. HINTON *(Bristol, UK)*

Contributors
JASMIN HURTADO *(Lima, Peru)*
STEPHEN M. CUSKEY *(Gulf Breeze, USA)*
DAVID TEPFER *(Versailles, France)*

Editor
F. A. SKINNER

INTRODUCTION

A general discussion on the survival and persistence of micro-organisms in natural environments was preceded by three formal presentations. **J. Hurtado** (*Peru*) provided a general survey of principles and relevant aspects of the subject; **S. M. Cuskey** (*USA*) considered the design of genetically-engineered micro-organisms (GEMs) with limited capacity to survive and **D. Tepfer** (*France*) discussed the construction of plant/micro-organism associations likely to benefit the plant.

Jasmin Hurtado (*Lima, Peru*) presented a general account of the survival, persistence and colonization by micro-organisms in the natural environment.

An organism introduced into new terrain will persist only if it finds a suitable 'niche' or set of favourable environmental conditions to which it is particularly well adapted.

RELEASE OF GENETICALLY-ENGINEERED
MICRO-ORGANISMS ISBN 0–12–677521–4

Factors that affect persistence in the natural environment include temperature, solar radiation, desiccation, soil type and nutrient status, salinity, pH, metal content as well as biotic factors, such as presence of microbial antagonists, parasitism and predation, production of bacteriocins and the physiology of the specific strain.

The environmental persistence of micro-organisms carrying foreign genetic traits should be considered but the environmental persistence of the foreign genes themselves must also be studied because a population of released micro-organisms that finds itself in a suitable environmental setting, may produce, evolve and transfer genetic material to other organisms in the environment.

GEMs may be evaluated for plasmid mobility by conducting mating experiments with laboratory strains as recipients and with members of the indigenous microflora recently isolated from environments to which the GEM is to be released. It is necessary to determine the effects of cell concentration and donor-to-recipient ratios and to evaluate conjugal mating techniques with regard to sensitivity and consistency for producing transconjugants. All successful DNA transfer processes require replication and recombination but bacteria in nature usually grow slowly and pass through fewer generations per year than they do in laboratory culture. DNA/DNA interactions and exchanges must therefore be extremely rare events. The importance of plasmid mobility in the environment should not be under-estimated but it is likely to be sporadic. Another important aspect is that moribund or dead GEMs may release DNA into the environment. Nucleic acids may absorb to soil colloids and so be protected from degradation. Modified DNA would then persist in the environment but, because of the ability of cells to discriminate against foreign DNA, would probably pass mostly into cells with homologous or related genomic sequences. Thus, while the potential of these transfer mechanisms to generate genetic novelty remains unchallenged, the extent to which this potential is realized in nature should not be overstated. Considerations regarding environmental release of GEMs can be eased if neither the organisms nor their DNA persists after their work is over.

Knowledge about the fate of released GEMs has been obtained from the study of natural micro-organisms that have been released to the environment: *Rhizobium* is an example.

Rhizobium strains inoculated into soils usually compete poorly with indigenous strains and, even in the absence of antagonistic strains, the introduced rhizobia grow slowly, if at all, in fallow soil. Thus, there is reason to believe that a GEM introduced into soil would probably not proliferate and would eventually die out unless it could occupy an ecological niche not already occupied.

Strong selection pressure can, however, permit a bacterial strain to persist in the soil environment. Thus, some mercury-resistant strains of soil bacteria can colonize mercury-contaminated soil. Such examples permit us to see how difficult it is to predict the fate of GEMs introduced into soil or the environment generally.

Thiobacillus ferrooxidans is a most important iron-oxidizing bacterium in mineral-leaching systems and some benefit would seem to be obtainable by genetic manipulation of this species to enhance its iron-oxidizing ability. In biomining operations in which *T. ferrooxidans* plays a part, other acidophilic organisms are also active and it is important to determine which members of this indigenous microflora are really the most important in the leaching process before deciding to improve it by genetic manipulation. *Thiobacillus ferrooxidans* might, in fact, not be the most important member of this mixed population so far as mineral leaching is concerned.

The use of conditional lethal genetic determinants for control of GEMs released to the environment was then considered by **Stephen M. Cuskey** (*Gulf Breeze, USA*).

Bio-control of released GEMs is not a popular field of study for several reasons. Some think that bio-control is not necessary either because the organisms will not survive or because the proposed releases are inherently safe. Another reason given for not studying bio-control mechanisms is that they will not work. Bio-control is fine for organisms grown in a controlled, laboratory environment but the real world is too complex and large for the control of released organisms. These objections may be valid but they are based on very little experimental evidence. The purpose of the work to be described is to test the second of the above objections, by determining whether it is possible to construct an organism which will effectively compete with the indigenous microbiota but which will automatically be destroyed after completion of its desired function.

Three main approaches to bio-control based on a conditional lethal phenotype have been developed. The first main approach made use of temperature-sensitivity to DNA damaging agents. This is not a conditional lethal phenotype, *per se*, but should render susceptible organisms less able to survive normal environmental stresses. Several *P. aeruginosa* mutants with cold-sensitivity to MMS were isolated. Tests showed that the back mutation rate to the wild type phenotype was relatively high, making this type of control mechanism less useful. In addition, chromosomal mutations are less useful in general because they must be generated anew for each released organism.

A second type of control mechanism is based on the accumulation of toxic metabolic intermediates which is due to inclusion of blocked or partial metabolic pathways. An example of this is mercury (Hg^{2+}) super-sensitivity. Cells containing the mercury regulatory gene (*mer*R) and the mercury transport gene (*mer*T) from the *mer* operon of TN*21* without the mercury reductase gene (*mer*A) take up mercury into the cell without de-toxification mechanism. These cells are sensitive to extremely low levels of normally sub-inhibitory levels of mercury. Cells with and without the super-sensitive cassette were incubated in sterile Escambia River water at a concentration of $10^3\,m^{-1}$. *Pseudomonas aeruginosa* strain PAOI was used in this study. After 2 days, various amounts of mercury, as mercuric chloride, was added. Cells with the super-sensitive genes were killed at a Hg^{2+} concentration of 0.5–1.0 $\mu g\,l^{-1}$, a concentration at which the control cells were unaffected. No survivors of the super-sensitive cell populations were found at sampling times of 1 and 19 days. This model, though not proposed for actual bio-control because of mercury toxicity, shows that populations can be totally controlled. Higher cell densities could not be tested with this system because of limitations in the mercury super-sensitive phenotype.

A third type of control may be the most useful. This is a self-contained system based on the use of normally plasmid-borne killing and protecting genes. A general system was designed with organisms for pollution control serving as models. A 'killing' gene was included which was constitutively expressed. A 'protecting' gene was under the control of a promoter which recognized the presence of a toxic waste. While the bacterium degraded the toxic waste, the action of the 'protecting' gene countered the action of the 'kill' gene and the cell survived. After complete biodegradation, the 'protecting' gene turns off and the cell dies. Examples of 'kill'/'protect' gene tandems are the *hok/sok* genes from plasmid R1 studied by Molin and co-workers in Denmark, *kil*A/*kor*A from plasmid RK2 studied by Figurski and Thomos and their co-workers and restriction endonuclease/methylase genes. Cuskey had used the *kil*A/*kor*A genes in a construct entitled pEPA93. Genetic control is provided by the OP2 promoter from the *TOL* plasmid, pWWO. This promoter, which usually requires the product of a positive regulatory gene (*xyl*S) as well as toluic acid, benzoate or a related aromatic acid for induction, was placed upstream from the promoter-less *kor*A gene. This provides for benzoate (or toluate) dependency in pEPA93-transformed cells. This construct is currently being tested to determine its effectiveness as a bio-control mechanism.

Benzoate-sensitive cells were also constructed by placing the OP2 promoter upstream from the promoter-less *kil*A gene. Transformed *E. coli* or *Ps. aeruginosa* cells did not grow in the presence of benzoate in a rich laboratory medium or a basal salts medium. These benzoate-sensitive cells are also being tested for stability of the construct and extent of population lethality after addition of benzoate to the growth medium.

It was concluded that use of normally plasmid-borne, natural 'suicide' genes with altered regulation of gene expression, may be a useful method for determining whether bio-control of released GEMs is feasible.

David Tepfer (*Versailles, France*) provided an account of nutritional mediators of plant–bacterium interactions. The work had been carried out jointly with Arlette Goldmann, Monique Maille, Brigitte Message and Veronique Fleury and with C. Boivin, C. Rosenburg and J. Denarie, at INRA-CNRS, Castanet-Tolosan, France.

Survival of a micro-organism depends on its nature, genetic modifications that may have been made to it and the physical and biological characteristics of the environments it encounters. Tepfer's group had tried to understand how certain soil bacteria proliferate in the soil surrounding certain plants.

Plant roots condition their immediate environment (the rhizosphere) by selectively excreting and assimilating a variety of substances. These include secondary metabolites that can have positive and/or negative effects on organisms living in the soil. In an effort to understand better plant–micro-organism interactions they searched for hypothetical plant secondary metabolites that were called 'nutritional mediators', by which substances are meant that, upon release into the soil by certain species of plants, can provide an exclusive energy source to certain soil bacteria. Such nutritional mediators would be positive mediators of plant–micro-organism interactions.

Root cultures were used to develop *in vitro* models for the rhizosphere; these have made it possible to study plant–micro-organism interactions under controlled conditions. Root culture in these models is facilitated by genetically transforming the roots with *Agrobacterium rhizogenes*. Transformed roots are capable of rapid proliferation, and they are both genetically and phenotypically stable. Most interestingly, they tend to over-produce secondary metabolites. In the work that was presented, these cultures were used to study the role of secondary metabolites liberated into the soil by plant roots in the maintenance of specific relationships between plants and soil bacteria.

The search revealed a family of previously undescribed compounds that were designated 'calystegins', since they were first observed in *Calystegia sepium* (a species of morning glory). Calystegins are abundantly produced by the roots of three species of the 105 species screened: two species of morning glory (*C. sepium* and *Convolvulus arvensis*) and by *Atropa belladonna*. Forty-four species of previously defined soil bacteria were tested for ability to degrade calystegins. Only one of these, *Rhizobium meliloti* strain 41, was able to catabolize calystegins and use them as a sole carbon and nitrogen source for growth. Calystegin catabolism in this strain is conferred by a large, previously cryptic plasmid (pRme41a).

A search has been made for other soil bacteria capable of catabolizing calystegins, both in the rhizospheres of plants that produce them and those that do not produce them. Calystegin-catabolic bacteria were found in the rhizospheres of plants which synthesize calystegins but not in the rhizospheres of other plants. It was suggested that the wild, calystegin-catabolizing strains may be beneficial to the plant.

DISCUSSION

Much experience has accumulated in connection with the inoculation of legume crops with *Rhizobium* spp. Here, each strain is specialized in that it will infect the roots of a limited range of hosts to form nitrogen-fixing nodules. Rhizobia are, however, saprophytes and can occupy other ecological niches in the soil itself and survive there, although, with the wide variability of the strains and the enormous range of soil conditions, it is rarely possible to predict the length of the survival period. This matter was raised by discussion of a hypothetical situation in which a crop of soybeans would be followed by fallow and then by peanuts, both crops receiving appropriate inoculants. The question of whether the peanut strain would establish itself in the soil more effectively after the fallow period than it would if applied immediately after the soybean crop, cannot be answered with certainty. It is a fact, however, that rhizobia usually do survive in soil after the legume crop has been lifted but not necessarily in numbers sufficient to benefit a succeeding crop of the same kind, so re-inoculation will be needed in the following season. This kind of result is, perhaps, not too comforting to those interested in ensuring the short-term survival of GEMs introduced into soil.

A further discussion centred on the two philosophies governing the introduction of organisms to soil. With useful organisms such as rhizobia, the concern is to ensure survival and persistence, but with GEMs the prime concern is for safety which implies that an introduced GEM does not proliferate abnormally, but dies out quickly. In fact, experience shows that although an introduced organism may survive it rarely remains established as a dominant member of the microflora.

It is especially difficult to estimate the number of cells of a particular microorganism in an environment such as soil when the number is small; indeed, if the number is very small the organism may be undetectable. Perhaps the actual level is not important; the important point is whether their presence is likely to cause trouble. A change in conditions could select for the organism and encourage its proliferation.

Suicide genes

Several important topics were raised in relation to S. M. Cuskey's talk on the introduction of suicide genes into bacteria destined for release or which might escape to the environment from a closed system. It was considered that such compromised bacteria could be of value when long-term survival would be inappropriate although there could be a danger that the bacteria might die too quickly if the concentration of pollutant, whose absence triggers the suicide process, falls too low. Studies so far have been confined to *in vitro* experiments and it is not known whether the suicide system described would operate in the natural environment. It is possible that the suicide genes could be transferred to other bacteria by transformation or transduction after death of the introduced organism. This might happen but is not likely to be a problem.

Other aspects of the topic discussed included the possible loss of the suicide gene by spontaneous mutation. If the rate of gene deletion were to be high, this could be important or even dangerous. It was also considered that the employment of the suicide gene strategy would have an important advantage in that it could allay public concern over the release of GEMs.

Calystegin production and utilization

Tepfer's study of calystegin production by morning glory and and its catabolism by a certain *Rhizobium* strain stimulated a general discussion on the complexity of the rhizosphere as a microbial ecosystem, especially in relation to the difficulty experienced in establishing introduced organisms in the soil. The plasmid-mediated ability of a *Rhizobium meliloti* strain to catabolize calystegins was regarded as an interesting example of a plant–bacterial interaction and illustrates a strategy which can be adopted by a bacterium to enhance its chances of survival and persistence in the environment. Such proliferation of a harmless saprophytic strain might benefit the plant by excluding other organisms including potential pathogens. In this field there is clearly scope for work on genetic engineering of bacteria and plants.

One aspect of the round table topics which was not discussed in any detail was that of 'colonization'. Clearly, this is a subject of considerable importance which deserves careful consideration.

17 Round Table 4: Genetic Stability and Expression

Chairman
PETER M. BENNETT *(Bristol, UK)*

Rapporteur
SIMON BAUMBERG *(Leeds, UK)*

Contributors
PETER M. BARTH *(Runcorn, UK)*
CARMEN SANCHEZ-RIVAS *(Buenos Aires, Argentina)*
ROBERT V. MILLER *(Naperville, USA)*

Editor
M. SUSSMAN

This round table dealt with phenomena that are neither predictable nor readily explicable but which are potentially of relevance to the behaviour of micro-organisms in the environment.

Peter T. Barth (*Runcorn, UK*) wished to dedicate his talk to Dr Bob Hedges who had died so tragically earlier in the year. It had been his enthusiasm, vast knowledge of the literature, hard work and genius that had contributed so much to the group working in Professor Naomi Datta's laboratory. They were the first group with Alan Jacob, Nigel Grinter and Peter Barth to recognize the phenomenon of drug-resistance transposition. It was Bob Hedges who recognized that certain puzzling plasmid analysis data could be explained by the transposition hypothesis. He invented the term 'transposon' to describe the, now familiar but then bizarre, concept of a distinct segment of DNA that could transpose itself into new sites on replicons. It had also been Hedges and Datta who found that there was a very large number of plasmids which conferred resistance to sulphonamides and streptomycin and it was they who had stimulated Grinter and Barth to investigate them. They discovered that these plasmids fell into a new incompatibility group, which they designated IncQ, and that they constituted a very uniform, widespread and broad host-range

RELEASE OF GENETICALLY-ENGINEERED
MICRO-ORGANISMS ISBN 0-12-677521-4

group of plasmids. R300B, which is indistinguishable from RSF1010 and R1162, is the prototype of this group and has become the progenitor of a large array of cloning and expression vectors used by many research workers around the world. IncQ vectors were used throughout the dehalogenase cloning project to be described below and it seemed fitting in this connection that Bob Hedge's indispensable contribution should be remembered.

Peter Barth described the cloning of a D-chloropropionate dehalogenase from *Pseudomonas putida* AJ1. This was sought because chemical synthesis of chloropropionic acid gives a 50:50 racemic mixture of D- and L-chloropropionic acid (D- and L-CPA) and only L-chloropropionate is the feedstock for the synthesis of several important compounds, including the ICI herbicide 'Fusilade'. The separation of the racemic mixture is difficult but a D-CPA dehalogenase would convert D-CPA to L-lactic acid which could then easily be removed.

Pseudomonas putida AJ1 contains stereospecific D- and L-CPA dehalogenases. A mutant of this strain, AJ1-23, is L-CPA$^-$ and contains only the D-CPA halogenase and thus might be a suitable source of the enzyme. Unfortunately, its expression is low, inducible by a costly inducer and the strain reverts to become L-CPA$^+$. It was, therefore, necessary to clone the D-CPA halogenase gene separately from that of L-CPA dehalogenase and to provide for high expression of the gene in a suitable host. The strategy for gene selection was by two means: (a) ability to grow on D-CPA as the sole carbon source and (b) hybridization with oligonucleotide probes to match the protein.

The D-CPA dehalogenase was purified and found to be a $\sim$30 kDa monomer which was active as its tetrameric form. The protein was sequenced and probes of two kinds were synthesized. These were *short* mixed probes of all possible codons and *long* unique (guessed) probes.

The initial cloning vector was pTB244, a derivative of the IncQ plasmid R300B. This has a broad host-range and is able to replicate in enterobacteria and in pseudomonads. Cloning was successful and yielded clones carrying genes for both D- and L-CPA dehalogenases. The former was separated by sub-cloning. Sequencing showed that the promoter was of the s^{54}, nitrogen status-regulated, type found for genes of nitrogen assimilation in *Klebsiella pneumoniae*.

The clones were unstable in *P. putida* and suffered deletions that sometimes removed the dehalogenase gene alone and at other times most of the plasmid markers. Colonies that expressed the dehalogenase were small, possibly because expression of the dehalogenase was deleterious and this may explain the selection for deletions. In addition, expression of the dehalogenase, unlike certain plasmid antibiotic-resistance markers, was not proportional to plasmid copy number, which could be varied by mutation. Such propensity for deletion and anomalous expression of cloned genes, of which this work was an example, could lead to anomalous results in untried conditions in the environment.

Carmen Sanchez-Rivas (*Buenos Aires, Argentina*) then described the generation of diploids after protoplast fusion and regeneration of differently genetically marked strains of *Bacillus* spp., in particular *Bacillus subtilis*.

Three kinds of progeny could be recovered from such experiments: (a) stable recombinants, (b) prototrophs and (c) biparentals. The latter two were initially unstable but could stabilize on subculture. Growth of prototrophs in non-selective media yielded further new recombinant types. Non-expressed markers could, however, still be shown to be present, by extracting DNA from the isolate and using it for transformation. Patterns of marker activity or inactivation accorded with a model for diploid genome formation by recombination between parental genomes around the origin and terminus of chromosome replication. This recombination was independent of *rec*E4, the *B. subtilis* equivalent of *rec*A. The inactivation could be either (a) reversible, while the types were still unstable and extracted DNA treated with proteinase K gave re-activation of markers, or (b) irreversible, when the types had stabilized and proteinase K had no effect. The work is described in detail by Sanchez-Rivas *et al.* (1982) and by Levi-Meyrueis and Sanchez-Rivas (1984). Similar results had been obtained in some cases with streptomycetes.

The results implied that genetic information in these groups of organisms could be cryptic—possibly in organisms from nature or laboratory organisms after other kinds of genetic transfer.

Robert V. Miller (*Naperville, USA*) discussed studies of gene transfer between micro-organisms in the freshwater environment. These had demonstrated that transfer by conjugation and transduction occurs at low but significant levels in such an ecosystem and had also provided new data which suggest that genetic instability and evolution of both the engineered sequences and their microbial hosts occurs at accelerated rates in natural environments. The significance of these preliminary findings may potentially far exceed those of gene transfer for the evaluation of risk of recombinant DNA in the environment as the consequences of genetic variation are significant even if transfer events are rare.

Prophage induction *in situ*

Environmental studies on transduction in the freshwater environment had demonstrated the major environmental reservoir of phages capable of mediating gene transfer by transduction is the induction of prophages from naturally occurring lysogens (Saye *et al.*, 1987). In laboratory simulations with water from a field site, the rate of prophage induction was found to remain constant at the spontaneous induction level. However, *in situ* dramatically increased levels of induction were observed in a periodic pattern (Fig. 1) which suggests that stresses capable of raising the rate of prophage induction operate in the environment. Since induction of lysogens is the most likely source of phages capable of transduction *in situ*, alterations in the rate of prophage induction may have a dramatic influence on environmental rates of gene dissemination.

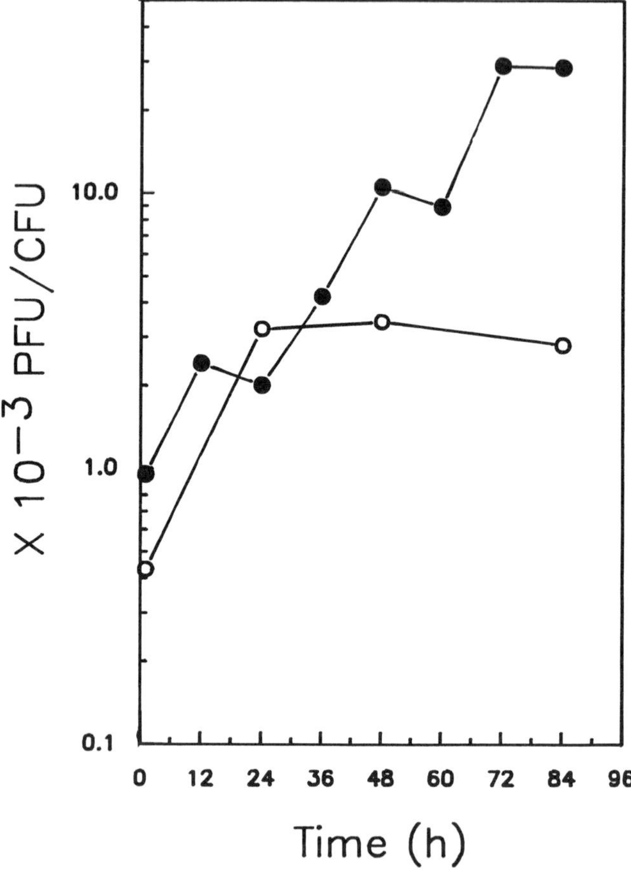

Fig. 1. Induction of prophage during long term incubations in laboratory simulations or *in situ* field trials. *Pseudomonas aeruginosa* RM296, an F116L lysogen, was washed to remove free phage and inoculated at a cell density of 10^4 CFU ml^{-1} into test chambers containing sterilized water from Fort Loudoun Lake, Knoxville, Tennessee, USA. Chambers were incubated either in laboratory simulations (●) designed to mimic *in situ* conditions or *in situ* in Fort Loudoun Lake (○). Samples were collected at the times indicated and assayed for bacteria and bacteriophages by methods described by Saye *et al.* (1987). A representative experiment of three is shown. Data from all experiments were qualitatively the same.

Increased mutation frequency *in situ*

During investigation of the transduction of chromosomal loci *in situ*, it was observed that reversion frequencies were raised from 50 to 1000-fold *in situ* above frequencies observed in laboratory simulations designed to mimic the conditions found at the field site. These data indicate that environmental factors may induce mutation rates which cannot at present be predicted from laboratory studies.

Increased rates of recombination and/or transposition *in situ*

During field trials on the conjugal transfer of plasmid DNA (O'Morchoe *et al.*, 1988), it was observed that the frequency of genetic alterations observed in transconjugants was much higher *in situ* than in laboratory simulations designed to emulate the *in situ* conditions. Approximately 50% of the transconjugants isolated from *in situ* matings showed at least one rearrangement or deletion in the transferred plasmid DNA. In laboratory simulations, the frequency of such alterations was less than 1%. It appears that genetic evolution of plasmid DNA mediated by the processes of transposition and/or recombination may be accelerated *in situ* compared with rates observed in the laboratory. Increases in the relative rate of either or both of these processes *in situ* would cause a significant increase in genetic instability in the environment.

During these environmental studies, preliminary evidence was obtained that suggests reassortment and modification of genetic material in natural ecosystems may far exceed the rates obtained in the laboratory setting. These include rates of mutation, gene reassortment by recombination and transposition, and prophage induction. Some of the factors were shown to be inducible by various forms of stress under laboratory conditions. Current practices for the assessment of risk are based on the assumption that levels of genetic stability in the environment are similar to those observed in the laboratory. Miller's results suggest that stress, operating *in situ*, may induce increased rates of genetic variability which cannot, at present, be adequately modelled in the laboratory.

DISCUSSION

In the general discussion that followed, **S. K. Farrand** (*Urbana, USA*) and **S. P. Mann** (*Cambridge, UK*) described instances of anomalously low expression of cloned bacterial genes in heterologous hosts. **G. J. Stewart** (*Tampa, USA*) described the increased levels of transformability of marine isolates, while **W. Wackernagel** (*Oldenburg, FRG*) described experiments on transformation on the surface of sand grains, which, rather as in Miller's experiments, showed higher levels of genetic transfer than in the laboratory and higher frequencies of deletion during transfer. Miller pointed out that in his experiments it was likely that genetic events occurred in organisms on the surface of particles in the water. **Roy Curtiss** (*St Louis, USA*) suggested that, perhaps, enterobacteria were best hosts for environmental release because sufficient was known about them to ensure that such unforeseen events did not occur.

The implications of the discussion were (a) that unexpected problems involving instability and/or low expression of cloned genes were not uncommon and (b) that various lines of work led to the suspicion that DNA transactions such as recombination, mutation, transposition and prophage induction could all occur at elevated levels in the natural environment. If this is confirmed, much further effort should be directed to solving this problem because of its considerable importance.

REFERENCES

Levi-Meyrueis, C. and Sanchez-Rivas, C., 1984. *Molecular and General Genetics* **196,** 488–493.

O'Morchoe, S., Ogunseitan, O., Sayler, G. S. and Miller, S. V., 1988. *Applied and Environmental Microbiology*, in press.

Sanchez-Rivas, C., Levi-Meyrueis, C., Lazard-Monier, F. and Schaeffer, P., 1982. *Molecular and General Genetics* **188,** 272–278.

Saye, D. J., Ogunseitan, O., Sayler, G. S. and Miller, R. V., 1987. *Applied and Environmental Microbiology* **53,** 987–995.

18 Round Table 5: Case Histories of Deliberate Release

Chairman
STUART W. GLOVER *(Newcastle-upon-Tyne, UK)*

Rapporteur
RAMON SEIDLER *(Corvallis, USA)*

Contributors
LIDIA WATRUD *(St Louis, USA)*
EDWARD WEDMAN *(Corvallis, USA)*
FAUSTINO SINEREZ *(Tucuman, Argentina)*

Editor
D. E. STEWART-TULL

Lidia Watrud (*St Louis, USA*) spoke about the considerations for the introduction of engineered organisms into the environment and stressed that part of her work was started before the regulations on the release of genetically-engineered micro-organisms (GEMs) was established in the USA. Nevertheless, the general safety criteria were established in the experimental design of laboratory and growth chamber tests, e.g. genes manipulated should be safe genes.

It was apparent that considerable advantages to the agricultural industry could be derived from the release of GEMs expressing toxin activities directed against particular insect pests: selectivity, improved efficacy, longer periods of control, reduced use of conventional chemical insecticides and cost effectiveness since the farmer would only have to use the GEM once or twice in a season. The criteria used for the selection of recipients for the recombinant gene product were that they be non-pathogenic, plasmid-free or cured, possess limited plasmid acceptance and able to maintain the introduced trait. Microbial isolates were identified which colonized plant roots at high population densities. Following sufficient taxonomic characterization to rule out known pathogens, selected isolates were characterized for biochemical and antibiotic sensitivity traits.

The organism used to receive the *Bacillus thuringiensis* ssp. *kurstaki* toxin

RELEASE OF GENETICALLY-ENGINEERED
MICRO-ORGANISMS ISBN 0–12–677521–4

gene encoding for the protein active against insects was *Pseudomonas fluorescens*. Using antibiotic markers, studies were carried out in the laboratory to assess environmental fate and toxicological effects of the *P. fluorescens* recombinant (Ps/Bt). To minimize concerns for the horizontal transfer of engineered genes, the delta-endotoxin gene from *B. thuringiensis* ssp. *kurstaki* was chromosomally inserted into the root colonizers.

In controlled microcosm studies, Ps/Bt was detected predominantly in the top 15 cm of the soil. Survival of the *Pseudomonas* root colonizer 112-12 in sewage was limited and generally a decreasing population of the GEM was found.

These engineered microbes have yet to be tested in the environment. However, regulatory support for the safety of the delta-endotoxin gene and for the host *Pseudomonas* root colonizers has recently been demonstrated. In June, 1987, the US Department of Agriculture (USDA) granted approval for a field test of tomatoes containing the delta-endotoxin gene; in October 1987, the US Environmental Protection Agency (EPA) granted approval for field testing the same *P. fluorescens* host organism engineered to express a non-pesticidal trait (lactose utilization).

The microbial release involving the Ps/Bt bacterium has not occurred. However, the Bt toxin gene was cloned into tomato plants and a field experiment was conducted with the recombinant plant.

There was no doubt about the efficacy of the toxin gene activity on insect control. Untreated tomato plants were stripped of all leaves whereas the recombinant plants containing the Bt toxin gene were strong and healthy with a complete coverage of leaves.

Ramon Seidler (*Corvallis, USA*) provided a carefully considered, responsible scheme for the measurement of dispersal of ice$^-$ recombinant bacteria released at California test sites.

Scientists from two organizations conducted the first spray releases of genetically-engineered ice nucleation deficient *Pseudomonas* species under the auspices of experimental use permits granted by the EPA Office of Pesticide Programs. These releases took place in California during April and May of 1987. EPA Office of Research and Development (ORD) scientists were responsible for designing a plan which would document the fate and dispersal characteristics of the viable recombinant bacteria by aerosolization during the spray events and on designated days following these events. The permit holders were responsible for extensive testing of plant, soil, water and insect specimens for recombinant bacteria. Field experiments were conducted at Brentwood, California, where scientists from Advanced Genetic Sciences, Inc. (AGS) conducted spray releases on strawberry plants on 24 April and 12 May 1987. On 28 May 1987, Dr Steve

Lindow of the University of California, Berkeley, sprayed recombinant *Pseudomonas* onto potato plants at a test plot in Tulelake, California.

P. syringae is the most common bacterial species responsible for plant frost injury due to production of ice nuclei active at temperatures warmer than −5°C. Scientists have learned that mutants of *P. syringae* and *P. fluorescens* may lose the ability to produce ice nuclei. When such mutants are present in high concentrations on plant surfaces, they can prevent frost from occurring down to about −5°C. Recombinant DNA technology has been applied to construct strains that have precise deletions affecting only their ability to produce ice nuclei. Such mutants are genetically stable and still possess all the other ecological attributes that allow them to grow and colonize leaf surfaces. Such recombinant deletion strains can be tested for their ability to compete with wild-type *Pseudomonas* species present on plant surfaces. When the recombinant strains are present in sufficient densities, they too can protect plant surfaces from frost as can conventional ice nucleation deficient mutants. The objectives of the experiments conducted by AGS and UCB were to evaluate, under field conditions, the ability of recombinant strains to control frost. The EPA considers these recombinant strains to be pesticides. The pests which might be controlled were the ice nucleation active, naturally occurring strains of *P. syringae* and *P. fluorescens*.

ORD scientists were involved in these efforts in order to provide documentation to regulatory office personnel within EPA on the releases, to verify proper functioning of the sampling equipment and provide recommendations for evaluating fate and transport of recombinant organisms for future field releases. The objectives of the EPA efforts were: (1) to design a sampling procedure to determine drift of recombinant *Pseudomonas* species during aerosol application and on subsequent days, (2) to determine distributions of the GEMs on and off the plot during the meteorological conditions encountered during the spray and post-spray periods and (3) to provide recommendations for evaluating fate and transport of recombinant organisms for future field releases.

The design selected by AGS consisted of a plot of 841 m^2 (0.2 acre), subdivided into a 4 × 4 Latin square and surrounded by a buffer zone 15 m wide (Fig. 1). Each square contained about 150 strawberry plants. The treatments applied to plants by AGS were: (1) control, water only, (2) an ice^- recombinant *P. syringae*, (3) an ice^- strain of *P. fluorescens* and (4) a mixture of a fungicide and bactericide. The purpose of the last treatment was to evaluate frost protection of plants that did not contain viable bacteria capable of ice nucleation. Thus, the effects of chemical elimination of wild-type ice nucleators was compared with the biological replacement by the recombinant *Pseudomonas* species. Each bacterial treatment contained about 200 million cells per ml in a volume of 10 l. Two releases were conducted by AGS, each was evaluated by EPA personnel. The releases took place on 24 April and 12 May 1987.

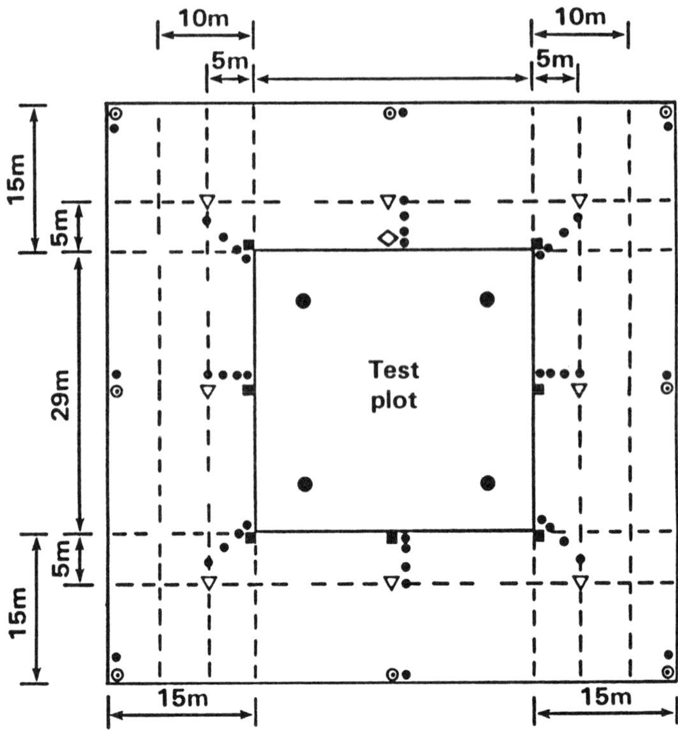

Fig. 1. Location of meteorological and bacteriological sampling devices on the Brentwood test plot: ◇, meteorological and bacteriological tower; ■, bacteriological sampling tower; ▽, sampling platform; ⊙, AGI-sampling pole; •, gravity plates and oat plants; ●, multiple sampling platform.

At the UCB site investigators utilized a randomized complete-block design with four replications of seven different treatments (Fig. 2). The plot planted with potatoes was about 1776 m^2 surrounded by a buffer zone 30 m wide. The treatments used by UCB were: treatment 1, untreated control, treatments 2–5 involved spraying either *P. syringae* Cit7dellb or TLP2dell recombinant strains, treatments 3 and 5 also received the wild-type *P. syringae*, treatment 6 was sprayed with a mixture of both wild-type strains. The application of wild-type strains of ice$^+$ bacteria ensured a challenge population would be present to evaluate the efficacy of frost control by the recombinant strains. Wild-type strains were applied only after the application of ice$^-$ bacteria and only after all monitoring devices for the detection of ice$^-$ bacteria were removed from the test plot. This procedure prevented the growth of wild-type strains on the agar media being used to detect the recombinant strains. Treatment 7 was analogous to

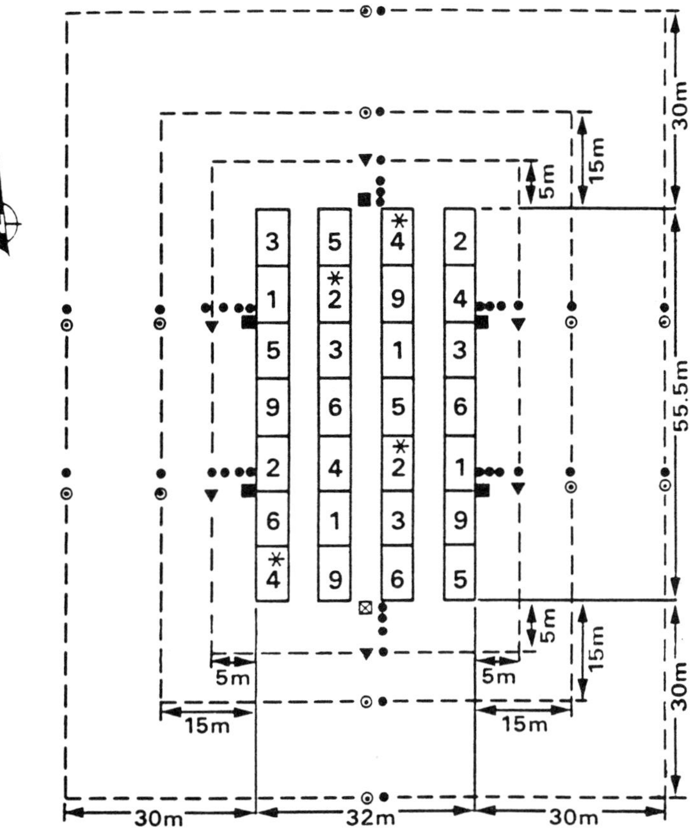

Fig. 2. Locations of meteorological monitoring and air sampling devices on the Tulelake test plot: □, meteorological and bacteriological tower; ∗, multiple sampling platform; ■, bacteriological sampling tower; ▼, sampling platform; ⊙, AGI-sampling pole; ●, gravity plates and oat plants.

treatment 4 of the AGS release. The Tulelake release evaluated by ORD personnel took place on 28 May 1987.

Five different types of sampling devices were employed by ORD to fulfil the study objectives. These were: (1) six-stage Andersen samplers, (2) Reynier slit samplers, (3) all-glass impinger samplers, (4) gravity plates (exposed petri dishes) and (5) sentinel potted plants. Although all samplers were found to have desirable attributes, results suggested that all-glass impingers (AGIs) and gravity plates (open petri dishes) were the most useful samplers of choice. Both were reliable, easy to set up and use and were relatively inexpensive.

Wind conditions were nearly ideal for minimizing drift during the release; they averaged only 0.1–0.2 m s^{-1} at 1 m above the ground. The records indicate air direction was highly variable during the releases; the aerosol movement did not favour significantly any single compass direction.

Reynier slit samplers provided a temporal description of the aerosol during the spray events. Results illustrated that the aerosol passed through most stations at Brentwood in 2–4 min, a time which approximates the spray period for a given Latin square. At the Brentwood releases, growth zones of the oxidase-negative *P. syringae* were clearly distinguishable from the oxidase-positive *P. fluorescens*. These strains were sprayed in succession and the Reynier results reflected this by the growth patterns on the agar plates. Usually aerosols passed through a station only once, even though wind direction was variable. Growth patterns on the Reynier samplers confirmed that there was occasional cross-contamination between the various Latin squares which received recombinant bacteria. At the Tulelake release, two zones of growth that represented the two recombinant strains Cit7dellb and TLP1dell appeared on most Reynier plates. The larger the growth zones the longer the period of exposure to the bacterial aerosols. Aerosol exposures recorded on the plates at Tulelake were typically of 6, 8 and even occasionally up to 10–14 min. The aerosols which took longer to pass through a station may have resulted from the larger plot size and thus the larger surface area being sprayed. Total spray times at all releases were about 20 min.

Based on sample calculations for the second Brentwood spray, less than 0.001% (or 15 million) of the total viable cells released entered the aerosol cloud at plant level and settled onto the 841 m^2 plot. Over 99.99% of the cells sprayed out were directly deposited onto the plants and soil. The total number of bacteria which settled onto the entire 15 m buffer was 1.2 million or 0.0001% of the total bacteria sprayed. However, 8% of the 15 million cells in the aerosol spray drifted out onto the buffer zone. Thus, once the microbes entered the aerosol, a relatively large proportion were able to drift away from the origin of application. A key factor in a release is to minimize the numbers of bacteria that appear in the aerosol cloud directly over the sprayed area.

Plots of average viable count per gravity plate versus distance from edge of sprayed vegetation were used to estimate the end-point of the aerosol drift. Such plots predicted that the extent of the detectable aerosol drift (0.1 CFU per gravity plate) on the day of spraying at Brentwood would be 19.2 m, while at Tulelake it was 35.2 m. There were only minor differences in the meteorological conditions during the two releases. Possible reasons for the further drift at Tulelake include: a slightly greater cell concentration (source strength) was detected on the Tulelake release; the plot was of a larger dimension, therefore, the aerosol covered a larger surface area; six-stage Andersen samplers revealed that the spray at Tulelake consisted of larger droplets; the droplets would not have evaporated for 30–50 s and, therefore, provided a longer period to drift before total evaporation of the droplets occurred. This phenomenon allowed a longer period

Table 1. Average colony-forming units of recombinant *Pseudomonas* spp. in aerosols detected by gravity plates at different locations on the test site (Brentwood event 2).

		Average CFU per m^2		Total CFU per area
Gravity plates located:*				
I.	Over the vegetation	1.8×10^4		1.5×10^7
II.	On the towers (0 m)	2.2×10^3		—
III.	3 m into buffer zone	1.1×10^3	0–3 m	6.1×10^5
IV.	5 m into buffer zone	1.7×10^2	3–5 m	1.9×10^5
V.	15 m into buffer zone	1.9×10^2	5–15 m	3.5×10^5

*Counts on plates at canopy level (I, II) or on the ground (III, IV and V). m^2 per buffer zone area: 0–3 m, 372 m^2; 3–5 m, 296 m^2.

of drift and may have maintained the viability of the bacteria for longer times.

On days 1 and 2 following the spray, recombinant bacteria were detected in some of the samplers. This resuspension was not detected on day 4. Wind movement greatly influences the extent of resuspension and the distance cells will travel.

Research will be conducted on the extent of resuspension of microbes from plant and soil surfaces.

The following major conclusions can be drawn from evaluating the dispersal of ice^- recombinant bacteria at California test sites:

(1) EPA scientists successfully designed, tested, and evaluated procedures for studying the air dispersal of recombinant bacteria released onto small test plots.
(2) All five types of samplers used in the study had desirable attributes under certain conditions. Results from this study indicated that AGIs and gravity plates were the samplers of choice due to their ease of use, low cost and satisfactory recovery efficiencies.
(3) Over 99.99% of the sprayed bacteria settled directly onto the soil and plants on the test plot. From an extrapolation of the data, the total number of bacteria which settled onto the 15 m buffer zone at the second Brentwood spray was 1.2×10^6 CFU or only 0.0001% of the total bacteria originally sprayed. During the spray events, viable cells in the aerosol drifted to the back of the buffer zone at concentrations of approximately 1 CFU per gravity plate.
(4) A greater drift in bio-aerosol particles occurred at the Tulelake event than at either of the Brentwood events. The larger area sprayed and the greater

source strength of cells applied at Tulelake may have significantly influenced aerosol drift.

(5) Research is required to understand quantitative aspects of re-entrainment since recombinant bacteria were detected in the air on days 1 and 2 following the spray events.

(6) Nozzle size is an extremely important factor influencing extent of drift and the numbers of recombinant bacteria present in each aerosol droplet. Research is needed to learn how nozzle-type influences the survival of aerosolized bacteria especially when numerous cells may be present in some droplets while only low numbers are in others.

Edward Wedman (*Corvallis, USA*) described the pathogenicity, immunogenicity and transmissibility of a recombinant vaccinia virus in calves.

A limited amount of research with recombinant viruses in livestock has been reported. These studies were undertaken at the Wallaceville Research Centre, Ministry of Agriculture and Fisheries, Upper Hutt, New Zealand, with the approval of the Ministry and the US Department of Agriculture.

Although Sindbis virus infection is widespread among livestock in certain geographic locations, it is not considered a pathogen and for that reason, the recombinant had no commercial value but was especially attractive as a recombinant virus model for studying live virus recombinant vaccines in livestock species.

The vaccinia virus used was the WR strain and the Sindbis virus was the HR strain. The recombinant vaccinia:Sindbis virus was constructed by Rice and others in 1985.

Six-month-old Holstein steers were examined daily for signs of illness throughout the experiment. Temperatures were taken daily for 1 week post inoculation. All animals were killed and incinerated at the end of each experiment. Tissue collected from the inoculation site, prescapular lymph node, liver, lung and spleen were ground using sterile sand, mortar and pestle. A 10% suspension in tissue culture media and whole plasma was inoculated onto monolayers of monkey kidney cells in 24-well plates to isolate virus.

Intradermal inoculation of ten calves with the recombinant virus produced localized nodular lesions at the site of inoculation without secondary or haematogenous spread. Reinoculation with the same recombinant resulted in new nodular lesions, despite detectable serum neutralizing antibody to both the recombinant and Sindbis virus. Pox lesions at the site of inoculation appeared as early as 2 days and persisted for as long as 16 days. Inoculated calves did not transmit virus to pen contact calves.

Another group of 14 calves was inoculated intradermally on the opposite side of the neck. Nodules produced by inoculation of the recombinant were removed periodically from these calves over a 17-day period and attempts were made to isolate the recombinant virus. From this same group of calves, two were killed on days 3, 5, 7, 9 and 12. The recombinant virus could not be isolated from the prescapular lymph nodes, plasma serum, spleen, liver or lung. Skin biopsies yielded the recombinant virus at the site of inoculation over the 17-day period.

In a separate experiment neutralizing antibodies against the recombinant virus and the Sindbis virus were detected in a few calves at 7 days, even though initially the levels were low; titres increased markedly after the second and third inoculations.

From these studies, it was concluded that the recombinant vaccinia virus was infective and immunogenic for calves. It caused pox lesions to develop at the site of inoculation. This occurred even with reinoculation made several weeks apart. It did not cause illnesses in calves nor was it transmitted to uninoculated pen contact calves.

The recombinant virus could not be passaged from one calf to another even by intradermal inoculation. Nor did it cause illnesses to develop even when inoculated intravenously in massive doses.

These findings should help allay concerns about the special danger of using recombinant viruses for animal vaccines. However, all live virus vaccines, whether modified by passage through alternate hosts, cell culture or the recombinant DNA process should be thoroughly tested for safety and efficacy before release to the professionals and public.

The paper of **Jose L. La Torre** (*Serrano, Argentina*), commenting on the field trial of a recombinant vaccine in Argentina was presented by **F. Sineriz** (*Tucuman, Argentina*). Argentina's sanitary authorities learned, in September 1986, that a field experiment to test a vaccinia–rabies recombinant vaccine, was under development in the county of Azul, Province of Buenos Aires, involving human beings and domestic livestock. This investigation stemmed from an exclusive agreement between the Wistar Institute (Philadelphia, USA) and the Pan American Health Organization (PAHO). The experiment started in June 1986 in the Experimental Farm, operated by the Pan American Zoonoses Centre (CEPANZO) of PAHO, in Azul. The correspondence is shown below:

27 May 1986
Carlyle Guerra de Macedo,
Director,
Pan American Sanitary Bureau,
Pan American Health Organization,
525 Twenty-Third Street, NW,
Washington, DC 20037

Dear Sir,
Enclosed is a copy of the letter of understanding between the Pan American Health Organization and The Wistar Institute which concerns Argentine field trials of a veterinary vaccinia recombinant rabies vaccine. The agreement has been signed on behalf of Wistar by its director, Dr Hilary Koprowski.

Sincerely yours,
Warren B. Cheston
Wistar Institute

5 June 1986
Dr Warren B. Cheston,
Associate Director,
The Wistar Institute,
36th Street at Spruce,
Philadelphia, Pennsylvania,
19104-4268

Reference: Letter of Agreement concerning Argentine Field Trials of a Veterinary Vaccinia Recombinant rabies vaccine.

Dear Dr Cheston,
Thank you for your letter of 27 May 1986 returning a fully executed original of the above-referenced agreement. We wish to acknowledge receipt of the agreement and to thank the Wistar Institute for its collaboration with the Pan American Health Organization.

Sincerely yours,
Hernan L. Fuenzalida
General Counsel,
Pan American Health Organization

19 May 1986
Dr Hilary Koprowski,
Director,
The Wistar Institute,
36th Street at Spruce,
Philadelphia, PA 19104-4268

Dear Dr Koprowski,
Following your conversations with Dr Mario V. Fernandes, Coordinator of our Program of Veterinary Public Health and Dr Joe R. Held, Director of our Pan American Zoonoses Center (CEPANZO), I am pleased to inform you that the Pan American Health Organization (PAHO) will carry out an experiment on the safety and immunizing properties of a vaccinia–rabies recombinant vaccine for cattle according to the following terms;

1. The experiment will take place in the field station of CEPANZO at Azul, in Argentina.

2. The experiment will be supervised by Dr Joe R. Held, with the collaboration of Dr Ana Maria Diaz (Chief of the Rabies Laboratory), Dr Elmo de la Vega (Chief of Pathology), and Mr Luis Lozano (Chief of the Azul Field Station), all from CEPANZO.

3. The experiment will be conducted as follows:

3.1 *Recombinant Vaccine*:

Vaccinia–rabies recombinant produced and provided by the Wistar Institute will be used at a concentration of 10^8 plaque forming units (pfu) of vaccinia-recombinant per animal.

3.2 *Animals*:

Forty head of adult milk producing Holstein cattle to be obtained by PAHO.

3.3 *Personnel*:

Four animal caretakers vaccinated against smallpox to be contracted locally by PAHO.

3.4 *Organization of the Trial*:

Ten animals will be vaccinated subcutaneously in the neck with 1 ml of recombinant vaccine. The vaccine will be administered from a 'multiple dose' syringe, similar to the ones used for routine vaccination of cattle. Ten other animals will be vaccinated with the recombinant vaccine by scarification of the previously shaved neck of the animals.

3.5 *Separation and contact of the animals*:

The ten cattle vaccinated suhcutaneously and ten vaccinated by scarification will be kept in separate enclosures, but each group will be kept in close contact with the ten non-vaccinated animals. The enclosures for each group of 20 animals (ten vaccinated and ten non-vaccinated) will be separated by a distance of $\frac{1}{4}$ mile and will consist of a fenced-in field, with a shed which will be constructed prior to the start of the trial. During the first months of the trial the animals will be kept in the shed constantly to ensure closest possible contact between the vaccinated and non-vaccinated cattle.

3.6 *Contact with humans*:

Two animal caretakers will be assigned to each of the two groups of cattle for the duration of the trial. They will milk the animals two or three times daily but will be instructed to avoid contact with the pustule that will most probably develop on the neck of cattle vaccinated by scarification.

3.7 *Clinical observations*:

All animals will be checked daily for development of skin or any other lesions, and the time of their appearance and eventual disappearance will be carefully recorded. The caretakers will be instructed to report immediately the development of any skin lesions, particularly on their hands.

3.8 *Material collected*:

Material from skin lesions developed in vaccinated or control cattle will be collected once or twice while the lesion lasts and preserved for isolation and characterization of the vaccinia virus.

3.9 *Collection of blood*:

Blood will be obtained from the animal caretakers and from *all* animals before vaccination and at 1, 3, 5 and 6 months after vaccination. Serum will be separated at CEPANZO laboratories and the presence of neutralizing anti-rabies antibodies determined at CEPANZO. Aliquots of serum will be sent to Rhone Merieux in Lyon, France, to determine the presence of anti-vaccine antibodies. Rhone Merieux will also receive material from skin lesions for isolation and characterization of vaccinia virus.

3.10 *Challenge of virus*:

If the vaccinated cattle develop anti-rabies antibodies, these animals and representative controls will be challenged at Azul with a lethal dose of rabies street virus of South American origin.

4. The financing of the project will be as follows. Upon signature of this letter, the

Wistar Institute will pay to PAHO US$70 568 (seventy thousand, five hundred and sixty-eight dollars), which represents the total cost of the proposed project. PAHO will administer these funds in accordance with its established procedures and regulations.

The budget for the project is as follows:

COST OF ANIMALS	*Australes*	*Dollars**
40 cows (A200 each estimate based on market price 15 Dec. 1985)	A8000	$10 000
FEED		
Hay 15 kg/animal/day (A0.061 kg)		
40 × 15 × 0.06 × 240	A8640	$10 800
Concentrate 3 kg/animal/day (A0.40/kg)		
40 × 3 × 0.40 × 240	A11 520	$14 400
Pasture per day (A0.40)		
40 × 0.40 × 240	A3840	$4800
SUPPLIES AND EQUIPMENT		
2 milking sheds (A2500 each)	A5000	$6250
Miscellaneous animal handling and milking equipment and veterinary supplies	A2400	$3000
LOCAL SALARIES AND BENEFITS		
4 persons (A10 per day)		
4 × 10 × 240	A9600	$12 000
10% overtime	A960	$1200
Administrative costs (13%)	A6494	$8118
TOTALS	A56 454	$70 568

*Based on rate of exchange at time of preparation of estimates.

5. Wistar agrees to indemnify, defend, and hold harmless PAHO from any and all losses, liabilities, costs, or expenses ('Indemnified Losses') it may suffer as a result of any action, suit, claim, or governmental or other proceeding or investigation arising out of or relating to the field trial, to the extent that such Indemnified Losses shall be caused by Wistar's negligence or willful misconduct or breach of its obligation hereunder.
6. The Wistar Institute and PAHO will consult prior to any publication of findings, reports, or other documents related to this project. In consultation with PAHO, Wistar will convene a meeting of Wistar, PAHO, and any other Organization that they agree to at the conclusion of the project to discuss the results obtained.
7. Any property purchased under the terms of this agreement shall, upon completion of the project and termination of the agreement, become the property of PAHO.
8. This agreement will be valid from the moment of its signature, and will be in effect for a period of one year.
9. This agreement can be modified, extended or rescinded by mutual agreement, and can be terminated unilaterally by one of the parts by notification 60 days in advance.

If you are in agreement, please sign this and the other original enclosed, and return one of them to PAHO.

Sincerely yours,
Deputy Director,
Pan American Sanitary Bureau

No permission or consent was requested from Argentina's authorities nor had any information become available about the organization of this investigation. The experimental protocol clearly shows that the aim of the investigation was to assess the possible effect of the recombinant viruses on vaccinated animals, as well as on those in contact with animals and on the human beings directly involved with the experiment. It is evident that this investigation, as it was planned and conducted, had a series of ethical, legal and scientific implications which called for a critical and thorough evaluation.

Ethical implications

Neither the national authorities nor the local scientific community were informed about the planning and performance of a rather critical investigation to be undertaken in their own country.

The Custom Office's franchises and the diplomatic status enjoyed by PAHO staff, under the UN–Argentina agreement on technical cooperation was apparently used for the introduction into the country of the recombinant viruses. The type of contents of parcels or luggage introduced under such franchises is not declared.

From available information, there was no involvement of Argentinians in the planning of the investigation. Participants in its implementation at the experimental farm were not informed about the risks and potential consequences of such an experiment.

Moreover, the four caretakers involved in the experiment were not under any kind of medical supervision. Furthermore, they were not vaccinated against smallpox before the experiment started. The only sign of previous immunization considered by the staff responsible for the investigation was the presence of scar or scars compatible with those left by a previous vaccination.

The milk from vaccinated cattle was consumed without pasteurization by the caretakers and their families, including children. The bulk of the milk was sent to the local market and distributed for selling after pasteurization.

No warning signs were placed near the experimental area, showing a clear lack of understanding of the risks involved or the intention of hiding the nature of the experiment.

Legal implications

Argentina, like most of the developing countries, has no specific regulations for the research, production or release into the environment of GEMs. However, there are two laws (No. 3959, of 28 December 1902, and No. 13636, of 31 January 1967) prohibiting the introduction into the country of exotic micro-organisms. Therefore, at least the spirit of these laws was violated by the PAHO staff who participated in the experiment. Consequently, in the event that human beings, domestic livestock or wild animals were affected by the recombinant virus, Argentina's health authorities would be compelled to denounce the case to the corresponding international forums and to take the corresponding legal actions against the people and/or institutions responsible for the implementation of this experiment.

Argentina, as CEPANZO's host country, has over the years substantially contributed funds and facilities in the establishment and operation of its headquarters and laboratories. This situation obviously implies that there is mutual responsibility regarding the operation of the centre. The way CEPANZO planned and implemented this experiment clearly violated the established agreement and also the normal relations between the PAHO's staff and their national counterparts.

For these reasons, it might well be that the corresponding national authorities would denounce the existing agreement and eventually interrupt CEPANZO's activities in Argentina.

Scientific implications

Argentina's scientific community, as well as the national sanitary and health authorities, are greatly interested in the new generation of vaccines for human beings and animals. The development of vaccinia recombinants created a great expectation in this area and no less than three local laboratories are fully committed to this subject. However, due to the fact that the use of attenuated vaccine strains, made by recombinant DNA or by standard procedures, requires very detailed studies to evaluate their behaviour and fate in the field, a cautious approach was adopted. In the particular case of recombinant vaccinia viruses it is clear that no satisfactory animal models were available for assessing their virulence and efficacy as a vaccine (Brown *et al.*, 1986). Therefore, before releasing vaccinia virus into the field it is imperative to develop and evaluate the corresponding tests, both *in vivo* and *in vitro*, for the theoretical determination of the virulence of vaccinia strains on man and animals. It will be important also to determine whether changes in host range or tissue tropism (Kriegler *et al.*, 1984–85) could result from the genetic modification of the virus and also by possible alterations in the viral envelope caused by the insertion of a foreign gene product (Brown *et al.*, 1986). Regarding the impact that the release of new vaccinia strains could have on the environment it should be pointed out that at present, little is known about the ecology of orthopoxviruses.

The way in which these viruses are maintained in nature and what is the role of wild-life reservoirs are open questions that require careful investigation (Baxby *et al.*, 1986).

This lack of information is important if vaccinia recombinant strains carrying genes of other immunizing antigens are to be widely used on human beings and domestic livestock. The possibility that such new strains may become established in nature, as vaccinia may have become in Indian buffaloes (Baxby *et al.*, 1986), and/or undergo recombination with other existing orthopoxviruses should be carefully considered.

For the above-mentioned reasons the Argentine authorities interrupted the experiment; the animals were killed and properly buried. Before they were killed, blood and milk samples from the animals were taken and stored under appropriate codification. These samples as well as others taken from the four caretakers, unvaccinated against rabies, and from 13 persons from the experimental farm, 12 of whom were previously vaccinated against rabies, were assayed for the presence of neutralizing antibodies against rabies virus. Preliminary results, performed by staff of the National Institute of Microbiology of Argentina which were never involved in this case, indicates that all vaccinated and contact animals had become seroconverted 6 months after the beginning of the experiment.

Samples taken by CEPANZO before the beginning of the experiment and confiscated by Argentine authorities were, in all cases, negative for rabies antibodies.

Analysis of the blood samples of three of the four caretakers indicates that one of them had become seroconverted for rabies virus antibodies; one was always negative. The other caretaker was not available for clinical or serological tests.

The results obtained with the 13 PAHO technicians indicated that 12 were positive for antibodies against rabies; all were vaccinated several times against rabies. The remaining one was negative and was not vaccinated before or after the experiment.

These results suggest that the recombinant virus passed from the vaccinated animals to all of their contacts and at least in one case to the human beings directly involved in the handling and milking of the animals. At present other possible animal contacts such as calves, rodents, etc., are being tested in order to determine the extent of the spreading of the virus.

In order to be absolutely objective and neutral, the Argentine authorities have requested international cooperation in order to confirm or dismiss the above results.

Concluding remarks

The way the field trial was conducted in Argentina clearly violated ethical and legal principles and could have consequences that will affect the development of new generations of vaccines not only in Argentina but also in the rest of the world.

It is clear that ecological organizations worried by genetic release will use this situation to protest against the development of rDNA products and their use in the environment. If they succeed in their actions, the development of new products, that could be extremely relevant for man, plants, animals and the environment, will be seriously endangered.

In this particular case in Argentina the problem was not the experiment itself but the way it was planned and implemented. Argentina's scientific community and government agencies are always open for cooperation and for scientific exchange with any country or scientific group around the world; there will not be any change in this attitude. Nevertheless, it is important that, in such cooperative efforts, ethical, legal and scientific principles are respected, whether or not written regulations exist.

DISCUSSION

Regarding questions on the Californian releases **Ramon Seidler** (*Corvallis, USA*) responded that a three-dimensional, meteorologically driven random walk model is under development at EPA. The model currently predicts the downwind movement of particles released over a small area such as the Californian release sites. The model will be improved when information becomes available on droplet evaporation and death kinetics of the microbe. In response to a question about the age of plants in the experiment it was explained that plants were sprayed when young to achieve a better colonization by applied bacteria. Leaves on young plants are not heavily colonized by bacteria and are more amenable to colonization by applied strains.

In view of the seroconversion of animal handlers in the field trial of a recombinant vaccine in Argentina, **Edward Wedman** (*Corvallis, USA*) was asked whether checks were made on people handling the animals. It seemed that there was no monitoring of the personnel connected with the release as all had been vaccinated with pox virus before the study began.

Many people expressed their concern over the field trial in Argentina. At the end of his presentation **Faustino Sinerez** (*Tucuman, Argentina*) expressed his opinion that a code is necessary to prevent such experiments being carried out again under similar circumstances. The main problem would seem to be that it is not known what experiments are being done throughout the world in those countries where there are no regulatory procedures. Responsible scientists want to earn the trust of the public and there was support for the idea of applying some pressure on countries who did not have regulatory controls. The OECD is meeting shortly to format a series of guidelines and the EEC are making efforts to draw together their guidelines—the ultimate aim would be to harmonize regulations worldwide.

The idea of greater international co-operation was followed up by a plea from **John Arbuthnott** (*Dublin*, *Eire*), who agreed to receive any recommendations for international action to oversee the future development of attenuated vaccines. Possibly, there is a need for an international agency on vaccine development because matters are developing rapidly and there is a disparity between what is being done at the research level and the practical use of GEMs. In 1983 WHO set up a special group to look at the development of vaccines against the five major diseases, dengue, poliomyelitis, hepatitis, tuberculosis and capsulated organisms.

There was agreement that the agency provides much useful financial support but it was still necessary to urge countries to support the work of the special group.

Concern was expressed that some regulated workers might export their GEMs to countries where there is little or no control. In the USA, before GEMs can be exported, an import licence from the country of destination is required. If the health authorities of that country has doubts, these must be carefully considered.

An interesting analogy was drawn by **Martha Gilliland** (*Nebraska*, *USA*), 'a regular engineer not a genetic engineer', who explained that we can, in clear technical terms, design safe structures but frequently ignore the human technical interface. The example discussed during the session indicated that the same sort of problem could occur in designing engineered organisms. **John Beringer** (*Bristol*, *UK*) stressed two points: (1) that any regulations and procedures must be 'user friendly' and (2) that guidelines are drawn up before the technology enters the commercial market-place. Therefore, it seems that in responsible countries real efforts have been made to ensure that as far as is humanly possible the human interface has been taken into consideration.

The discussion ended with some of the economic difficulties experienced by developing countries where vaccines are urgently required. **Fred Brown** (*Wellcome Biotechnology*) reminded the meeting that the recombinant vaccinia virus would cost pennies.

SUMMARY

The purpose of this session was to provide information on previously conducted or planned environmental releases of recombinant organisms. The nature of the products discussed varied from bacterial pesticides to recombinant animal vaccines. Some level of controversy accompanied the research endeavours of these early experiments. It is clear that all experimental procedures discussed resulted from very careful evaluation of major release issues.

In the first presentation it was learned that advantages to the chemical agricultural industries can result from the development of Bt toxins delivered by new bacterial strains or through recombinant plants to control insect pests. Extensive studies have been conducted to evaluate ecological ramifications of introducing a *Pseudomonas* root colonizer containing Bt toxin. These studies were only conducted in controlled laboratory or microcosm tests. Although there were regulatory constraints on the release of the engineered microbial pesticide, the Bt toxin has been sucessfully field tested in engineered plants. Results demonstrated field efficacy including protection of tomato plants from insect pests. A non-pesticidal form of the microbial host organism engineered with *lac*Z and *lac*Y genes has also been successfully field tested during the last year.

Results from air monitoring of the first two environmental releases in the USA were presented. Many types of monitoring equipment were used to evaluate field performance in anticipation of future activities. However, it must be emphasized that the complex array of equipment tested was for experimental purposes. Results clearly demonstrated that petri-dishes can be satisfactorily used to estimate the direction and distance of movement of recombinant bacteria in aerosols. Automated samplers can be used to provide more quantitative evaluations as well as timing of aerosol movements. Evidence illustrated the extent of viable cell movement ceased at the edge of the buffer zones surrounding the release sites.

The recombinant vaccinia:Sindbis virus was used as a model for studying efficacy of live virus vaccines in cattle. This experiment was conducted at a research station in New Zealand with the full knowledge and approval of the authorities. Results demonstrated that nodular lesions were found at the site of inoculation only and that no transmission of the virus occurred to control animals penned with inoculated cattle.

Antibody titres to both the recombinant virus as well as to the Sindbis virus was documented. The recombinant virus was not isolated from major body organs of the experimental animals. Virus recovery was only recorded for a transient period from the initial nodular lesions at the site of inoculation. Antibody titres increased markedly after repeated injections.

Ethical, legal and scientific implications were raised about another animal vaccine in a field test of a recombinant vaccinia–rabies virus. Planning and implementation of this experiment were not the major problems. Concern centred on the lack of information provided to appropriate government authorities in the country where the test was conducted.

Concerns were expressed about the spread of the experimental virus from the test animals to all the contacts involved in the test, including human beings. Concerns were also expressed about the possible establishment of the recombinant virus in the natural indigenous animal populations.

It was clearly emphasized that the government and scientific communities of the country are always open for scientific collaboration with any country or scientific group provided that communication and open consultations are appropriately emphasized by all concerned parties.

19 Round Table 6: Use of Microcosms

Chairman
E. P. GREENBERG *(Ithaca, USA)*

Rapporteur
N. J. POOLE *(Bracknell, UK)*

Contributors
H. A. P. PRITCHARD *(Gulf Breeze, USA)*
J. TIEDJE *(East Lansing, USA)*
DENNIS E. CORPET *(Toulouse, France)*

Editor
F. A. SKINNER

H.A.P. Pritchard (*Gulf Breeze*, *USA*) described the characteristics of microcosms.

Many types of research questions which evolve from a consideration of biotechnology risk assessment can be addressed with natural environmental samples and their associated microbial communities. In general, these samples are placed in some convenient container, incubated under some test condition or conditions and a result recorded. Depending on the framework of the questions asked, the results can easily provide new insights or verify certain hypotheses. This type of natural microbial community testing is simple, straightforward and in many cases, environmentally relevant. For example, to determine whether gene exchange occurs between an added donor and the indigenous microflora, a flask containing soil, plant material, sediment or water can be inoculated with the donor, incubated and transconjugants or transductants can be sought. The results lead to certain conclusions about gene exchange in natural microbial communities and provide more relevant information than pure culture and sterile environmental components. The great array of possible studies of this type including the appropriate controls, suggests that they should not be constrained by procedural criteria, definitions and/or standardization. Thus, it is wrong to

RELEASE OF GENETICALLY-ENGINEERED
MICRO-ORGANISMS ISBN 0–12–677521–4

refer to them as microcosm studies. If they are, this will dilute the importance of the term.

So what is a microcosm? It is an analogue to the field or a field within the laboratory. It is a laboratory system created and used because of two important considerations.

First, if one is interested in determining whether the results obtained in the type of tests described above are the same under the conditions in which the natural physical and biological integrity of the environmental sample are maintained, then it will be necessary to set up a microcosm. Physical integrity in the case of aquatic environments could be the sediment–water interface, a highly active zone of microbial activity that might be important in certain types of experiments or research questions. Biological integrity might be the maintenance of interactions between trophic levels, such as bacterial phytoplankton and grazing animals.

Second, if it is desired to extrapolate research results in a quantitative manner to the environment, then a microcosm is necessary. Extrapolation in this case means relating the laboratory data to the site where the environment samples were taken.

If neither of these considerations is pertinent to the research question being asked, then it is argued that a microcosm study is unnecessary. If it is felt that these considerations are meaningful, then it is possible to define a microcosm as a laboratory system containing an intact piece of the field which has been shown to behave, ecologically, like its counterpart in the actual field. The 'intact' aspect is brought out to accommodate the integrity consideration. The physical design of the microcosm will be controlled by this need for intactness, the ecological behaviour aspect is designed to accommodate the extrapolation consideration. The behaviour assessment, a process called field calibration, will be a function of the ecological parameters selected for comparing microcosm and field responses. Table 1 lists several types of functional and structural

Table 1. Field calibration measures/toxic effects parameters.

Functional	Structural
Thymidine uptake	Lipid and fatty acid profiles
Heterotrophic activity	Discriminant analysis of mixed measures
Oxygen uptake/carbon dioxide production	DNA/RNA sequencing
Hydrolytic activities	
Reduction of tetrazolium salts	
Litter decay rates	
Inorganic nutrient cycling; nitrification/denitrification	
Biodegradation of xenobiotic chemicals	

parameters that could be measured to determine the difference or similarity between the microcosm and a field site. Field calibration does not imply that the microcosm should be an exact analogue to the field; instead, it allows one to determine where the microcosm stands relative to the field. Depending on the calibration, a decision can be made regarding the environmental significance of the microcosm and ultimately the results it will produce.

A variety of other types of laboratory test systems designed to model the environment are also available. For example, synthetic communities, with known numbers of selected organisms placed in sterile media or environmental components and jar ecosystems in which natural samples are incubated over long periods in the laboratory to evolve a sustaining ecosystem which is unique relative to natural ecosystems, have been used. These systems obviously are not appropriate for examining research questions that deal with the integrity of ecosystem components or extrapolation. They are, however, very useful for the generic application of complex laboratory test systems and for the screening of genetically-engineered micro-organisms (GEMs) which might cause adverse effects when introduced into these systems. It must be remembered, however, that screening results cannot be extrapolated to the field. Instead a data base from the test system must be developed so that relative evaluations of new information can be made.

In the EPA biotechnology risk assessment programme, microcosms as defined above, are used for two purposes. First, they are used to verify the environmental significance of data generated in a variety of simpler experiments. For example, if the survival and colonization pattern of a GEM that is derived from shake flask or test-tube studies has to be verified under more environmentally relevant conditions, the organism is introduced into a microcosm and its resulting survival colonization pattern is compared with the previous results. If it turns out that the microcosm gives the same pattern, then to some extent, the simpler test data have been verified and can be used in place of microcosms in future experiments. In this type of verification study, it is clear that microcosm studies alone are not sufficient—considerable information from other experiments and test is mandatory.

Second, the microcosms are used to detect potential ecological effects that result from the introduction of a micro-organism. The principle here is that the microcosms are black boxes of many interacting processes and it is assumed that interference with one process will affect others which will then be reflected in certain integrative measures of ecosystem structure and function. In many cases, these measures are the same as those used in field calibration and, thus, Table 1 reflects the type of measures that could be taken. If potential ecological effects are to be investigated in this manner, considerable research is required in the areas of sensitivity, signal-to-noise ratios from ecological background data on the microcosms, and extrapolation may be necessary.

Next, the various types of microcosms were described by **J. Tiedje** (*East Lansing, USA*).

For the purpose of this subject it is useful to distinguish two types of microcosm: those that are meant to represent a subunit of nature, brought into the laboratory, with its native biotic and abiotic components relatively undisturbed, and those that are simpler but which share the feature of allowing the study of some principle of a micro-organism(s) in its natural habitat.

The first type is more commonly the type meant when macro-ecologists or eco-toxicologists speak of microcosms. This type has been of value, for example, in studies on bio-accumulation of chemicals, and has also recently been used successfully in studies on the fate and gene exchange of GEMs. It is not at present suitable for use in determining ecological effects because it is too insensitive and one does not know what to measure, or in risk assessment terminology, the end points are not clear. Therefore, it is premature for submitters of requests for the environmental release of GEMs to use results from microcosms as evidence to justify a 'no ecological effects' conclusion.

The second type of microcosm—the simpler one—is often the type envisaged by microbial ecologists. This type have a major role to play in obtaining risk assessment information. It is often the most relevant laboratory approach to obtaining data on competitiveness, survival, gene exchange and certain types of dispersion. These microcosms should be designed to answer specific questions and their limitations should be known. For example, when used to study gene exchange the conditions used have a dramatic impact on the results, e.g. exchange frequencies vary from as high as for filter mating to non-detectable, depending on method of introduction and moisture content.

A typical soil microcosm of this simpler type is a soil core. In contrast to an aquatic microcosm, or a leaf phyllosphere, a soil core is not difficult to maintain in the laboratory under conditions not greatly different from those in the field. To represent the soil community reasonably, the core should contain not only soil but roots, detrital particles, anoxic microsites and the micro- and mesofauna typically found in soil cores, and the soil should not be air-dried, nor sterilized. The design of the core often depends on the purpose of the study so that the key features being tested are not compromised. For example, if one is studying transport, an intact core of reasonably large dimensions is necessary if it is hoped to approach predictability in the field. If one is studying gene exchange or competition a more uniform distribution of the organism may be necessary to obtain interpretable results and, thus, a mixed core may be more appropriate. Consequently, the design must be versatile to accommodate various needs for set up as well as for sampling.

In the design of soil core microcosms, several considerations are important:

(1) Organism addition. This includes how organisms are physically added

(i.e. injected, mixed in, spread on surface), the physiological state of the inoculum (washed, not washed, starved), the density of the cells, and whether the organism is added in a carrier (e.g. vermiculite). These features will have a major impact on the results and must be carefully evaluated, selected and controlled. Density is mentioned in particular, because of the too frequent tendency to add more organisms than the ecosystem can support. The added concentrations should be in the range of the carrying capacity of that system and the intended dose range for the organisms.

(2) Sampling scheme. Possibilities include horizontal or vertical sub-cores in larger soil core microcosms for temporal monitoring, leaching and total (destructive) sampling.

(3) Designs appropriate for treatments. Considerations include a sufficient volume for root growth for the desired length of experiment, maintenance of moisture and aeration status, leaching designs that simulate rainfall, temperature control and later organism additions.

(4) Analytical methods for monitoring populations or processes. For population studies, plating, immunofluorescence and gene probes are the possibilities. The considerations are how much sample is needed, how many replicates and the capacity for sample analysis. The spiral plater and laser counter are a major advancement in capacity and accuracy for plate analysis.

Tiedje's group had used a soil core system divided into vertical sections so that different organisms can be mixed into separate layers. The sections are stacked and sealed by tape. After a period of organism adaptation to soil, some of the organisms are displaced downward into sections below by simulated rainfall and leaching. Thus, competition and gene exchange can be studied in a more natural way since the movement and physiological state are realistic in nature.

The gut of gnotobiotic animals as microcosms for the assessment of the safety of GEMs was considered by **Dennis E. Corpet** (*Toulouse, France*).

Microcosms are laboratory systems which mimic field ecosystems. They have been used to study the interactions between pollutants and bacteria. A few microcosm studies have been done on the survival of recombinant bacteria (Pritchard and Bourquin, 1984; Armstrong *et al.*, 1987). However, the most dreaded complication of the use of GEMs would be the establishment of the GEM, or the transfer of its DNA, in the digestive tract of humans. A special type of microcosm is proposed here, based on germ-free animals maintained in isolators and associated with complex floras.

The *isolator* is an airtight plastic container in which gnotobiotic animals are raised and which prevent the entry and escape of bacteria. Food, bedding and water are sterilized by irradiation or autoclaving, and introduced into the isolator through a sterile transfer system, with peracetic acid as sterilant. The air goes into and out from the isolator through absolute membrane filters and animals are handled with rubber gloves which are sealed to the isolator wall. These experimental systems are safe, because the bacteria used in the experimental animals cannot contaminate the outside. These systems yield reproducible results because their flora can be fully determined, and bacterial contaminations from the outside are avoided.

The *gut microflora* is a complex ecosystem, in relation to its animal or human host, and harbours a high density of bacteria (more than 10^{11} CFU g^{-1}) which are mainly strict anaerobes, belonging to hundreds of different species (dominant populations are of *Bacteroides*). The main ecological parameters in the gut are the flow rate due to intestinal transit, the very low oxygen content, and the low concentration of substrate remaining for bacterial growth.

Human-flora-associated mice

The complex faecal flora from humans can be transferred to germ-free mice and, provided they are isolated, the gross composition of the flora remains similar to that of the donor's flora for at least several weeks (Hazenberg *et al.*, 1981; Corpet, 1987).

Colonization resistance

The indigenous microflora is one of the major defence mechanisms which protect the body against colonization by invading bacteria. For example an *E. coli* strain, even if recently isolated from stools, cannot be implanted into the gut flora of other individuals of the same host species (Ducluzeau *et al.*, 1970). Results from *in vivo* and *in vitro* studies and mathematical models suggest that the antagonistic effect of the microflora against bacterial invaders is due to competition for limiting substrates and for adhesion sites (Freter *et al.*, 1983a). Besides, the microflora of healthy humans prevents the establishment of R-plasmid bearing enterobacteria in the gut, since these resistant strains, which are found in most people, are cleared from the stools of volunteers given a sterile diet (Corpet, 1988).

Survival of GEMs *in vivo*

The *E. coli* strain Chi-1776, which is certified for recombinant DNA experiments, fails to establish in germ-free mice (Wells *et al.*, 1979). This strain, and *E.*

coli DP50supF, died rapidly in conventional mice, such that their excretion in the faeces was reduced to 1 and 5% of the inoculum, respectively, after an oral dose (Freter *et al.*, 1983a).

Plasmid transfer *in vivo*

In germ-free mice associated with two isogenic clones of *E. coli*, one of which was harbouring an R-plasmid, one plasmid rapidly spread from the resistant strain to the plasmid-free strain (Corpet and Lumeau, 1987), while the other did not. Plasmids of this last class were a disadvantage to the bearing strains, which were slowly eliminated from the gut of the disassociated mice, mainly because of the detrimental effect of the plasmid on the growth rate (Corpet, 1986). However, antibiotic levels ranging from 1 to 10 μg ml^{-1} of drinking water, were enough to give an ecological advantage to the R-plasmid bearing strains in the gut of gnotobiotic mice (personal unpublished data).

Besides, even a transient strain, which cannot colonize the gut, can transfer its plasmid to resident bacteria, if the number of viable cells passing through the gut is high enough. However, the resulting transconjugant populations are often too small to be detected (Freter *et al.*, 1983b).

Possible *in vivo* studies and criteria for a safe GEM

(1) GEMs should not colonize germ-free mice.
(2) GEMs must be eliminated from the gut of human-flora-associated mice.
(3) GEMs must not transfer DNA to resident intestinal bacteria:
  (a) even under an antibiotic selective pressure or
  (b) in the presence of a conjugative plasmid (co-transfer).

The *in vivo* genetic transfer could be assessed with a DNA probe (Holben *et al.*, 1988). It would be the *in vivo* counterpart of the *in vitro* conjugation techniques which are used to measure the genetic stability in GEMs (Walter *et al.*, 1987). One should keep in mind that up to now, most GEMs are designed and tested for use in plant husbandry, but in the near future GEMs will be administered to humans (e.g. through fermented milk), and to cattle (e.g. probiotics, living bacteria added in feedstuffs to promote animal growth).

DISCUSSION

Microcosms

The three papers on microcosms for aquatic, terrestrial and animal systems led to a lively discussion on '*What is a microcosm?*'

The term 'microcosm' seems to mean all things to all men, ranging, for example, from 'synthetic' to 'natural' systems, the latter containing an intact piece of the field which has been shown to behave ecologically like its counterpart in the actual field. It was agreed that arguing over definitions was a pointless exercise but that workers who use microcosms should understand the uses, limitations and bias of their particular systems and ensure that people using their results are also aware of these facts. To discuss microcosms and their value in studying GEMs was difficult enough because of generalities; it was agreed that this discussion was made more difficult by people raising hypothetical examples such as assessing the risk of inserting "botulinum genes in lactobacillus which is then used in the manufacture of yogurt", because they were unrealistic examples of what people are trying to achieve with GEMs and have confused the debate.

Microcosms and decision making

A microcosm is just one of the experimental techniques available in a continuum from gene sequencing to field evaluations, all of which can provide *quantitative* information required for risk evaluation. A decision on the risk associated with the release of a GEM should *not* be based solely on the results from a laboratory microcosm; other information is essential. If designed correctly a microcosm can provide valuable information on questions concerning ecological implications, the survival and spread of GEMs, or, most importantly, in determining the competitiveness of a GEM when compared with its 'wild' counterpart. It should be remembered, of course, that many potential uses of GEMs will involve strains with increased persistence.

It is unrealistic to believe that there can be one 'generic-microcosm' which could be used to evaluate the release of all types of GEMs. Each GEM will have its own peculiar characteristics; however, it will be possible to develop general guidelines.

Microcosm design

Though it might be considered axiomatic, it is essential that the questions to be asked with the microcosm, and the experimental design to be used, are carefully thought out. The results, once obtained, must then be considered while bearing in mind the limitations of the experimental design. Quantifiable results and the careful use of mathematical models will help in refining the microcosm. These challenges, and examples of failure to consider them, were discussed. Particular emphasis was placed on the need to consider the level of complexity, sensitivity, stability and reproducibility, the need for field calibration and the parameters measured, for example, the survival of GEMs added to a soil microcosm will be strongly influenced by its mode of application, the concentration added and the sampling method. In aquatic microcosms it is essential to model the physical characteristics, for example, water turbulence.

Conclusions

The participants came to the following consensus view on microcosms.

Recommendations

(1) The particular microcosm should be carefully defined.
(2) It should be ensured that the user of the results understands the limitation of the microcosm.
(3) Microcosms should be quantitatively calibrated to the field conditions.
(4) Microcosms should be designed to answer specific questions.
(5) Gnotobiotic animals should be used to screen GEMs wherever appropriate.
(6) Appropriate soil, water or animal microcosms should be used where appropriate.

Main uses of microcosms

(1) To screen for effects of GEM characteristics which are not desirable.
(2) To provide quantitative data for modelling, and hence refining the microcosm.
(3) To provide one of the tools of microbial ecology, and the results to form one component in decision making.

REFERENCES

Armstrong, J. L., Knudsen, G. R. and Siedler, R. J., 1987. *Current Microbiology* **15,** 229 232.
Corpet, D. E., 1986. *Journal of Antimicrobial Chemotherapy* **18,** Supplement C, 127–132.
Corpet, D. E., 1987. *Antimicrobial Agents and Chemotherapy* **31,** 587–593.
Corpet, D. E., 1988. *New England Journal of Medicine,* (in press).
Corpet, D. E. and Lumeau, S. 1987. *Zentralblatt für Bakteriologie und Hygiene* **264,** 178–184.
Ducluzeau, R., Bellier, M. and Raibaud, P., 1970. *Zentralblatt für Bakteriologie und Hygiene* **213,** 533–548.
Freter, R., Brickner, H., Fekete, J. *et al.*, 1983a. *Infection and Immunity* **39,** 686–703.
Freter, R., Freter, R. R. and Brickner, H., 1983b. *Infection and Immunity* **39,** 60–84.

Hazenberg, M. P., Bakker, M. and Verschoor, B. A. 1981. *Journal of Applied Bacteriology* **50,** 95–106.
Holben, W. E., Jansson, J. K. and Tiedje, J. M., 1988. *Applied and Environmental Microbiology* **54,** 703–711.
Pritchard, P. H. and Bourquin, A. W. 1984. *Advances in Microbial Ecology* **7,** 133–215.
Walter, M. V., Porteous, A. and Seidler, R. J., 1987. *Applied and Environmental Microbiology* **53,** 105–109.
Wells, C. L., Johnson, W. J. and Balish, E., 1979. *Zentralblatt für Bacteriologie und Hygiene*, Supplement 7, 197–200.

20 Round Table 7: Fate and Effects

Chairperson
ROSA GRIGOROVA *(Sofia, Bulgaria)*

Rapporteur
J. GWYNFRYN JONES *(Ambleside, UK)*

Contributors
R. A. PRINS *(Haren, Netherlands)*
SONJA SELENSKA *(Sofia, Bulgaria)*

Editor
M. SUSSMAN

This session dealt with some of the least understood and least predictable aspects of the release of genetically-engineered micro-organisms (GEMs).

The proceedings were opened by **R. A. Prins** (*Haren, Netherlands*) with a discussion of joint work with R. Hengeveld of the Research Institute for Nature Management, POB 201, 6800 HB Arnhem, Netherlands. They considered that the parallel between the future release of GEMs and the so-called 'biological invasions' by higher organisms is difficult to avoid. If general principles could be derived from a study of the numerous examples of biological invasions, such principles might be of value in considering the fate and potential effects of the release of GEMs.

Biological invasions

Recently there has been renewed interest in the mechanisms underlying biological invasions as is apparent from an increasing number of publications (Joenje *et al.*, 1986; Programme on the ecology of biological invasions, Scientific Committee on Problems of the Environment [SCOPE]). Hengeveld (1988) has

RELEASE OF GENETICALLY-ENGINEERED MICRO-ORGANISMS ISBN 0-12-677521-4

recently carried out a major study of the mechanisms of biological invasions and the major conclusions will be described.

Classical views

Biological invasions are spectacular phenomena characterized by the rapid spread of a species over considerable areas. Invasions are known in different aspects of population biology, though often by different names and emphasizing different aspects. In epidemiology where spatio-temporal aspects of disease incidence are studied, invasions are known as epidemics, whereas in palynology invasions are called migrations. Ecologists speak of 'species introductions' or 'colonizations' and in population genetics 'waves of advance' and 'the spatial spreading of genes' are discussed. Yet, all these terms pertain to more or less the same process.

In ecology, populations are often considered spatially stable and the view is held that communities consist of both qualitatively and quantitatively co-adapted species that mutually control their abundance. Foreign species would have difficulty in penetrating established communities but, if successful, they would disrupt the structure of the community until the point is reached when the established species can control the density of the invaders and thus establish a new multi-species equilibrium. 'Closed' communities would be less invadable than 'open' ones but, by circular reasoning, predictions of the degree of 'closed-ness' or 'open-ness' depends on their invadability. Not all species would be good invaders, some species having properties that enable them to invade foreign communities more easily than others. What these properties are is not clear, although there have been many suggestions.

Mobility of the species, especially of animals, is important in the colonization of islands or new land, such as young polders or volcanoes. At first much land remains unoccupied, resulting in low competition pressure both among the established species and with new arrivals. Later an equilibrium is formed between the number of immigrants and that of the established species which may eventually die out because of too high a competition pressure. Such a competition pressure relates negatively with the available surface area and that might explain the positive relationship between the number of species and island size.

This theory of species equilibrium has greatly stimulated spatial analysis in ecology and has recently resulted in the theory of patch dynamics (Pickett). In patch-dynamics natural systems are seen as mosaics of local, suitable patches, colonized by various species at different times. Eventually these species die out to give individuals of the same or of other species the possibility of colonization. As various patches differ both in quality and age, highly intricate, spatially-dynamic systems can result and these determine the species composition of the area as a whole.

New insights

Analysis of various cases of quite different types of invasion, such as holocene tree invasions, the spread of early European man, the invasion of New Zealand by red deer and by thar, the spread of cholera and measles in North America and Iceland, etc., all demonstrate that individualistic behaviour rather than coherence of structural inertia of communities characterizes colonization. Invasions usually start with the local introduction of individuals into an environment not occupied by the species. When, in this 'empty' environment, an individual produces on average more than one new individual, a population will build up. In the course of time, this population will spread in a wave-like fashion.

Rates of expansion

An important question is how fast the expansion will take place. Skellam was the first to investigate this problem by means of a diffusion model originally derived by R. A. Fisher for the spread of advantageous genes. Several mathematically-inclined ecologists applied such diffusion models for the analysis of invasions. Unfortunately the Fisher–Skellam model is valid only under fairly limited conditions. Firstly, an individual is assumed to move at random throughout life. It is, however, well-known that many species settle down permanently after their juvenile period. Secondly, all individuals are assumed to be equal in the sense that their reproduction and death rates depend only on their local environment and not, for example, on their age.

A model called the *advancing wave model* was, therefore, developed by Hengeveld (1988) based on a diffusion component, combining spatial diffusion with logistic growth of local populations. It was shown that for long-distance diffusion and for normal Gaussian dispersion and for the situation that N_0 is negligibly small relative to N_t, the expansion rate of the area invaded becomes:

$$A = 4\pi D r t^2$$

Where A is the area, D is the diffusion rate, r is the growth rate and t is time. The square root of the area occupied, that is the radius of the area multiplied by the constant $\sqrt{\pi}$ increases linearly with time: changes in expansion rate reflect changes in growth rate. A period of neighbourhood diffusion has to be distinguished from the subsequent long-distance diffusion. In the case of population densities that are too low, a curvilinear relationship exists between $\sqrt{A}$ and time t.

Structured populations

Van den Bosch *et al.* (1988) have discussed this general model for the spatial spread of newly introduced species, by applying the model to the well-recorded invasions of the Collared Dove (*Streptopelia decaocto*) and of the muskrat (*Ondatra zibethicus*).

They used mechanistic explanatory models that depend on the concept of structured populations (Metz and Dieckmann, 1986). This departs from the notion that not all animals are equal. For example, the body size of mammals may vary and this explains the differences in energy requirements, which relates to input–output considerations of the environment. When the frequency distributions of the body surface area of these mammals are measured, it can be calculated what size classes can persist in the system. Although considerable simplifications are made, a model developed by Belowski (1987) provides a first approximation of the necessary population sizes and habitat area for averting extinction over some time period.

The advantages of the model developed by Van den Bosch *et al.* (1988) is that no assumptions are made about survivorship, reproduction and dispersal of individuals and this overcomes the problem that accurate information about the life history characteristics are usually not available. The theory (Dieckmann, 1978, 1979; Thieme, 1977, 1979) starts from two types of life history characteristic at the individual level. The first type, the demographic characteristics, are the well-known life-table statistics. The second type, the dispersal characteristics, describes the distribution of individuals around their place of birth. With these characteristics it is possible to calculate the number of individuals at any time at any place and in this way it is possible to study the spread of the population.

Sonya Selenska (*Sofia, Bulgaria*) then discussed the fate of newly-constructed *Rhizobium* strains after introduction into the soil. This depends on ecological factors, such as the type of soil and its previous use, nutrient availability, mineral composition, pH, moisture, salinity, temperature, the effects of bacteriophages, epiphytic bacteria and mycorrhizal association, indigenous rhizobial populations, production of bacteriocins, etc. (Dowling and Broughton, 1986).

Indeed, many soils contain numerous indigenous Rhizobium strains which may be poorly effective or even ineffective but well-adapted to their environment and may often be highly competitive against newly introduced inoculants. In connection with this it would probably be reasonable to deal with strains previously selected for competitiveness in the particular environmental conditions into which they would be re-introduced. Also important would be the traceability of the strains.

For experiments on this subject a convenient mutant of *Rhizobium meliloti* 41 was obtained by random Tn*5-mob* mutagenesis (Simon, 1984). This mutant contains a Tn*5-mob* inserted into the plasmid which, according to several authors (Bromfield *et al.*, 1985; Toro, 1986), participates in competitive relations of *R. meliloti* and is known to be more effective in symbiosis than the parental strain. Thus, it was possible to determine the viability of the bacteria, their maintenance of symbiotic activity and to detect possible transfer of plasmid sequences of migration of Tn*5* among rhizobia and/or other micro-organisms, as well to study the effects of such genetic changes, if they take place.

Another important factor in symbiotic nitrogen fixation is the host-plant genome. It is well known that about 25–30 proteins are specifically expressed within nodules and an even larger number of host factors induce and control nodule formation (Long, 1985). It should then be reasonable to manipulate both the macro- and micro-symbionts in such a way that nodulation can be established within a specific legume–*Rhizobium* combination. Several strains of *R. meliloti* with an increased symbiotic activity were obtained by introduction on non-oncogenic variants of Ti plasmids and opine secretion by nodule cells induced by transconjugants was detected. Since Selenska's strains utilize opines, the secretion makes it possible to colonize the rhizosphere of the host plant leading to an improvement in their competitiveness with indigenous strains. However, it is important to answer the following questions.

First, is it possible that opine secretion creates conditions for the accumulation of pathogenic strains of *Agrobacterium tumefaciens* around the cells of transformed nodules, which in turn would be harmful for the environment?

Second, is it to be expected that any pathogenic effect results from *R. meliloti* containing Ti plasmid or is it possible that a conversion of non-pathogenic *R. meliloti* to a pathogen such as *A. tumefaciens* occurs? According to Heuman (1984), such conversions are possible.

Third, does the Ti plasmid transfer to other soil bacteria?

So far results have only been obtained relevant to the second question. Pathogenic effects by the transconjugants have not been detected on *M. sativa* even when a wild-type Ti plasmid was introduced into *R. meliloti*. But questions still remain.

It would be useful to investigate the behaviour of newly constructed strains, first, in model ecological systems similar to the real surrounding conditions and only after that to release them into the environment.

DISCUSSION

Among the comments made in general discussion, there was a re-assertion that macro-ecology mathematical models should be applied to microbial ecology. However, lack of knowledge about species-composition may hamper progress towards a theory of microbial ecology. With regard to invasions, it was suggested that, perhaps, there was an absence of evidence for invasion by organisms that were neither pathogenic nor commensal. The model may, therefore, be of limited use in the study of GEMs and their dispersal. However, it might have greater relevance to released virus populations.

Philip F. Entwistle (*Oxford, UK*) doubted whether the dispersal of free-living micro-organisms follows any known pattern but that of pathogens, or at least the disease that they cause, certainly does. For instance, the early spread of nuclear polyhedrosis virus disease in populations of pine sawfly (*Neodiprion sertifer*), from any initially small epicentre, disperses along an indented curve. This can be normalized by 'logging' the units of distance (x) and of disease proportion (y) ($\log y = a - b \log x$): the gradient of this curve ($-b$) appears to be a constant. For instance, it was almost exactly the same for an example from Scotland and one from the FRG. Other systems could be quoted. Of course, this type of solution was classically observed and analysed by van der Plank in South Africa and P. H. Gregory in the UK for the primary spread of plant diseases. Van der Plank, given to pithy descriptions wrote something like:

"For any given host–pathogen association, the rate of spread ($-b$) is a constant but the more propagules generated at the epicentre, the further the disease spreads."

Eventually the primary situation matures to a wave form travelling out from the epicentre; but this itself breaks down and spread then appears rather chaotic. This is because disease does not only spread continuously but also discontinuously. Chance events can lead to a degree of 'jump spread' leading to the creation of new epicentres. These grow and interact with the original centre and with each other resulting in the state of apparent chaos one finds in a generalized epidemic, epizootic or epiphytotic.

One can imagine that all this might happen with organisms larger than pathogens; for instance, insects and even smaller vertebrates may be subject to discontinuous 'jump spread'.

J. Gwynfryn Jones (*Ambleside, UK*) then broadened the discussion by suggesting that although little was known about the survival of micro-organisms, nothing was known about their effects in the natural environment. Survival of both the organism and the gene should be considered, particularly since very closely-related strains of the same species of bacterium, all containing the same plasmid, show widely differing survival curves. Perhaps, further consideration should be given to the information already available about the survival of organisms in the field, as, for example, the survival and dispersal of organisms applied to agricultural land in the form of slurry.

The question was then raised as to which effects should be considered. While pathogenicity would be high on a priority list for risk-assessment, possible long-term effects on major geochemical cycles should not be ignored. It was proposed that microbial ecology was moving too slowly for biotechnology, which was already prepared for large field trials. The next question would be whether the environment has a buffer capacity to cope with GEMs and that, therefore, deleterious effects would not be detected. It was generally considered that survival and transfer of a foreign gene was more important than the effect it might have in the environment, since the former precede the latter. The scale of introduction of a GEM into a particular environment should also be borne in mind, since threshold concentrations are important with regard to effects. In the closing stages of the discussion it was suggested that, since there does not appear to be any concern about the release of organisms obtained by conventional enrichment, where gene transfer has almost certainly occurred, there should not be any real worries about the release of a GEM.

REFERENCES

Belowski, G. E., 1987. In *Viable Populations for Conservation*, Soule, M. E. (ed.), pp. 35–57. Cambridge: Cambridge University Press.

Bromfield, E. S. P., Lewis, D. M. and Barran, L. R., 1985. *Journal of Bacteriology* **164,** 410–413.

Dieckmann, O., 1978. *Journal of Mathematical Biology* **6,** 109–130.

Dieckmann, O., 1979. *Journal of Differential Equations* **33,** 50–73.

Dowling, D. N. and Broughton, W. J., 1986. *Annual Review of Microbiology* **40,** 131–157.

Hengeveld, R., 1988. *Mechanisms of Biological Invasions*. London: Chapman and Hall. In press.

Heuman, W., 1984. *Molecular and General Genetics* **197,** 425–436.

Joenje, W., Bakker, K. and Vlijm, L., 1986. *Proceedings of the Dutch Academy of Science, Series C, Biological and Medical Sciences* **90,** 1–80.

Long, S., 1985. In *Nitrogen Fixation Research Progress* Evans, H. J., Bottomley, P. and Newton, W. E. (eds), pp. 87–93. Dordrecht: Martinus Nijhoff.

Metz, J. A. J., and Dieckmann, O., 1986. *The Dynamics of Physiologically Structured Populations*. Springer Lecture Notes in Mathematics 68. Berlin: Springer-Verlag.

Metz, J. A. J., de Roos, A. M. and van den Bosch, F., 1988. In *Proceedings of the Falsterbo Workshop on Size and Age Structured Populations*, Eberman, B. and Persson, L. (ed.), In press.

Simon, R., 1984. *Molecular and General Genetics* **196,** 413–420.

Thieme, H. R., 1977. *Journal of Mathematical Biology* **4,** 337–351.

Thieme, H. R., 1979. *Journal of Mathematical Biology* **8,** 173–187.

Toro, N., 1986. *Molecular and General Genetics* **202,** 331–335.

van den Bosch, F., Hengeveld, R., Metz, J. A. J. and Verkaik, A. J., (1988). *Journal of Biogeography*, submitted for publication.

21 Round Table 8: Pre-release Considerations

Chairman
GILLES PELSY *(Paris, France)*

Rapporteur
MONICA RILEY *(Stony Brook, USA)*

Contributors
B. J. J. LUGTENBERG *(Leiden, Netherlands)*
HISAO UCHIDA *(Tokyo, Japan)*

Editor
C. H. COLLINS

Early in the planning stages of a 'release' project, many matters should be considered. The list of these should include, information on toxonomy, toxicology, survival, genetic stability, genetic exchange, host range, efficacy, field design and monitoring plans.

Clearly, many factors enter into design of the genetic delivery system, and each factor presents choices. The list of choices is long and only a few were mentioned. There is the choice of the host and the vector system. Choices can be made about parameters of the gene expression system, about the range of genetic transferability of the gene and the persistence of the organism or the gene in the target environment. One can plan for either a long- or a short-lived system. In addition, a marker trait can be chosen for incorporation into the construct to permit tracking of the fate of the organism or the gene in the environment.

Other factors that could receive pre-released consideration are characterization of the target ecosystem in both biological and physical terms. Consideration could be given to the potential interaction of the introduced organism with indigenous populations at the site and to the potential effects of physical factors and climatic conditions on the fate and interactions of the introduced organism.

To a large extent, it is the individual investigator who defines the list of pre-release considerations for a given project. In many countries, guidelines of

RELEASE OF GENETICALLY-ENGINEERED MICRO-ORGANISMS ISBN 0-12-677521-4

considerations are provided by governmental bodies. Factors of concern are listed, relevant information and evaluation is required to establish the relative safety or level of hazard of a planned release. All such information is a part of *pre-release considerations*.

In the first of two talks, **B. J. J. Lugtenberg** (*Leiden, Netherlands*) described a proposal for engineering a safe marker system for Gram-negative host micro-organisms. As an alternative to, for instance, antibiotic resistance markers, he proposed that an outer membrane protein of *E. coli* be modified by splicing in a surface protein derived from a *Rhizobium* strain. The hybrid protein would extend beyond the surface lipoprotein molecules and might then serve as a surface antigen for that modified host organism. According to this scheme, specific antibodies to the surface-located hybrid protein would be used to trace the fate of the organism and perhaps also would allow recovery of some of the released organisms in antigen–antibody complexes.

When a genetically-engineered micro-organism (GEM) is released it is necessary to follow its fate. Marking a bacterium with a transposon would be very convenient, but as we have heard in other discussions, would arouse considerable opposition. New marker genes should, therefore, be developed, and ideally they would have the following characteristics: (1) specificity, (2) safety, i.e. occur naturally in soil, (3) easy detectability, (4) general applicability and (5) they should be scientifically rewarding. Requirements (1)–(4) point to surface-exposed protein as a marker. The use of antibodies would then allow the organism to be recovered while it is growing in the soil or the rhizosphere. The new chemical make-up could then be used as a *reporter* to elucidate the nature of the growth limitation under the environmental conditions being studied.

Specificity

The cell surface is subject to attack by, for example, chemicals, enzymes, bacteriocins and bacteriophages. Adaptation of cell surface components to ecological niches is, therefore, likely to be an important contributory factor to differences between individual strains. Indeed, profiles of membrane proteins, lipopolysaccharides (LPS) and C-polysaccharides (CPS) have been used to characterize individual strains within species or biovars, e.g. of *Pseudomonas* and *Rhizobium*. Synthesis of LPS, flagella, etc., requires many genes and to transfer all specific LPS information into other strains is very complicated. Cell surface proteins are, however, shielded from exogenous antibodies by O-Ag-containing LPS and most strains of soil bacteria contain O-Ag. Strain specificity for a single protein is hard to imagine, e.g. PhoE protein is strongly conserved in Enterobacteriaceae.

PhoE protein has the following properties: (1) it is induced by phosphate limitation; (2) it occurs in *all* Enterobacteriaceae; (3) its predicted amino acid sequence is strongly conserved (the *Pho*E gene can be functionally expressed in *Pseudomonas* as well as in Enterobacteriaceae; (4) its topology is known in great detail (e.g. there are many surface-exposed loops (Fig. 1)); (5) antigenic peptides have been introduced into a predicted loop, giving a determinant, that is poorly surface-exposed in *E. coli* markers. However, the strong conservation of PhoE may be a useful property to introduce into the outer membrane of any Gram-negative (soil) bacterium of choice as an *anchor*.

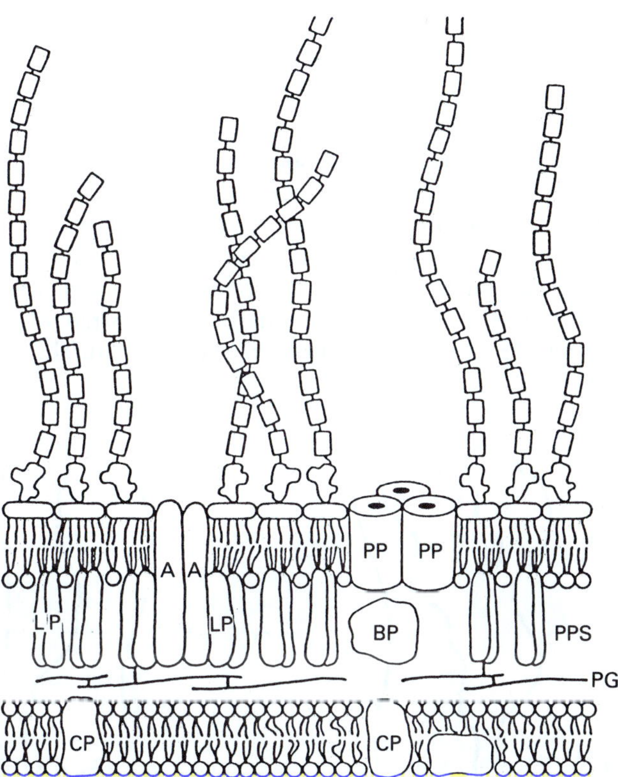

Fig. 1. Molecular architecture of the cell envelope of Enterobacteriaceae. The cytoplasmic membrane (CM) consists of a phospholipid bilayer in which proteins, including carrier proteins (CP) are present. The outer membrane consists of an asymmetric lipid bilayer with phospholipid and lipoprotein forming the inner leaflet. The outer lipid leaflet is formed by the lipid part of LPS probably together with those of enterobacterial common antigen (ECA) and capsule. Three types of outer membrane proteins, namely pore protein (PP), OmpA protein (A) and lipoprotein (LP) are indicated. The two membranes are separated by a peptidoglycan layer (PG) and a periplasmic space (PPS) which contains oligosaccharides and proteins, among which are binding proteins (BP). After Lugtenberg and van Alphen (1983).

There remain the problems of surface-exposure and specificity. The surface-exposed loops may offer opportunities to solve these problems, by constructing a hybrid protein as a novel surface-exposed marker gene.

Production of a protein that is surface-exposed in O-Ag-containing cells

To produce a surface-exposed protein: (1) PhoE protein is used as an *anchor* by introducing a large piece of protein into a surface-exposed loop as a tag; (2) since the protein must be exported by the export machinery of the cell, it must contain all the signals required for export and lack those preventing export. It should be noted that not every protein can be used as a tag.

NRPX protein

This is a 50 kDa protein produced by a soil bacterium, i.e. it is 'natural' and safe. It is secreted by some *Rhizobium leguminosarum* strains but only after induction, as determined by immunological screening of many soil bacteria.

As a *secreted* protein, NRPX would not be absolutely specific for the 'released' bacterium, but by converting it into surface-exposed protein, in a loop of PhoE protein, it will become a specific marker for the strain to be introduced when whole cells are used for detection or when the natural *nrpx* inducer is absent (Fig. 2).

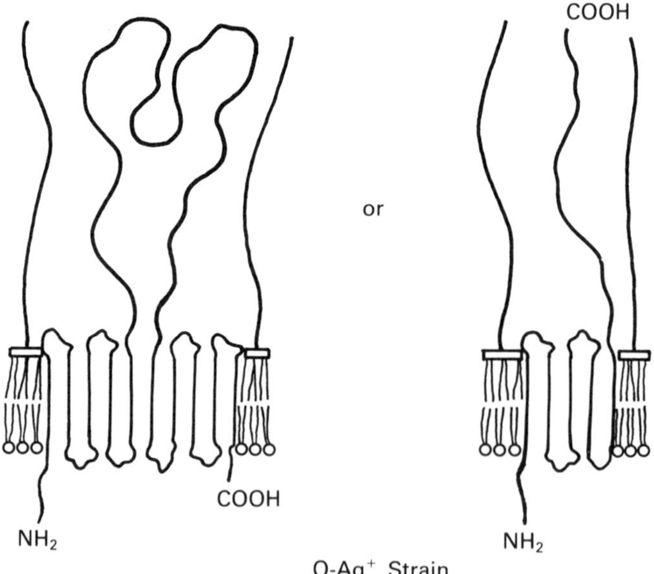

Fig. 2. Introduction of NRPX protein in a surface-exposed loop of PhoE protein.

To obtain constitutive expression of a new *PhoE*-NRPX-*Pho*E marker, a promoter which is constitutively expressed in *E. coli*, *Pseudomonas* and *Rhizobium* is used.

To apply the new *Pho*E-NRPX-*Pho*E marker in a bacterium of choice, the construct is introduced in a transposon devoid of antibiotic-resistance and transposase. The genes of the latter are cloned in a plasmid that can be maintained only under selective pressure. This makes its use in the laboratory possible. Derivatives suitable for release purposes arise spontaneously in the absence of selective pressure.

In summary, the requirements of specificity and safety are satisfied. Specificity is achieved by changing the cellular compartment and regulation, while safety is obtained by using components that occur in soil; *Pho*E in *Enterobacter* and NRPX in *Rhizobium.*

Hisao Uchida (*Tokyo, Japan*) discussed the present status of pre-release considerations in Japan.

In 1986 the Ministry of Agriculture, Forestry and Fisheries (MAFF) of the Japanese Government published, for public comment, their first draft guidelines for agricultural and environmental applications of biotechnology and in May of that year the OECD made the report of the *Ad hoc* Group on Safety and Regulations in Biotechnology publicly available. Within the framework of that report, 'Industrial Guidelines' were issued in June 1986 by the Ministry of International Trade and Industry of the Japanese Government (MITI). The Industrial Guidelines were formulated by the Recombinant DNA Advisory Committee to MITI which has been reviewing industrial proposals since then and rDNA committees have reviewed the experiments under the aegis of the Ministry of Education, Science and Culture, and the Science and Technology Agency of the Japanese Government. As yet, however, agricultural and environmental guidelines are not officially available in Japan, and it is still difficult to talk about releases of GEMs to the open environment.

Hisao Uchida reviewed some features of the MAFF draft, though he had not been involved in drawing this up. He then proceeded to describe recent studies by Yoshimi Okada of Tokyo on a live vaccine that cross-protects tobacco plants against tomato mosaic virus (TMV) infection. Finally, he expressed concern about attempts to formulate a universal scheme to assess the risk of environmental releases.

The main principles of the MAFF draft are 'step-by-step' testing and the 'case-by-case' judgements. As one step of testing recombinant plants, the MAFF draft guidelines introduced the Good Outdoor Practice (GODP) concept. This is a restricted and controlled test area with measures to avoid the spread and transmission of the test rDNA plants and their genetic material to plants outside the area. No details were given about the construction of these facilities.

The draft also mentions a controlled model environment, which is "a restricted area where the spread of rDNA micro-organisms and their genetic material are minimized".

The MAFF draft then lists various considerations in the assessment of safety. These are more or less similar to those employed by the Industrial Guidelines, except for additional descriptions of 'weediness' for recombinant plants. Although not mentioned in the draft, a weed is a plant difficult to control when grown on cultivated lands. 'Weediness' of a plant can be checked, according to the MAFF experts, by observing (1) asynchrony of germination, or duration of dormancy after appropriate germination stimuli are given to seeds, and (2) the ability of a part of a plant, such as its stems, leaves, roots, etc., to regenerate the whole plant.

Micro-organisms were classified into five categories by the same system as that adopted by the Industrial Guidelines (Table 1). Further 'case-by-case' considerations may be necessary, however, for certain deliberate releases such as microbial pesticides and live vaccines.

Table 1. Classification of recipient micro-organisms according to MITI, Japan

(1)	GILSP Non-pathogenic. Either to have (a) an extended history of safe industrial use or (b) restricted growth except under special cultural conditions
(2)	Category 1 Non-pathogenic, but not included in GILSP
(3)	Category 2 Pathogenicity to human is not deniable, but infection will not cause serious diseases. Availability of effective preventive and therapeutic measures
(4)	Category 3 Human pathogens not included in Category 2, but effective preventive and therapeutic measures are available
(5)	Special class Human pathogens without known preventive or therapeutic measures

A distinct feature of the MAFF draft guidelines is the proposal that once a GEM has cleared the safety assessment, on 'case-by-case' reviews, it is exempt from rDNA regulations.

The GEM can then be used in an 'outside GODP' environment.

In the 2 years since the draft was published for public comment, there has been no further discussion of it in Japan. In fact, there has been no pressure for 'Guidelines' because there have been no proposals to release GEMs. Compared with developments in the fermentation industries, activities involving agricultural or environmental application of rDNA technology have been scanty: historically, Japan has been a very conservative agricultural country.

Nevertheless, researches in the field of plant molecular biology are now gradually progressing in Japan. Although a broad range of potential applications of rDNA techniques in agriculture has been suggested, it is likely that the first practical case in Japan will be construction and use of transgenic plants showing enhanced resistance to chemicals and disease.

A specific example is the recent work by Yoshimi Okada and his colleagues in Tokyo. They observed strong cross-protection in transgenic tobacco plants, expressing a mild strain of tobacco mosaic virus. Cross-protection of plant viruses is a phenomenon in which a plant, infected with one strain of a virus, is protected from super-infection by a second, related strain. The phenomenon is well-known and has been used successfully for practical control in a few viral diseases. For example, a mild mutant of TMV-L, named TMV-L11A, was isolated in Japan in 1971 by conventional mutagenesis and has since then been used to protect tomato plants from infection by TMV. TMV-L11A is a stable strain giving practically no symptoms in its host plants. As it has a long history of safe agricultural use, it may be considered as a Good Industrial Large Scale Practice (GILSP) equivalent or, more appropriately, as a plant equivalent of live vaccines for animals. Okada integrated into the chromosome of the tobacco plant, cDNA of the whole TMV-L11A genome. The plasmid also has a kanamycin resistance marker. By means of the Ti-plasmid vector system, the cDNA was integrated into the chromosome of tobacco plant (*Nicotiana tabacum* cv. Samsum) and the transformed plant showed 1:3 segregation of the ability to produce the mild virus. A transgenic plant was isolated containing a single copy of the cDNA per cell, as analysed by Southern blotting. The transformed plant thus obtained showed efficient cross-protection against infection by the wild type TMV-L (Fig. 3).

Fig. 3. Cross-protection of plants transformed by TMV-L11A cDNA against TMV-L infection. (Courtesy: Yoshimi Okada.)

Two strategies have been known to produce virus-resistant plants. One is the expression of the viral coat proteins as developed by Roger Beachy of Wisconsin. The other is the expression of the satellite RNA as reported by B. D. Harrison and W. L. Gerlach. Compared with these two methods, expression of the mild strain by infection from within has the definite advantage of showing higher resistance to viral infection. On the other hand, the transgenic system has several drawbacks; the necessity for a stable mild strain, the possible decrease in yield and quality of crops by mild viral production and possible back-mutation of the mild mutation. These drawbacks can be overcome by introducing appropriate deletion mutations which may also help to reveal the mechanisms of the cross-protection.

The work of Okada's group raises interesting questions, at least from regulatory points of view. Is it necessary to obtain further data within GODP areas, even though the cross-protection by the mild strain has been used safely for a long time? Is it necessary to delete a part of the viral cDNA to abolish the production of infective viruses? What happens when the host plant has been changed from tobacco to tomato? The success of this work suggests that more cases of rDNA engineering of useful plants will be reported from Japan before long.

Pre-release considerations

Hisao Uchida said that there is no theoretical need to introduce new regulatory measures to control rDNA organisms just because they were modified by rDNA technology. Conventional methods were new when the technique was first introduced many years ago. Moreover, rDNA organisms should pose only a very small 'incremental' problem over non-rDNA organisms to assess the risks, because, as stated in the OECD/CSTP Considerations: "genetic changes from rDNA techniques will often have inherently greater predictability compared with traditional techniques, because of the greater precision that the rDNA technique affords to particular modifications". On the other hand, because of over-population of the country, the Japanese public in general reacts unfavourably to the idea of 'environmentally releasing' GEMs into the open air.

It may be pertinent here to recall past experiences of environmental release of microbial pesticides in Japan. Although a number of microbial pesticides have been registered in Japan, and outdoor experiments and commercial sales of preparations have been permitted, the use of microbial pesticides is limited and evaluation of its cost-benefit is still controversial. The first case of a microbial pesticide was *Bacillus thuringiensis*. These bacteria were first isolated and described in Japan in 1901, as a potent pathogen of the silkworm, and this fact has been and still is greatly affecting its evaluation as a microbial pesticide, although variants of the bacteria with low toxicity for the silkworm have been isolated. Until about 50 years ago silk was the main item of export from Japan. When the possibility of using *B. thuringiensis* as microbial pesticide was explored in the US, the Japanese Government prohibited its import to protect sericulture. Although *B. thuringiensis* preparations were registered as a Biological-Biorational Pesticide by EPA in 1960, its import was banned until 1971. Representative *B. thuringiensis* preparations have been produced in the US, France and the USSR. In the US, *B. thuringiensis* preparations are exempted from tolerance requirements and are applied at recommended doses even up to the day of plant harvest. In Japan, many companies applied for registration of *B. thuringiensis* preparations during 1974 but this was only granted in 1982, after another 8 years. Therefore, much information has accumulated about the safety and persistence of *B. thuringiensis* preparations, paying particular attention to their effects on the sericulture. In the meantime, there were advances in techniques of sericulture, including disinfection procedures, and *B. thuringiensis* is no longer a threat to silkworms. However, *B. thuringiensis* is listed in the current rDNA guidelines as an organism requiring P2 physical containment when used as a DNA donor. It is understandable that most companies diminished their interest in developing or improving microbial pesticides, in spite of the fact that a number of chemical pesticides are still causing trouble, and regulations will not be relaxed. This case is an example that shows the difficulty of changing a public concept once it has been established.

This is one reason why transgenic plants will be permitted in Japan before the environmental application of engineered micro-organisms. Obviously, 'agriculture' cannot exist without genetic breeding to improve useful plants and cattle and releasing them to cultivate and breed. To most Japanese, agriculture is part of the environment. It is interesting to note that most of them feel relieved by the sight of well-cultivated rice-fields because the scenery is 'natural' to them.

The basic philosophy of the risk assessment of the environmental releases now envisaged in Japan is to localize or contain the effect of the release within a certain area, in order to minimize the effect upon the 'ecosystem'. However, the time will soon come when an active application of GEMs is necessary to preserve 'nature'.

In conclusion, concerning *risk analysis approaches for environmental releases of GEMs*, safety can best be endorsed by history and experience, but not easily by a theory or by an analytical logic, nor even by 'experiments' within GODP or in a model environment. It is important to have a theory, but without knowing what to control and what controls what, formulation of a universal logic scheme is next to impossible. Hisao Uchida was not convinced by deductive approaches which attempt first to formulate generic guidelines in order to assess risks associated with a particular proposal. It is true that recombinant DNA techniques have been used for more than a decade, and experimental or industrial guidelines have been useful, but a wide spectrum of environmental diversity as well as application protocols almost preclude any generalization unless generic guidelines become extremely restrictive, or of very limited applicability. It might be more prudent to consider carefully and thoroughly each specific proposal by adopting the case-by-case approach, and to accumulate enough experience on environmental introduction of modified organisms before we state anything general.

In the discussion that followed there was disagreement on the merits of giving advance consideration to interactions between the released GEM and populations in the target environment. Some thought it would be helpful to incorporate principles and considerations of microbial ecology into the planning process. Others thought there are no useful *a priori* principles and no general rules to follow, so each project should be assessed individually, case-by-case.

Regarding B. J. J. Lugtenberg's proposal, some discussants were sceptical that such a hybrid protein would be accepted by the outer membrane structure. Some thought the three-dimensional structure would be altered unpredictably, possibly erasing antigenic characters. There was wide agreement that safe alternatives should be found in place of antibiotic resistance markers. Many believed it inadvisable to introduce antibiotic resistance markers into the environment.

One discussant suggested that every engineered organism approved for release should be placed in a public repository for possible further study in the future. **Roy Curtiss** (*St Louis*, *USA*) pointed out that in the US, most such organisms would be registered for a patent. A requisite part of the patenting process is placement in a public repository.

It was recommended that scientists be more forthcoming about possible problems on drawbacks where warranted, to increase their public credibility. Unqualified optimism was seen to strain the faith of the public.

Many discussants saw a need for better public education to allay irrational fears. Several working programmes were described but difficulties in communicating effectively were experienced by others. There was general agreement that efforts towards public education should be continued.

22 Closing Address

EDWARD A. ADELBERG

Department of Human Genetics, Yale University School of Medicine, New Haven, CT 06510, USA

I will begin, as have many of the speakers before me, by thanking the organizers of this meeting: in my case, not only for inviting me but also for the title of my talk—which leaves me free to discuss anything I like. I shall, nevertheless, talk mostly about this meeting.

I believe that I am expected to summarize the conference proceedings, but I cannot possibly summarize 13 lectures and eight round tables in the time allotted. Rather than a comprehensive summary, then, I shall present a highly personal view of this meeting: essentially one man's set of take-home lessons. At the end, I shall air some thoughts on the debate (or debates) referred to by Hans Kornberg in his opening address.

First of all, I see this meeting as an impressive account of exciting scientific progress: progress in the design of micro-organisms for beneficial use, progress in the engineering of micro-organisms for safety, and progress in our ability to monitor both cells and DNA sequences for persistence in the environment. I will say a few words about each of these developments.

DESIGN FOR USE

Julian Davies gave us an excellent overview of the rapidly developing technology employed in the production by genetically-engineered micro-organisms (GEMs) of specific proteins: including hormones, interferon, interleukins and protectively immunizing antigens. Other speakers gave us progress reports on specific projects: Daryl Dwyer, for example, reported on *Pseudomonas* constructs that degrade substituted aromatic compounds, troublesome pollutants in effluents of treatment systems from wood-processing industries and chemical

RELEASE OF GENETICALLY-ENGINEERED MICRO-ORGANISMS ISBN 0-12-677521-4

plants. Another example, which for me was one of the highlights of the meeting, was the report by Enzo Paoletti on the engineering of poxviruses to serve as safe, polyvalent vaccines. Although vaccinia virus has been safely employed as a vaccine by itself for over 200 years, and has been engineered to carry heterologous antigens, Dr Paoletti's group has gone one step further by using fowlpox as the carrier—thus producing an effective live vaccine made safe by its inability to replicate in mammalian cells. In the case presented, fowlpox virus carrying the rabies virus glycoprotein gene was used to protect dogs and cats against a challenge with live rabies virus.

ENGINEERING FOR SAFETY

Dr Paoletti's paper was thus an exciting example of the successful engineering of a micro-organism for both use and safety. Several other speakers have also addressed the issue of engineering for safety. I was particularly impressed by Roy Curtiss's attenuation of *Salmonella typhimurium* by deletion of the cyclic AMP system: the Δ*cya* Δ*crp* strain retains the ability to colonize the mouse gut while losing its virulence, and can thus be expected to serve as a safe carrier of genes for any desired immunizing antigens. Other attenuation strategies were discussed by Alf Lindberg.

A highly innovative approach to safety was provided by Stephen Cuskey's report, during Round Table 3, of a 'suicide plasmid' designed to allow the controlled destruction of a GEM when its job is done. In the example he gave, the plasmid carries a gene for the degradation of a pollutant, a killer gene, and a protector gene linked to an operator that is switched on by the pollutant. When the host micro-organism has degraded the pollutant below a threshold concentration, the protector gene is switched off and the killer gene product destroys the cell.

PERSISTENCE OF GEMs IN THE ENVIRONMENT

Many papers at this meeting have addressed the problem of the persistence of GEMs or of the foreign DNA sequences that they carry. Stuart Levy, for example, showed us that R-plasmids can spread globally in amazingly short periods of time, provided that selective forces exist. In the absence of selection, however, other reports at this conference have indicated that most introduced micro-organisms have a difficult time competing with their indigenous counterparts. This was illustrated, for example, by John Beringer's data for Rhizobia, and by Steven Lindow's data for ice$^-$ strains of *Pseudomonas syringae.*

DETECTION

In studies of persistence, sensitive and specific detection systems are required. Rita Colwell gave us a fine overview of the progress being made in the development of such systems, particularly by the use of fluorescent antibodies and radioactive probes. I found especially exciting the report by Ron Atlas, as part of Round Table 2, that a specific DNA sequence, if available as a probe, can be detected at the level of one copy per gram of soil, and that the detection of one copy per kilogram should be feasible. In this procedure, which can be fully automated, total DNA is extracted from the sample, a specific sequence is amplified by many orders of magnitude using primers and DNA polymerase, and that sequence is then detected with the specific probe. This technique should obviously detect any probe-able sequence, such as an introduced foreign gene, with exquisite sensitivity, and can be used to detect cells of a particular microbial strain, such as an introduced GEM, if a specific marker sequence is available. Progress has also been made in the insertion of synthetic marker sequences into GEMs, so that they can be detected with probes, as well as in the use of biochemical markers such as LAC, as presented by David Drahos.

THE DEBATE

Time does not permit me to comment individually on the other excellent papers that have been presented during the past 3 days. Rather, at the risk of making my talk controversial, I would like now to step back and take a broad view of the debate that has been raging in recent years. This debate—about the safety of GEMs in the environment—is not, to my mind, between molecular biologist and ecologists. Rather, it is a debate between those, whatever their background, who consider the benefits of specifically proposed releases to far outweigh the risks, if indeed a plausible risk exists, and those who demand proof of safety even when no specific potential harm has been defined. There are, I believe, ecologists on both sides of this debate.

If I had to apply labels to the contending factions, I might call them 'conservatives' and 'progressives', to borrow from the political lexicon. On the right are the conservatives, and on the far right are the reactionaries, who see all genetic engineering as dangerous, if not downright immoral. Interestingly, there is no extreme left wing; no-one, that is, who proposes wild schemes without consideration of the environment.* There are, in my experience, only moderates in the progressive camp; those who are ready to listen to scientifically legitimate

*I am indebted to Dr Bernard Davis for pointing out this asymmetry to me.

concerns and respond to them by appropriate risk assessment and by the design of biological containment systems.

The debate has gone on at two levels: general principles, and specific cases. Let me emphasize that general principles are important. It has been said again and again that we cannot generalize, that every proposed release must be analysed 'case-by-case'. Yes, of course—but at some point we *must* rely on general principles, if only to tell us when we have enough data. For we cannot test all possible events in all possible environments.

Let me illustrate the importance of this concept. If scientific experience and scientific reasoning were to suggest that most GEMs would spread in the environment and cause unwanted effects, then the burden of proof would have to be on the proponents of specific releases to demonstrate safety by exhaustive experiments. But if scientific experience and reasoning suggest that most GEMs will either not compete in nature or will do no harm even if they persist, then the burden of proof should be on the opponents of the release, who should be required to produce plausible scenarios not only of persistence in the environment but also of harmful consequences.

I am convinced that the principles of microbiology, microbial genetics and ecology all support the latter, moderate view, as persuasively argued in the literature by Bernard Davis. (I refer you, for example, to his recent articles in *Science* and in *Genetic Engineering News*.) Let me recite three of those principles.

First, the conservative stance has been that the release of GEMs is closely related to past introductions of established microbes, plants and animals into new environments. The great majority of these have been beneficial, but a few have created pests: e.g. kudzu vine, chestnut blight and the gypsy moth in the United States and prickly pear in Australia. As Davis cogently argues, however, this is the wrong analogy (and has done great disservice by implying that the burden of proof rests with the proponents of release). The pests in question were highly evolved to compete in nature; their transplantation to new regions of the globe, where certain natural checks were absent, allowed them to spread out of control. In contrast, GEMs are more analogous to domesticated organisms: microbes, plants and animals that have been bred for man's use, usually to the detriment of their ability to compete in the wild. Even escaped feral animals are no more fit than their wild ancestors.

As a second principle, we cannot expect to add significantly to the enormous amount of variation that is continually occurring in microbial populations in nature. It is only when we change the environment, introducing new selective conditions, that we cause major population shifts.

A third principle is that, for a GEM to become a pest, it must out-compete its indigenous relatives (and then go on to produce an unwanted effect). The small amounts of foreign DNA introduced by genetic engineering, however, are unlikely to increase fitness, which requires evolutionary co-adaptation of the entire, balanced genome.

For lack of time, I must refer you to Davis's excellent articles for a full articulation of these and other relevant scientific principles.

I said earlier that the debate has been going on at two levels: general principles and specific cases. But in fact there have been very few of the latter—the conservatives have come forward with very few plausible scenarios of harm associated with specific proposed releases. We heard one such concern during the question period following Dr Paoletti's talk: it was asked whether inserting the gene for a foreign antigen, such as the rabies virus glycoprotein, might alter the tissue tropism of the poxvirus with possible harmful consequences. This was a reasonable question that deserved an answer, which indeed was given (no such changes have been observed). I'm sure that there are other, scientifically legitimate concerns about specific release proposals, but so far they have been rare.

I would urge the moderates among us, then, to do three things:

(1) Educate the general public about the scientific principles that lead to a presumption of safety, in the absence of specific, plausible scenarios of harm;

(2) respond to scientifically legitimate concerns by undertaking the appropriate risk assessment experiments; and

(3) in the absence of scientifically plausible scenarios of harm, press for permission to 'proceed with caution', to use Roy Curtiss's words.

CONCLUSION

Let me return now to the conference which is just ending. I congratulate the organizers on bringing together such an impressive group of speakers and discussers: the work they have reported here, and the seriousness with which they are responding to legitimate ecological concerns, convinces me that the release of genetically-engineered micro-organisms into the environment for beneficial purposes will be overwhelmingly successful and free of significant risk.

REFERENCES

Davis, B. D., 1987a. Bacterial domestication: underlying assumptions. *Science* **235,** 1329.

Davis, B. D., 1987b. Point of view: is deliberate introduction ecologically any more threatening than accidental release? *Genetic Engineering News* October, p. 4.

Davis, B. D., 1988. Exposing fatal flaws in some arguments against recombinant release. *Genetic Engineering News* February, p. 20.

Index

GEM(s) = Genetically-engineered micro-organism(s)